Innovation and International Corporate Growth

Alexander Gerybadze · Ulrich Hommel ·
Hans W. Reiners · Dieter Thomaschewski
Editors

Innovation and International Corporate Growth

Editors
Professor Dr. Alexander Gerybadze
Center for International Management and Innovation
Center for Research on Innovation and Services (FZID)
University of Hohenheim
Schloss-Osthof-Nord
70599 Stuttgart
Germany
gerybadze@uni-hohenheim.de

Hans W. Reiners
BASF SE
EV - E100
67056 Ludwigshafen
Germany
hans-w.reiners@basf.com

Professor Ulrich Hommel
Strategic Finance Institute (SFI)
European Business School (EBS)
International University
Schloss Reichartshausen
Wiesbaden/Rheingau
EBS Campus Rheingau
65375 Oestrich-Winkel
Germany
ulrich.hommel@ebs.de

Professor Dr. Dieter Thomaschewski
University of Applied Sciences
Ernst- Boehe-Str.4
67059 Ludwigshafen
Germany
Dieter.thomaschewski@fh-ludwigshafen.de

ISBN 978-3-642-10822-8   e-ISBN 978-3-642-10823-5
DOI 10.1007/978-3-642-10823-5
Springer Heidelberg Dordrecht London New York

Library of Congress Control Number: 2010921282

© Springer-Verlag Berlin Heidelberg 2010
This work is subject to copyright. All rights are reserved, whether the whole or part of the material is concerned, specifically the rights of translation, reprinting, reuse of illustrations, recitation, broadcasting, reproduction on microfilm or in any other way, and storage in data banks. Duplication of this publication or parts thereof is permitted only under the provisions of the German Copyright Law of September 9, 1965, in its current version, and permission for use must always be obtained from Springer. Violations are liable to prosecution under the German Copyright Law.
The use of general descriptive names, registered names, trademarks, etc. in this publication does not imply, even in the absence of a specific statement, that such names are exempt from the relevant protective laws and regulations and therefore free for general use.

Cover design: WMXDesign GmbH, Heidelberg

Printed on acid-free paper

Springer is part of Springer Science+Business Media (www.springer.com)

Contents

Introduction: *Managing Innovation in Turbulent Times* 1
Alexander Gerybadze, Ulrich Hommel, Hans W. Reiners, and
Dieter Thomaschewski

Part I   Innovation and International Strategy

**R&D, Innovation and Growth: Performance of the World's
Leading Technology Corporations** 11
Alexander Gerybadze

**Technology Strategy for the Corporate Research Center
of a Diversified Global Enterprise** 31
Reinhold Achatz and Hans Jörg Heger

**R&D Internationalization in Multinational Corporations: Some
Recent Trends** 47
Heike Belitz

**The Chemical Industry Drives Innovation also in Times
of Crisis: BASF as an Example** 67
Andreas Kreimeyer

The Innovation Premium: Managing for High Return on Innovation . . 81
Tom Sommerlatte

**Innovation in the Interplay of Organization and Culture: The
TRUMPF Story** 91
Christian Koerber, Gabriela Buchfink, and Harald Völker

**Growth and Internationalization: Renewable Energy
and New Technology-Based Firms** 113
Christian Schwens, Holger Steinmetz, and Rüdiger Kabst

Part II   Efficiency of Innovation Processes in International Enterprises

Determinants for Failure and Success in Innovation Management . . . 127
Dieter Thomaschewski and Alexander Tarlatt

Innovation Generating and Evaluation: The Impact of Change Management 151
Dietmar Vahs, Verena Koch, and Michael Kielkopf

Implementing Change Management Successfully – Reinventing an Innovative Corporation: The Bayer Case 175
Alexander Moscho, Lydia Bals, Matthias Kämper, and Stefan Neuwirth

Ambidextrous Leadership in the Innovation Process 191
Kathrin Rosing, Nina Rosenbusch, and Michael Frese

Innovation Through Market Pull and Technology Push in the Heavy Equipment Business: The Voith Case 205
Bertram Staudenmaier and Michael Schürle

Principles of Collaborative Innovation: Implementation of Open and Peer-to-Peer Innovation Approaches 219
Gerhard Satzger and Andreas Neus

Managing Open Innovation Networks in the Agriculture Business: The K+S Case 239
Alexa Hergenröther and Johannes Siemes

Part III   Capital Markets, Finance and Innovation Performance

Modern Valuation Approaches for Corporate Innovation Activities ... 263
Andreas Krostewitz and Martin Scholich

Value-Based Management of the Innovation Portfolio 281
Ulrich Pidun

Innovation Performance Measurement 299
Peter Schentler, Frank Lindner, and Ronald Gleich

Accounting for Innovation: Lessons Learnt from Mandatory and Voluntary Disclosure 319
Thomas Günther

Integrated Financing Strategies for Innovation-Based Growth 333
Ervin Schellenberg

Corporate Venture Capital 349
Malte Brettel

Knowledge-Based Financing Strategies for Innovation Output 359
Thomas Rüschen and Frank Rohwedder

The Value Impact of R&D Alliances in the Biotech Industry 371
Hady Farag and Ulrich Hommel

Are Family-Owned Businesses Better Innovators? 393
Katinka Wölfer

Bibliography 417

Editors 441

Authors 443

Introduction: *Managing Innovation in Turbulent Times*

Alexander Gerybadze, Ulrich Hommel, Hans W. Reiners, and Dieter Thomaschewski

Innovation is the engine of growth and corporate restructuring. During phases of persistent growth (e.g. during the period from 1990 to 2008), this message is simple and straightforward. But what happens to innovation when the economy is characterized by "engine failure", and how does this effect the management of R&D and the role of innovation during a severe crisis? This book on *Innovation and International Corporate Growth* was planned just before the financial shock that turned world business upside down. Who will still be interested in a new book on innovation when firms and financial institutions are going bankrupt and managers are most concerned with reducing costs and with keeping their business running?

The authors are convinced that innovation and sustainable restructuring should still be on the agenda, but that it is necessary to rethink the optimistic scenario, where more investment in R&D automatically leads to successful products and new business models, resulting in rising spirals of wealth. More innovation does not necessarily mean smart innovation. Persistent innovation in times of discontinuity places emphasis on efficiency and problem-solving, on specific types of innovation better suited to periods of distress, while at the same time sowing the seeds for the next wave of business renewal.

Part I: Innovation and International Strategy

Smart innovation is built on effective competence for renewal and superior innovation management capabilities. *Alexander Gerybadze's* introductory survey on R&D innovation and growth analyzes the *Performance of the World's Leading Technology Corporations* over a ten year period just before the financial crisis began. The innovation success of high performers is built on excellence in idea generation processes, project development and portfolio management as well as on superior product launch and business scale-up capabilities. Effective innovation routines lead to

---

A. Gerybadze (✉)
University of Hohenheim, D-70599 Stuttgart, Germany

persistence in innovation even during periods of serious restructuring and distress. High performers, so the argument, are also promising candidates for managing the next round of innovation, and we can learn a lot from the description of their advanced managerial practices and organizational models.

Large corporations in the electronics industry went through a period of deep restructuring and had to invest persistently in R&D and innovation. The BCG Senior Executive Innovation Survey (2008) ranks General Electric, Sony, Samsung and Siemens as top performers, based on breakthrough products and innovative processes. Innovation management at *Siemens* is described by *Reinhold Achatz* and *Hans Jörg Heger*, and this chapter emphasizes corporate strategy, technology management and the role of the corporate research center. In contrast to the general opinion of analysts arguing for conglomerate discount, and despite the reduced role of corporate research in many other companies, Siemens has still maintained a strong corporate technology base. The company acts as an integrated conglomerate that builds on effective synergies between different businesses. A coherent strategy emphasizes trendsetter projects where advanced R&D is combined with de-facto standardization processes and a strong patent portfolio. This article also provides a very interesting account of how this company effectively manages a global network of research centers and describes how research units work together with business units.

The *Globalization of R&D* and increased offshoring of innovation activities is an important topic also addressed in Part I of this book. *Heike Belitz* analyzes recent trends in the globalization of R&D in multinational corporations from the U.S., Europe and Asia. Today, German companies spend an average of 30% of their overall R&D expenditures abroad, and foreign investors account for an increasing share of the national R&D base. The article analyzes the pros and cons of R&D offshoring in different industries and provides useful recommendations for management as well as for innovation policy.

Global R&D and the restructuring of innovation is probably most pronounced in the chemical and pharmaceutical industry. BASF, the world leader in the chemical business represents a very interesting example of persistence and continuous change. The description of how the *Chemical Industry continues to drive Innovation* even in times of crisis provides a very interesting account of how this company deals with globalization, breakthrough innovation and long-term megatrends. Even though the chemical industry has been strongly affected by the recent world crisis, BASF has continued to invest in long-term research and major platform technologies in order to remain at the forefront of technological change and global industry restructuring. The article by the chief technology officer, *Andreas Kreimeyer*, gives insights into managerial challenges during times of deep structural change in the world's chemical industry.

Tom Sommerlatte analyzes the value-creation potential of high-performing innovators. The *Innovation Premium* is measured by total shareholder return over the last five years. Premium companies distinguish themselves from other participants in their industry in that they direct their innovation efforts much more explicitly and rigorously to creating customer benefit, market success and company value.

Sommerlatte argues for a value-based innovation management and close integration between corporate innovation strategy, effective innovation processes and an innovation-friendly innovation culture. The article provides several case examples of high-performing innovators in Europe, the U.S. and Japan.

While most studies on high performers in innovation concentrate on large, blue-chip corporations, there are very few detailed case studies on family-owned mid-size companies. This group of firms represents a strong pillar for growth, export performance and innovation. The Trumpf story is an excellent example of a very dynamic firm which effectively deals with *Innovation in the Interplay of Organization and Culture*. *Christian Koerber, Gabriela Buchfink* and *Harald Voelker* provide a detailed account of the evolution of this machine tool company with its successive steps of diversification into laser cutting and welding as well as into the field of medical business. A strong, innovation-minded corporate culture still influenced by the majority owner and the persistent management of innovation projects are the main keys for sustainable growth and internationalization in this highly successful company.

The role of high-technology based start-up firms and support in their favour have been major concerns for governments in Europe, and this topic is still considered to be critical for sustainable innovation. The chapter on innovation strategies in small and medium enterprises (SME) by *Christian Schwens, Holger Steinmetz* and *Rüdiger Kabst* describes the growth and internationalization patterns of firms in four different industries: *Renewable Energy, Microsystems, Biotechnology* and *Nanotechnology*. Innovation management and internationalization strategies in SMEs differ distinctly from those in large multinational firms, and success is built on effective network relationships with customers, suppliers and foreign distributors. Effective project promotion programs as well as appropriate forms of venture finance are of critical importance for successful international growth and will therefore be addressed in Part III of this book.

Part II: Efficiency of Innovation Processes in International Enterprises

Innovation projects contribute to growth and competitive advantage. However, innovation can be risky and result in massive financial losses. Innovation projects usually tie up long-term resources and call for large investments. One of the main challenges facing innovation management is to reduce the probability of failure thus activating the determinants of success. This means that the function of innovation management is ultimately to perceive opportunities in the innovation process and reduce risks. Some very important challenges and opportunities are considered in Part II of this edition.

Hence, the major challenge facing innovation managers is to avoid failure and strengthen the determinants of success. *Dieter Thomaschewski* and *Alexander Tarlatt* reflect on the innovation process with all its needs as regards planning, directing and performance monitoring. Successful innovation follows a structured approach with three vital elements: systematic and consistent preparation, boundary-less and multi-party execution and the efficient coordination of all activities and operations. Analyzing deficits regarding effective and efficient innovation management, they address the five different perspectives that lead to satisfying innovation results. The authors describe critical requirements which must be fulfilled in order to achieve success. Successful innovation is sound, hard, sustainable and consistent work, directed by ten (mandatory) commandments that should govern the innovation process.

Change management is a fundamental need and vital for any innovation. *Dietmar Vahs, Verena Koch* and *Michael Kirchhoff* describe the need and ability to change and innovate. In order to succeed, researchers suggest understanding the new product development (NPD) project as a process which must be based on an innovative corporate culture. Emphasis is put on the stage-gate process as a management tool, and on the continuous evaluation of the new product development process. The stage-gate process divides the NPD-project into different stages, which enable the company to assess the development from idea generation to launch. By defining evaluation criteria for the various stages and through a continuous monitoring process, management is kept informed about deviations and can take the appropriate measures in order to secure innovation success.

Change management is a very special challenge for large companies, demanding the development of a very special culture. *Alexander Moscho, Lydia Bals, Matthias Kämper* and *Stefan Neuwirth* describe the C*hange Management Process at Bayer*, one of the most successful chemical companies. This article illustrates how the company has dealt with challenges during the last five years and has reinvented itself successfully. One key component that enabled the company to manage this fundamental process was paying close attention to thoughtfully designed change management activities. How Bayer managed to do so and how this ability has become – in itself – a capability, is highlighted by taking the example of the Schering Integration.

The ability to achieve a balance between exploration and exploitation is called ambidexterity. Ambidexterity is important within the innovation process. While idea generation builds mainly on exploration, project implementation requires exploitation. *Kathrin Rosing, Nina Rosenbusch* and *Michael Frese* argue that the chaotic nature of innovation processes leads to a necessity to switch flexibly between exploration and exploitation. Effective leaders are able to combine the two by being open to new ideas, while still maintaining a high degree of efficiency in their routine business activities. The authors term this ability ambidextrous leadership. They provide a practical example and discuss pathways to the effective implementation of ambidextrous management.

Investment in R&D is an absolute necessity for manufacturers of advanced investment goods. What drives R&D, whether market pull or technology push, has been the subject of discussion for some time. *Bertram Staudenmaier* and *Michael Schürle* present the Voith case and describe how the two forces are combined. Voith remains one of the largest family-owned companies, which is active in the paper, energy, mobility and services sectors. The authors conclude that market-pull and technology-push strategies are of almost equal importance, both requiring that customers must receive a significantly higher added value. This means that employees have to be able to judge developments from the customer's perspective. The Voith case demonstrates that customer orientation and technology orientation can work together effectively and need not exclude each other.

New developments in information and communication technology (ICT) are an unprecedented success story, slashing the high costs of connecting and coordinating people and information. *Gerhard Satzger* and *Andreas Neus* describe models and the implementation of open and peer-to-peer innovation approaches. These open networks expand the reach of contributors to innovation and value creation processes far beyond the confines of any single organization. The authors analyze new modes of *Collaborative Innovation* which build on openness of information flow, common ownership, and peer-to-peer structures. Finally, these dimensions come with their own set of principles, which are markedly different from those of traditional "closed-shop" innovation.

The agricultural business is rather traditional and has existed for thousands of years. One might therefore deduce that innovation is a minor part of such a basic sector. *Alexa Hergenröther* and *Johannes Siemes* from the K + S AG, one of the leading companies in the fertilizer business, describe new dynamic forms of agricultural innovation. The authors discuss the need and pressure for innovation as well as the significance of open networks in this environment. Fields of opportunities become even wider by promoting open networks. The adoption and diffusion of innovation in agriculture are greatly supported by these new open and collaborative forms. Even if agriculture is torn between tradition and progress, the transfer of knowledge offers many opportunities, which should not be neglected.

Part III: Capital Markets, Finance and Innovation Performance

High-performing companies are able to generate marketable innovations on a consistent basis. They do so by integrating their investments in general R&D and innovation into their performance measurement systems and by utilizing innovative financing solutions offered by alternative investors and organized capital markets. Part III of this volume addresses various aspects in this context ranging from how to measure the value of corporate innovation activities right up to their performance impact. *Andreas Krostewitz* and *Martin Scholich* set the stage with their chapter on *Modern Valuation Techniques for Corporate Innovation Activities*. Academics and practitioners alike are still struggling to identify workable approaches for the pricing of future cash flow streams resulting from the adoption of innovation. The authors discuss how to apply traditional valuation techniques for these types of investments and study the accompanying limitations. They then turn their attention to the real option and the risk compound valuation approaches as alternative pricing methodologies. While resolving some of the more fundamental problems associated with the application of mainstream valuation techniques, these are not completely free of downsides and limitations. Krostewitz and Scholich have put together a tour de force providing the reader with a very up-to-date review of relevant knowledge on this topic.

The essence of modeling the economic performance of innovation investments is to grasp the risks associated with such projects. In his chapter on the *Value-based Management of the Innovation Portfolio*, *Ulrich Pidun* explains how innovative simulation techniques can be used to analyze the volatility dimension of corporate innovation activities from a portfolio perspective. He argues that companies need to optimize the trade-off between return and risk, between the short and the long term, as well as between exploitation and exploration. Ulrich Pidun proposes an integrative portfolio-based approach, which encompasses strategic decision-making, the actual selection of projects and the allocation of scarce financial and managerial resources. This chapter presents the key requirements for an effective value-based innovation portfolio management and describes the different techniques that can be combined to support an integrated approach to innovation management.

The *Performance of Innovation Activities* must be monitored on an ongoing basis as part of sound management practice. *Peter Schendler, Frank Lindner* and *Ronald Gleich* present an up-to-date review of relevant techniques from a performance management perspective. They argue that existing measurement systems do not satisfy the requirements of companies. In order to establish successful innovation monitoring covering the discrepancies of systems typically found in corporate practice, the authors propose a novel concept covering all levels of innovation performance and outline its practical implementation.

Innovation activities must be integrated into a company's accounting system and should be disclosed to investors in a meaningful form. *Thomas Günther* explains in his chapter how companies should organize the *Accounting for Innovation* in order to minimize the gap between the relevance of innovation for actual and potential investors and stakeholders on the one hand, and the actual disclosure of information on innovation capital on the other. He explains that current national (German GAAP) as well as international (IFRS) accounting limits the possibilities of disclosing innovation capital in a broader sense, and that it is restricted to marketable, controllable development projects in accordance with IAS 38 (Intangible Assets) and IFRS 3 (Business Combinations). While purchase price allocations have a tendency to reveal intangible assets, innovation capital accounting still plays a minor role. Günther explains why companies still hesitate to disclose information related to their innovation activities voluntarily, mainly due to the restricted measurability and objectivity of information as well as to fears concerning the external effects of disclosing information to customers and competitors.

Financing Innovation-based Growth Strategies put special demands on financial budgets and often involve discontinuous jumps in the company's asset base. *Ervin Schellenberg* explains in his chapter why companies in such situations should rely on integrative financing strategies, which fully utilize the entire range of available capital market instruments. He shows that growth can be financed on or off the balance sheet and that the ultimate selection of an appropriate structure will depend on the individual company's characteristics.

Many multinationals have been frustrated by their inability to generate innovation endogenously and have subsequently turned to the creation of *Corporate Venture Capital* units that invest in high-tech start-ups and fund spin-offs of the parent company. *Malte Brettel* discusses the evolution of this sub-market of the venture capital industry and explains the attractiveness of moving innovation activities to separate legal entities from a corporate perspective. His exposition, however, also shows that performance effects have been very mixed, leading many companies either to sell off their investment portfolios in later phases or to re-integrate corporate venture capital subsidiaries back into the parent organization.

The asset-backed securities industry has been one of the key drivers of financial market growth in recent years. While the mainstream market mostly covers standardized financial claims (receivables, consumer loans etc.), a market niche has been developing in recent years with the focus on the acquisition and grooming of patents. *Frank Rohwedder* and *Thomas Rüschen* provide readers with an overview of this highly innovative market segment in their chapter on *Knowledge-based Financing Strategies for Innovation*.

Innovation is often the result of a collaborative effort, and biotechnology is arguably the industry which is most reliant on utilizing partnerships with other companies to generate innovative growth. *Hady Farag* and *Ulrich Hommel* provide a detailed account of *Collaborative Value Creation in the European Biotechnology Industry*. This chapter represents the first inquiry into the value of European biotech alliances in general and is the first contribution to analyze the value of alliances across different stages of the lifecycle.

Academic finance literature has always placed a strong focus on the link between ownership structure and corporate performance. Recent years have seen a rising number of studies, which specifically analyze whether family ownership leads to superior financial performance. In her chapter, *Katinka Wölfer* reviews the existing evidence on this issue and specifically addresses the question of *whether Family-owned Businesses are Better Innovators*. The author is able to show that family ownership can provide the basis for more rewarding innovation investments. She argues that widely-held companies may benefit from adopting the management styles of well-run family corporations, which often focus on sustainable development rather than on the maximization of short-term shareholder returns.

Part I
Innovation and International Strategy

Part I
Innovation and International Strategy

R&D, Innovation and Growth: Performance of the World's Leading Technology Corporations

Alexander Gerybadze

Contents

1 R&D, Investment and Growth 11
2 When does Innovation Lead to Sustainable Growth? 14
3 Which Companies are the Top Performers in Innovation? 17
4 Managing Relational Dynamic Capabilities 21
    4.1 Standard-Setting Excellence 22
    4.2 Intellectual Property Management 23
    4.3 Managing Complementary Assets 23
5 Managing the Dual Cycle of Innovation 25
6 New Forms of Integrated Innovation Management 26
7 Integrated Design and Corporate Intellectual Property Management 28
References 29

1 R&D, Investment and Growth

Research and development (R&D) and Innovation are the drivers of change and the key determinants of growth in many industries and service sectors. The Industrial R&D Investment Scoreboard commissioned by the European Union provides data for the 2000 largest R&D spenders in Europe, North America and Asia.[1] World industrial spending for R&D has reached a level of € 373 billion and is expected to grow continuously, in spite of the financial turmoil and restructuring of the world economy after 2008.

A. Gerybadze (✉)
University of Hohenheim, D-70599 Stuttgart, Germany

[1] European Commission (2008), The 2008 EU Industrial R&D Investment Scoreboard, Seville and Luxembourg.

Table 1 The World's leading R&D industries 2007

Industry	R&D expenditures 2007 (Mio €)	R&D as % of sales 2007	R&D growth over last 4 years (%)
1. Pharmaceuticals & Biotechnology	71430	16.1	33.0
2. IT-Hardware & Equipment	68191	8.5	23.6
3. Automobiles & Parts	63234	4.2	8.8
4. Software & Computer Services	26624	9.7	27.7
5. Electronics & Electrical Equipment	26094	4.1	10.2
6. Chemicals	16428	2.8	1.5
7. Aerospace & Defence	15109	4.4	25.8
8. Industrial Engineering	11004	2.7	24.1
9. General Industrials	8129	2.1	11.7
10. Healthcare Equipment & Services	6552	6.5	32.3

Source: R&D Scoreboard (BERR, 2008), INTERIS Database University of Hohenheim 2009

As can be seen in Table 1, R&D expenditures are strongly concentrated within a few technology-intensive industries, with the top-five sectors accounting for more than two thirds of global R&D spending: (1) Pharmaceuticals and Biotech, (2) IT Hardware and Equipment, (3) Automobiles and Parts, (4) Software and Computer Services and (5) Electronic and Electrical Equipment. Within each sector, a few large firms with powerful R&D portfolios and strong innovation management capabilities account for the lion's share of resources. R&D and innovation activities of large firms address three types of strategies in mixed combinations: incremental innovation, dynamic growth strategies and industry creation.

1. *Incremental, piecemeal innovation.* In many industries, considerable R&D is required just to keep business going, to continuously renew products and processes and to defend market shares. This is particularly the case in well-established and mature industries. European firms in manufacturing industries often follow such incremental, more defensive types of strategies.
2. *Dynamic growth strategies.* Other industries are characterized as dynamic, fast-growth and high-tech. Firms in these industries need to adapt their portfolio of products and master breakthrough innovation persistently. As a result, the largest part of investment is directed towards new activities, mainly R&D and product development.

3. *Industry creation.* A third component of R&D spending is directed towards the creation of new industries, often a matter of long-term, high-risk investment involving venture capital and corporate diversification. Some of the fastest growing companies among the top R&D spenders did not exist before 1990, and the type of business they are in has been "created from scratch".[2]

European corporations have most strongly emphasized structure-enhancing, incremental R&D activities in established industries such as Automobiles, Chemicals, Electrotechnical Equipment and Industrial Engineering. These are often not the dynamic sectors and not the ones characterized by large increases in R&D spending. Still, these sectors are being transformed continuously and corporations need to be smart in adapting to new technologies, which are often generated in other, more dynamic high-tech sectors.

The strongest increase in R&D spending over the last ten years has occurred in *dynamic, high-tech growth industries.* Growth rates can attain 15–30% per annum, with above-average profitability. In order to participate in these dynamic industries, corporations must invest between 10 and 20% of annual revenues for R&D and the name of the game is speed in product development and efficient innovation management. As a result, the following dynamic industries have made rapid increases in R&D spending and have attracted the awareness of financial investors.

- Pharmaceuticals and Biotechnology,
- IT Hardware and Equipment,
- Software & Computer Services,
- Electronics & Electrical Equipment,
- Healthcare Equipment and Services.

Investments in these dynamic, R&D-intensive industries were mostly dominated by U.S. corporations, and were also targeted by corporations in Asian countries. In many of the most dynamic fields, European investors, with a few exceptions, were not among the high performers and have often lost out to their American and/or Asian rivals. This was particularly the case in Pharmaceuticals and Biotech, in IT Hardware and Equipment, in Semiconductors, as well as in Consumer Electronics.

European corporations have concentrated their R&D efforts in traditional manufacturing industries including Automobiles and Parts, Chemicals, Industrial Engineering and General Industrials. R&D intensities in these sectors tend to be in the range of 2–5% (R&D as percent of revenues). R&D and innovation activities tend to be less dynamic, and annual average rates of growth of R&D are more or less in the range of output growth. Temporary exceptions have been noticed

---

[2] Take as a prominent example Google, a start-up firm established in the late '90s. This newly created firm is today one of the most valuable American corporations and No. 59 on the list of the world's largest R&D spenders.

in Automobiles and Machinery, where firms had to increase their R&D spending for the absorption of Advanced Electronics, IT and Software. In the fields of Pharmaceuticals, Biotechnology and Healthcare, European firms have increased their efforts, but R&D investments have been concentrated increasingly in North America.

2 When does Innovation Lead to Sustainable Growth?

High-performing innovators effectively manage the full cycle of idea generation, project selection and execution, and they effectively address growth targets in their existing industries or in new, more dynamic market environments. *Innovation excellence* is more than innovation management, involving constant rejuvenation and effective market creation activities, year after year. Some companies have been successful in introducing new products to the market in one generation, but were unable to remain at the leading edge over a longer period of time.[3] Only those companies that continuously invest considerable amounts in R&D, that persistently expand their base of technological capabilities, and that remain at the forefront of new product introductions for successive generations will attain stable growth and strong financial performance. They need to maintain a repetitive cycle of innovation as described in Fig. 1. High margins and above-average returns are fed back into the pipeline, to invest more for R&D than rival firms, and to manage the new product development pipeline effectively year after year.

Innovation-excellent companies have a strong track record of turning innovation inputs (R&D expenditures, ideas and managerial inputs) into strong *innovation competence*. However, innovation competence is a necessary, but non sufficient

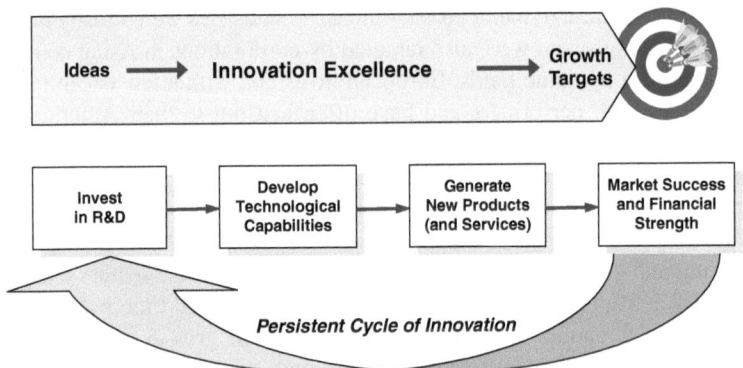

Fig. 1 Innovation excellence: Managing the dynamic cycle of innovation

---

[3] The term "Innovation excellence" was coined by Arthur D. Little to describe corporations that are persistent high performers in innovation management within their industry. See ADL (2005a,b).

condition for long-term growth and for persistent market success and financial performance. Some important complementary factors, such as strategic direction, organizational capabilities and dynamic interaction with other firms as well as standard-setting activities, are required to effectively attain strong and lasting innovation performance. The whole set of complementary factors can be defined as *dynamic capabilities*, and these are described more precisely below.

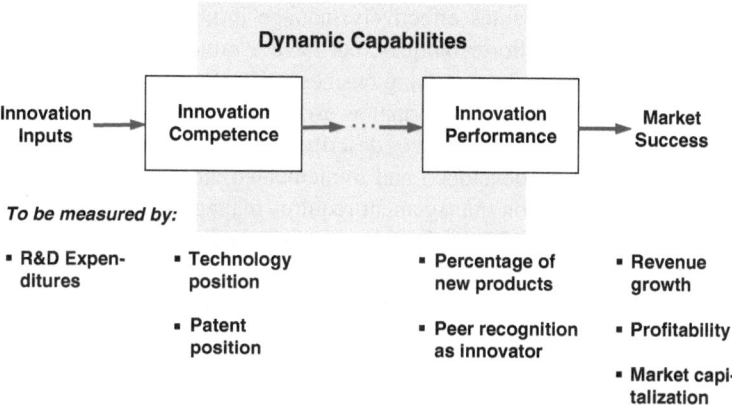

Fig. 2 Dynamic capabilities help to transform innovation competence into market success

Dynamic capabilities within innovation-excellent companies build on strong managerial capabilities at the corporate as well as the business unit level.[4] Innovation is strongly emphasized by corporate strategy and investment policies support expensive and often risky R&D projects. Superb performers have developed effective innovation routines, a strong new product pipeline and the ability to evaluate and absorb risks better than rival firms. Dynamic capabilities must be supported through innovation-enhancing organizational structures, allowing for the effective integration of corporate research as well as R&D performed within the major business units. Corporate capabilities are directed towards promising growth targets and this requires a balancing of priorities between short-term (often financial) objectives and long-term projects. Effective *intra-corporate innovation management* builds on excellence at three simultaneous stages:

1. Excellence in the idea generation and selection process, i.e. the continuous transformation of promising ideas and concepts into sound projects.
2. Excellence in project development and execution and the appropriate balancing of a large number of diverse projects, often attained through effective portfolio management.

---

[4] See ADL (2005a, 2006), Teece (2007) and Eisenhardt and Martin (2000) for an excellent description of dynamic capabilities within top innovation performing firms.

3. Excellence in turning projects into commercial ventures, in managing business scale-up activities and in generating strong new business units based on internally-generated innovation projects.

Some companies may be strong in idea generation and selection, but they often miss the subsequent stages. Due to failures in project development and portfolio management, they are unable to focus, thus taking on too many projects at a sub-critical level. Other companies effectively manage project development and use sophisticated R&D portfolio techniques, but have a rather weak track record for turning completed projects into growing business units. Strong launch and business scale-up capabilities are just as important as excellence in R&D and product development. All three types of competences described by the shaded boxes in the upper part of Fig. 3 need to be developed and implemented simultaneously. Excellence in intra-corporate innovation management requires managing the full cycle of idea generation and selection, R&D project development and portfolio management, as well as a systematic business development and scale-up process.

The most admired innovation performers have implemented an integrated innovation process as an effective routine. Stage-gate processes and firm-specific innovation management routines have been developed extensively over the last ten years. These techniques have often become "standard operating procedures" for many companies and are highly promoted by many management consulting firms.[5]

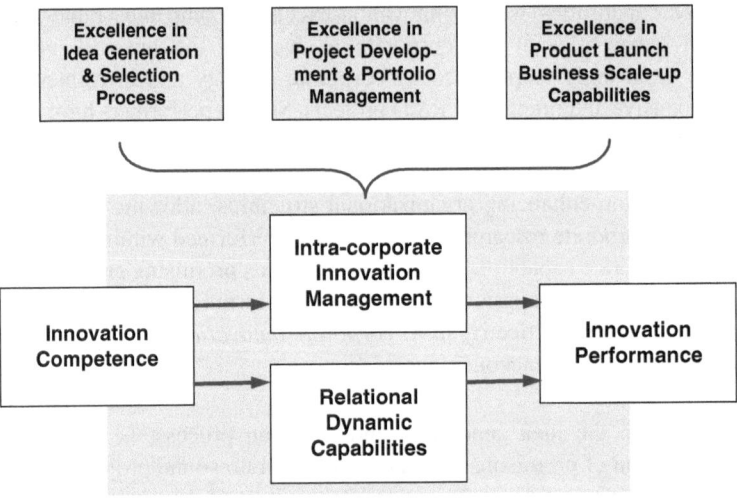

Fig. 3 Intra-corporate dynamic capabilities based on internal innovation management

---

[5] In a private conversation, Mr. Jaworski, managing director of 3 M in Germany, pointed out "that you can find at least 10 consultants in the neighborhood of Düsseldorf that are ready to implement a new structured innovation process within six months".

As a result, these techniques of innovation process management have turned into a "base competence", a type of "hygiene factor" required to stay in business, but they rarely serve as a differentiator for explaining above-average innovation performance.[6] Managing external relationships and relational dynamic capabilities are often much more difficult to master, as will be described in Sect. 4.

3 Which Companies are the Top Performers in Innovation?

Top performers in innovation are those corporations that manage a persistent stream of new products and services over a longer period of time. They often push the frontier and are considered innovation leaders in their industry. However, the balance between financial performance indicators and innovation-oriented investment projects needs to be mastered. Overstretched R&D budgets can be as bad as underinvestment in new product development.[7] Unfortunately, there are not many systematic studies on the relationship between R&D investment, corporate growth and financial performance. Some specialized consulting companies as well as university-based research centers have developed assessments of corporate innovation performance, however these are often incompatible with respect to the chosen evaluation technique.

- The Boston Consulting Group (BCG) has developed a senior management survey which is published annually. As part of the survey based on subjective assessments, the company asks top managers: "Which three companies do you consider the most innovative and why". As one of the results, BCG publishes a list of the top 50 most innovative global companies.[8] This is based exclusively on extensive in-house research attempting to unveil the innovation-to-cash process.
- Arthur D. Little (ADL) has promoted the concepts of Innovation Excellence and the Innovation Premium. In an earlier study on the Innovation Premium, Jonash and Sommerlatte (1999) analyzed the link between innovation performance and financial performance. ADL has since then published several studies on Innovation Excellence, including a ranking of the most admired innovators (ADL, 2005a,b).
- Booz & Company publishes an annual survey of the 1000 largest R&D spenders, has developed a detailed performance metric and identifies the high-performers among the large R&D spenders in the world. See Jaruzelski and Dehoff (2008, 2007).

---

[6] In their study on successful breakthrough innovation in 32 companies, Cotterman et al. (2009) come to a similar conclusion: Stage-gate processes are important but they do no longer serve as a differentiating factor.

[7] See Knott (2009, 2008) for a more recent study on the ambivalent influence of R&D on financial performance indicators.

[8] For the latest ranking see BCG (2009a,b). This study as well as earlier versions (BCG 2008, 2007) can be accessed via the internet.

- Research at the Center for International Management and Innovation at Hohenheim University has led to innovation auditing and innovation performance measurement for the world's largest R&D spenders. The Center tracks the evolution of the Top 100 R&D spenders over time and analyzes the relationship between R&D, sales growth and market capitalization.

In the following, we will analyze R&D investments and innovation performance for the period 1997 to 2007. The R&D expenditures of the top 100 R&D corporations have been increased consistently at an average annual growth rate of 7% over the last 10 years. R&D growth was considerably higher than revenue growth during that period, and was not seriously affected by the recession in 2001/2002. The top 100 firms spent $ 327 billion on R&D in 2007, representing 60% of total business expenditures on R&D.

During the period under investigation, firms from very dynamic high-tech sectors were responsible for large increases in R&D spending, e.g. Software and IT services (386% compound growth between 1997 and 2007), Semiconductors (+291%), Consumer Electronics (+170%) and Pharmaceuticals and Biotechnology (+158%). Firms from these dynamic sectors have significantly improved their position in this ranking. As an example, Microsoft is now No. 1 on the list of leading R&D spenders with an $ 8.2 billion R&D investment budget in 2007. Several dynamic firms which were much further back on the list in 1997, now appear among the top 100. Dynamic Schumpeterian competition drives the quest for R&D and innovation. Large incumbent firms such as ABB, AT&T, Saint Gobain, 3 M and Xerox are no longer on the list of the top 100 R&D spenders in 2007. New entrants in newly created business (Internet services, Network equipment and Biotech companies) have risen to the top, including Amgen, Cisco, Google and Yahoo.

Innovation performance cannot be based solely on growth in R&D spending, however. Reliable evaluations of corporate innovation performance need to take the following into account:

- Long-term and stable investments into most promising R&D projects,
- the build-up of unique pools of knowledge and strong patent positions,
- stable and above-average growth in revenues,
- high percentage shares of new products (introduced during the last 5 years),
- growing market shares and above-average profit margins,
- and, last but not least, considerable increases in market capitalization.

Unfortunately, only a few consistent econometric studies linking these variables have already been published. Investment analysts and industry-specific research organizations certainly use data uncovering these relationships, but exclusively in-house or for their client relationships.[9] As a result, publicly accessible rankings of

---

[9] Probably the most extensive benchmark studies on R&D performance and "pipeline studies" are carried out in Pharmaceuticals.

the world's most innovative corporations are often based on subjective evaluations, and are typically results of surveys among top managers. These tend to be biased in favour of highly visible firms, often high-tech consumer good producers and, last but not least, on US-based corporations. As an example, the BCG 2008 list of the most innovative corporations in the world contains 35 companies that can be considered as manufacturing firms with heavy involvement in R&D. In addition, the BCG list also contains 15 companies which are service providers investing in innovation, but rarely for dedicated R&D projects. Surprisingly, this last group contains

Table 2 Ranking of the World's 100 largest R&D corporations

	Company	R&D expenditures 2007 ($ million)	CAGR 1997–2007 (in %)	Former rank R&D 1997		Company	R&D expenditures 2007 ($ million)	CAGR 1997–2007 (in %)	Former rank R&D 1997
1	Microsoft	8 164	14.0	32	51	Denso	2 505	7.1	59
2	General Motors	8 100	−0.1	1	52	BT	2 492	16.1	109
3	Pfizer	8 089	13.9	31	53	NTT	2 435	−0.4	19
4	Toyota Motor	7 974	7.8	6	54	Philips Electronics	2 345	1.4	28
5	Nokia Finland	7 721	20.3	68	55	Hyundai Motor	2 343	24.7	201
6	Johnson & Johnson	7 680	12.3	26	56	Fujitsu	2 274	−1.6	13
7	Ford Motor	7 500	1.5	2	57	Texas Instruments	2 155	3.1	41
8	Roche	7 325	12.5	29	58	SAP	2 132	15.1	112
9	Volkswagen	7 198	10.3	20	59	Google	2 120		
10	Daimler	7 147	7.7	8	60	Procter & Gamble	2 112	4.6	53
11	Sanofi-Aventis	6 671	5.6		61	BASF	2 046	3.4	46
12	Samsung Electronics	6 489	21.8		62	Volvo	2 029	5.8	61
13	GlaxoSmithKline	6 461	11.8	34	63	Sun Microsystems	2 023	8.5	78
14	Novartis	6 414	8.8	18	64	Delphi	2 000		
15	Intel	5 755	8.5	21	65	AMD	1 847	13.3	110
16	IBM	5 747	2.6	4	66	Qualcomm	1 829	20.4	178
17	Robert Bosch	5 205	10.0	36	67	LG Electronics	1 802		
18	Matsushita Electric	5 175	4.0	7	68	EMC	1 767	20.8	187
19	AstraZeneca	5 042	7.9		69	Takeda Pharmaceuticals	1 730	10.9	98
20	Honda Motor	4 940	8.9	30	70	Nortel Networks	1 723	−2.0	
21	Alcatel-Lucent	4 924	9.4		71	Infineon Technologies	1 709		
22	Siemens	4 921	0.7	3	72	STMicroelectronics	1 705	10.8	
23	Merck	4 883	10.1	38	73	Sharp	1 699	5.3	70
24	Sony	4 869	7.6	24	74	United Technologies	1 678	3.2	58
25	BMW	4 597			75	Nestle	1 656	10.8	102
26	Cisco Systems	4 499	12.7	57	76	Merck	1 642	12.8	117
27	Motorola	4 429	4.4	12	77	Fuji Film	1 584	9.5	94
28	Ericsson	4 256	3.0	10	78	NXP	1 547		
29	Nissan Motor	4 161			79	Daiichi Sankyo	1 527	17.2	167
30	EADS	3 949			80	Astellas Pharma	1 503	8.0	
31	Bayer	3 867	5.2	23	81	Honeywell	1 459	11.3	113
32	Boeing	3 850	6.5	33	82	Novo Nordisk	1 457	12.4	127
33	Hitachi	3 693	−0.5	5	83	Caterpillar	1 404	9.3	104
34	Hewlett-Packard	3 611	1.4	9	84	Broadcom	1 349		
35	Renault	3 600	8.2	42	85	DuPont	1 338	−5.9	16
36	Toshiba	3 527	2.9	17	86	France Telecom	1 307	3.5	73
37	Eli Lilly	3 487	8.7	49	87	Dow Chemical	1 305	4.7	81
38	Canon	3 296	8.7	50	88	Safran	1 297		
39	Bristol-Myers Squibb	3 282	8.1	47	89	Medtronic	1 275	14.7	161
40	Amgen	3 266	16.1	89	90	Unilever	1 269	3.2	72
41	Wyeth	3 257	7.6		91	Continental	1 231		
42	Peugeot (PSA)	3 032	10.5	66	92	Telstra	1 231	36.2	
43	General Electric	3 009	6.6	43	93	Lockheed Martin	1 206	3.9	80
44	NEC	2 995	1.0	14	94	Royal Dutch Shell	1 201	5.5	87
45	Schering-Plough	2 926	11.9	75	95	Yahoo!	1 195	59.6	
46	Finmeccanica	2 858			96	Mitsubishi Electric	1 188	−1.9	44
47	Oracle	2 741	15.6	97	97	Valeo	1 155	11.7	137
48	Fiat	2 545	6.8	54	98	Electronic Arts	1 145	25.5	
49	Boehringer Ingelheim	2 529	10.5	76	99	Applied Materials	1 142	5.6	90
50	Abbott Laboratories	2 506	6.1	51	100	UCB	1 142		

Table 3 BCG Ranking of the Most Innovative Global Companies

Company		Rank R&D expenditures 2007	Primary reason for selection
1	Apple	124	Breakthrough products
2	Google	54	Unique customer experiences
3	Toyota Motor	4	Innovative processes
4	General Electric	43	Innovative processes
5	Microsoft	1	Breakthrough products
6	Tata Group	295	Breakthrough products
7	Nintendo	272	Breakthrough products
8	Procter&Gamble	60	Innovative processes
9	Sony	24	Breakthrough products
10	Nokia	5	Breakthrough products
11	Amazon	114	Unique customer experiences
12	IBM	15	Innovative processes
13	Research in Motion	261	Breakthrough products
14	BMW	25	Unique customer experiences
15	Hewlett-Packard	34	Innovative processes/new business models/Unique customer experiences
16	Honda Motor	20	Breakthrough products
17	Disney		Unique customer experiences
18	General Motors	2	Breakthrough products
20	Boeing	32	Breakthrough products
22	3 M	138	Breakthrough products
26	Samsung Electronics	12	Breakthrough products
27	AT&T	107	Unique customer experiences
29	Audi		Breakthrough products
31	Daimler	10	Breakthrough products
35	Cisco	26	Breakthrough products
38	Siemens	22	Breakthrough products
42	Exxon Mobil	135	Innovative processes
44	BP	173	Innovative processes
45	Nike		Unique customer experiences
46	Dell	166	New & differentiated business models
47	Vodafone	205	New & differentiated business models
48	Intel	15	Breakthrough products

Source: BCG Senior Executive Innovation Survey (2008, 21)

strong innovators based on the opinion survey completed shortly before the financial turmoil (such as Goldman Sachs, Bank of America, ING and HSBC).[10]

Almost two thirds of the manufacturing firms in the BCG 2008 list also appear on the list of the top 100 R&D spenders. Typical names that tend to have "high visibility" as innovators include Apple, Google, Nokia and 3 M. Big R&D money also

---

[10] Maybe the quest for innovation in financial derivatives has led some of these firms to accept risks that later resulted in the financial domino game. The most recent ranking published by BCG in April 2009 has thus excluded Goldman Sachs, Bank of America and ING from the list of the most innovative corporations.

supports high innovation as is the case for GE, Toyota, Procter & Gamble, Boeing, Samsung and a number of other big R&D spenders. A few German corporations with brand names are mentioned (BMW, Audi, Daimler and Siemens) on the BCG list. To summarize, this list tends to focus on US-based corporations and a few Asian firms with strong inroads into the U.S. market.

4 Managing Relational Dynamic Capabilities

Successful innovators are often active in turbulent market environments and need to be strong in their adaptive skills as well as in their abilities to deal effectively with other market participants. They must develop and maintain strong external as well as internal capabilities.

> Dynamic capabilities include difficult-to-replicate enterprise capabilities required to adapt to changing customer and technological opportunities. They also embrace the enterprise's capacity to shape the ecosystem it occupies, develop new products and processes, and design and implement new business models. It is hypothesized that excellence in these 'orchestration' capacities undergirds an enterprise's capacity to successfully innovate and capture sufficient value to deliver superior long-term financial performance (Teece, 2007, 1319f).

The power to shape the ecosystem of innovation and the ability to orchestrate and deal effectively with other innovation partners,[11] will be summarized under the term *relational dynamic capabilities*. Innovation excellence builds strongly on these relational capabilities, including the ability to sense, anticipate and influence trends and investment patterns. David Teece (1986) originally emphasized these external or relational innovation success factors, and recently elaborated on this framework (Teece, 2007). Three major factors are most critical for attaining benchmark performance in highly dynamic market environments.

1. The ability to participate and actively influence standard-setting processes and the major evolutionary pattern of a new product, a new technological field or a new business model.
2. The ability to absorb and control intangible assets to create value in environments where these assets are co-produced and distributed. Managing the dynamics of the appropriation game for intellectual property (IP), and effectively coordinating with other owners of tangible assets have become most critical for success and high performance in new product (or service) markets.
3. Finally, success in markets is often dependent on a number of complementary goods or complementary assets. Strategic control of complementary assets and

---

[11] This includes the whole set of relevant "co-producers" of a complex innovation, including lead customers, innovative suppliers, service providers, regulators, competitors, research centers and universities.

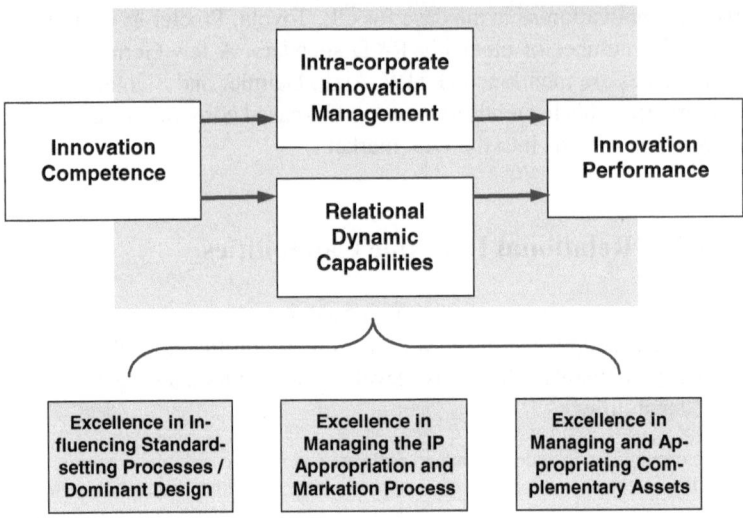

Fig. 4 Success factors related to external dynamic capabilities

the ability to leverage relevant assets owned by other firms often makes the crucial difference between successful and non-successful innovators.

4.1 Standard-Setting Excellence

Most high-tech markets are strongly dependent on the formation of norms and on agreement on standards. In the computer industry, in mobile telecommunication and in factory automation, innovation performance is critically dependent on the ability to influence product standards, interfaces and system configuration agreements. Standards and dominant designs evolve through complex social and political processes. Firms that may be excellent in R&D and technological innovation often lose out by "betting on the wrong horse". This has often been the case for European firms in semiconductors, computers and consumer electronics that were unable to influence the formation process of a strong standard, which, by contrast, was more stringently promoted by corporations in the U.S. or Asia.

Standard setting and R&D activities must be seen as strongly interrelated activities. R&D project selection should be based on criteria such as "likelihood of addressing the winning standard". Successful innovators are thus often forerunners and opinion leaders in informal as well as formal standard-setting consortia. Those who influence winning standards for world markets will attain a much higher market potential and larger sales volumes and will benefit from economies-of-scale, thus enabling themselves to concentrate expensive product-development projects on highly elastic markets allowing for a much higher return on investment.

4.2 Intellectual Property Management

The role of intellectual property in various forms has increased in importance over the last ten years, and markets for intangible assets will be the most important battlegrounds in the future. Pharmaceuticals and Biotechnology are the most outstanding examples where control of patents is often synonymous with blockbuster markets and extremely high profitability. In many other high-tech markets, patents may not be such a powerful weapon, but related forms of IP protection such as trademarks, brands, trade secrets etc. often serve as useful mechanisms for attaining differentiation and innovation success. Various novel forms of IP protection and markation such as labels, domains, internet practices, 3-D brands etc. have been introduced during recent years. As a result, new types of differentiation strategies have evolved and these have extended our repertoire of effective separation mechanisms.

In parallel, Intellectual Property (IP) regulation and litigation in different countries has become a very complicated and often cumbersome process. IP professionals are increasing in numbers (patent attorneys, trademark attorneys and consultants), and this has resulted in secondary markets for IP protection and advice that need to be addressed in parallel to R&D and technology development directed at product markets. As a result, technology-intensive companies must pursue smart IP strategies in combination with their product development efforts. Excellence in managing IP appropriation and markation processes has become the most critical success factor influencing strong innovation performance.

4.3 Managing Complementary Assets

A third, very important approach for managing relational dynamic capabilities builds on interdependencies between products and complementary goods. Successful innovators are smart in leveraging the dynamics of complementary goods. In information technology industries, different building blocs such as servers, PCs, software, various peripheral components as well as networks must be effectively integrated. Firms such as Cisco have grown effectively as solution providers for integrating a whole array of components and complementary assets often provided by other market participants. Another example is the market for Industrial Automation and Manufacturing. The integration of machine-tools requires effective coordination between factory automation systems, hardware, controls, field buses, software, sensors and many other components and technologies. Companies with strong relational capabilities are able to shape the manufacturing ecosystem in their industry.

Successful innovators understand the intricate interdependencies between co-specific assets and are able to benefit from high-margin revenue streams. Classical examples of winning combinations are razors and blades or printers and supplies (ink or toner), where high profit margins are based on (temporary) monopolies for highly specific complements. Firms like Apple, Microsoft and Cisco are extremely strong in exploiting the dynamics of complementary assets. Apple has been most effective in promoting new combinations of audio, video and mobile communication. It has entered the audio market and the mobile phone business, and has been a forerunner for new combinations of so far unrelated market segments. According to the BCG survey, the company ranks as the most admired innovator in the world.

Cisco was established around twelve years ago as a "bridge-builder", offering new and fast combinations as a problem-solver for customers around the world.[12] The company has recently announced that it will enter the market for servers and challenge companies like IBM, Dell and Hewlett-Packard. Following the strategy of combining and leveraging markets for complementary assets, the most successful corporation is probably Microsoft. This company is one of the most valuable corporations in the world and now heads the list of the world's largest R&D spenders. Microsoft is extremely strong in combining software, network solutions and the internet, up to the point where anti-trust authorities become concerned about complementary goods monopolies.

The dynamics of innovation in complementary goods markets in European companies often follow different routes from those pursued by North-American high-tech firms as just described. European firms often have a stronghold in low- or medium-tech sectors (food, machinery, transportation etc.). Still, these more traditional sectors are being transformed constantly through the application of high-tech based complements, as well as through new system configurations and business models. Siemens has effectively combined new solutions in factory automation, controls, software and digital signalling.[13] Orchestrating new developments in complementary goods markets can consist of integrating diverse components into an effective new business ecosystem. New combinations of goods and services can also exploit upstream as well as downstream complementarities. Leveraging upstream complementary assets is a typical strategy that can be observed in solar power and wind energy. While world markets for solar cells and modules become more and more competitive, special machinery and equipment companies in Germany have expanded successfully into a viable global solar supplier market. On the other hand, many companies in the Machinery industry, in Transportation and Medical equipment are effectively promoting new business models for exploiting complementary assets in downstream activities, building on service strategies, customer solutions and integrated lifecycle management.[14]

---

[12] Cisco's logo depicts an artificial variant of the Golden Gate bridge as a symbol for its entrepreneurial role as a bridge builder.

[13] See, for example, the case of Siemens, described by Achatz and Heger in this volume. In the field of Industrial Automation, Siemens focuses its R&D activities on trendsetter projects, for which the corporation can actively influence and shape its ecosystem.

[14] An excellent example can be found in the chapter on innovation at Trumpf by Körber, Buchfink and Völker in this volume.

5 Managing the Dual Cycle of Innovation

The model of innovation outlined above (in Figs. 2 and 3) needs to be complemented by a dual cycle of innovation that we often observe in highly innovative companies. Inducements to innovation come from demanding customers and complementary goods manufacturers as well as from service firms, which often stimulate corporations to think about and develop appropriate new solutions. Certain firms have attained a strong reputation as innovative solution providers in a specific line of business, and they are persistently involved in co-production or in co-innovation activities. Typical examples include

- Cisco, that was founded on the idea of connecting and developing working solutions within the network business. Over the years, the company has developed a unique brand position as innovative problem-solver for large companies worldwide.[15]
- Tetrapak has followed a similar strategy as innovative solution provider to the food and drink business worldwide. This company has developed new packaging solutions and service concepts for the milk and soft-drink business and has consistently expanded into packaging machinery and other lines of business across the world. Tetrapak has also been ranked as a most admired innovator in the ADL (2005) survey.
- A similar strategy was followed by SGL, a specialty carbon and graphite manufacturer from Germany that has gained a good reputation as solution provider to steel and aluminum producers worldwide. In addition, the company builds on core technologies in carbon fiber composites and is constantly developing new technical solutions in automobile companies, for the semiconductor industry as well as for photovoltaic firms.

This dual-cycle strategy of co-producing innovation builds on joint work with demanding lead customers. Pilot projects are used to build a strong reputation as innovative solution providers for similar problems in related industries. Successive projects help to generate a sequence of products and service solutions and a broad base of knowledge about customers and specific applications worldwide. Over the years, the company develops a comprehensive repository of knowledge about customers, problem functionalities and workable solutions.

In Fig. 5, this "secondary" cycle of innovation is outlined in the upper part. Innovation is triggered through projects with demanding customers and the evolutionary path of firms involves (1) reputation building as problem-solver, (2) joint work with a series of lead customers, (3) the continuous generation of new solution and application capabilities used to (4) feed and extend a company-specific repository of knowledge. While this is the predominant innovation cycle for professional service firms, manufacturing-based corporations still rely on the classical

---

[15] See the case study on Cisco in Jennewein, Durand and Gerybadze (2007) and Jennewein (2005).

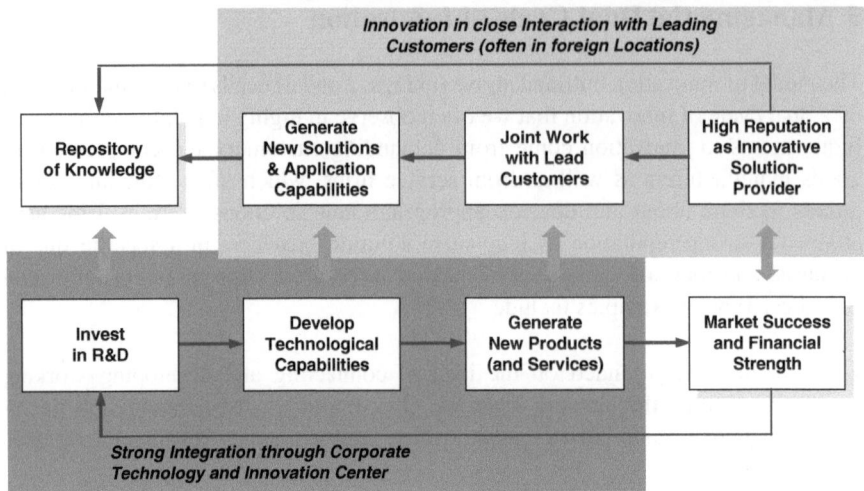

Fig. 5 Effective management of the dual cycle of innovation

cycle of innovation management as outlined in the lower part of Fig. 5, but even these manufacturing firms increasingly emphasize the dual cycle of innovation as a complement.

6 New Forms of Integrated Innovation Management

One of the key themes in most studies on innovation-excellent companies is corporate coherence and strong integration and dedication towards renewal and creative business development. Leading innovators maintain an effective integration between corporate strategy and technology strategy. They strongly emphasize the corporate innovation mission, not just as a statement or lip-service, but as a sustained effort to "push the innovation frontier". And this strong mission must be implemented with similar efforts at all levels of the corporation (corporate units, business units and major product groups). Most companies considered as the admired innovators in their industry follow *one* stringent innovation mission, which is coherently implemented at all levels of the firm.

This innovation mission may often be in conflict with efficiency and with capital market priorities. Financial investors and stock markets put pressure on quarterly earnings and short-term profits, and the CEOs of large corporations are under extreme pressure to optimize financial performance and to focus on balance-sheet optimization. As a result, finance-minded managers and hardcore cost-cutters often rise to the top and leave more long-term oriented projects for innovation and business renewal by the wayside. Sustainable innovation leaders, however, need strong CEOs who actively support the mission for renewal and long-term growth without neglecting their capabilities to deal with short-term capital market pressures. Not surprisingly, some of the most admired innovators are still led by owner managers who can emphasize corporate mission and who serve as an integrator within the company. Take as an example Microsoft, Apple, Samsung, Cisco and Dell, where family owners still play a dominant role. In a similar way, mid-size companies with strong innovation performance in Europe are often influenced by the sustained efforts of a dominant entrepreneurial figurehead.[16]

While CEOs and CFOs have to follow mainly financial missions and need to be supportive of innovation, the prime responsibility rests with the Chief Technology Officer (CTO) or the Senior Vice President of R&D. Most large R&D-intensive corporations have introduced the role of a CTO as a new position. However, CTOs are often not in the center of the power game. There are significant differences with respect to the role, responsibilities and the strength of the CTO across large corporations.[17] After a phase of implementing this CTO role, many companies have even undermined their position, and in a number of large firms, the CTO is no longer a member of the Executive Board. This tendency is in stark contrast to the increasing role played by R&D and the innovation mission. Corporations that remain at the forefront of technological change and innovation need to have a strong CTO, who serves as the network node between the CEO and the major R&D units and innovation projects. Most innovation-excellent companies, such is our hypothesis, need to place strong priority on the CTO function and should select a strong leader to serve as a corporate-wide integrator.

Large investments in R&D are critical requirements for success, but they are not a sufficient condition. R&D budgets are often diluted over too many activities and it is critical to manage an effective mechanism for selecting and implementing large strategic projects.[18] Strong innovators select a limited number of "corporate projects" or "top-priority projects", and these need to be managed and governed appropriately, with top management support, strong project managers and an effective stage-gate process leading to the effective launch of large and growing new business units. A typical form of how effective innovation project management can be implemented within the corporate hierarchy is described in Fig. 6.

The multitude of projects pursued in many R&D-intensive companies and the need to permanently adapt projects to changing technological and market priorities requires the implementation of a strong steering committee as a governance body. Most innovation-excellent companies have one corporate-wide steering group overseeing the major projects and initiatives. In very research-intensive companies, we often find a differentiation between a research committee and an additional group responsible for major business development activities. Corporate research

---

[16] Take, as an example, the innovation success story of Trumpf and the strong role of the majority owner, Mr. Leibinger.

[17] See the Global Benchmark Survey of Strategic Management of Technology that provides a survey of the role CTOs play in American, European and Japanese firms in Roberts (2001).

[18] See Cooper (2009) and ADL (2005a), who emphasize the role of "Strategic buckets" and "top-priority projects".

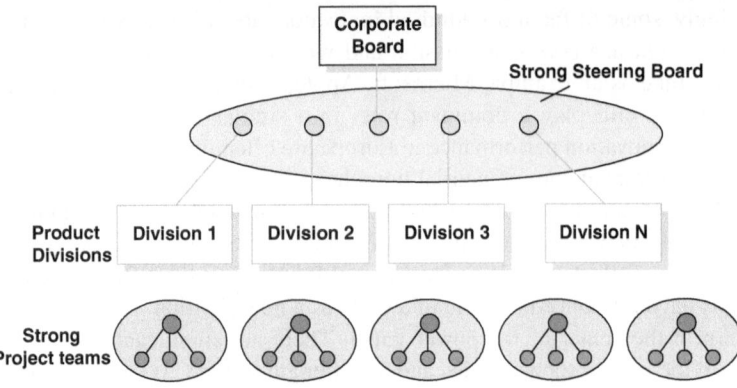

Fig. 6 Organizing for innovation

committees are typically headed by the CTO and involve the major members of the large R&D laboratories. For projects with a more direct commercial orientation, the corporate steering committee is often headed by the CEO himself, or alternatively by the CTO. The general managers of the major business units are regular members of this committee and they will ensure that strategic projects are transferred appropriately to the manufacturing as well as marketing and sales organizations.

7 Integrated Design and Corporate Intellectual Property Management

Another characteristic of integrated innovation management involves design and the integrity of form and content. Many of the leading corporations in product innovation are also leaders in design. An annual contest for industrial design takes place in different countries, and the International Forum on Design (IFF) has published a ranking of the Top 100 corporations, in terms of product design. Many of the corporations considered as leaders in industrial design such as Apple, Samsung and Sony also rank high in terms of R&D spending and innovation excellence. There seems to be a reason behind this phenomenon: top-level innovators have a clear mission and this mission is communicated from top to bottom and across the world: we develop and manufacture new products that customers like and that are impressive at first sight. Such corporations with integrated design gain a strong reputation, earn above-average returns, and these are used to fund the next generation of breakthrough products.

Corporate design is linked to the industrial design of products which again is linked to well-formulated technical and esthetic design attributes. Advanced corporations follow a "gestalt principle", a coherent logic, which applies similarly to the design of office buildings, the layout of equipment in factories and the form of products. Finally, this corporate design idea extends to chosen colors and symbols, which are used in brochures, presentations, business cards etc. This "gestalt principle" is extended to the R&D function and many of the leading innovators develop new products that can instantly be recognized as products of this particular company. Think of Apple or Sony, where similar design ideas are applied for quite a diverse set of products. Similar principles apply for the use of colors, e.g. blue for IBM or magenta for Telekom. This integrated view of innovation and design has become a typical characteristic of large innovation pioneers from the U.S. and Asia, and it is also a typical strategy for medium-sized innovators from Europe, who follow very intelligent design principles.[19] Design and branding help to create uniqueness and company specificity.

Just as technical functions are connected to esthetic functions, different forms of intellectual property, such as trademarks, copyrights, labels, logos, etc. are mobilized together with more technical IP rights (product and process patents). A new field of integrated intellectual property management emerges, which is effectively developed by leading innovative corporations. In contrast to more traditional approaches, where firms employed patent attorneys in their patent department, trademark specialists in marketing, and a number of other specialists in the legal department dealing with diverse aspects of intellectual property, this has now become a much more integrated managerial function that needs to be located very close to headquarters. Corporations like IBM, Siemens or Samsung that manage large patent portfolios are developing a more integrated approach to intellectual property management, and this is becoming a rather large headquarter function, often under the leadership of the CTO. As effective intellectual property management is becoming as important as R&D management, this organizational capability is developing into a major factor for explaining advanced innovation performance.

References

ADL (2005a). Global Innovation Excellence Study 2005. Innovation as Strategic Lever to Drive Profitability and Growth, Arthur D. Little, Rotterdam.
ADL (2005b). Global Innovation Excellence Survey, Arthur D. Little in Collaboration with VNONCW, ADL Rotterdam.
ADL (2006). *Innovation Excellence. Erfahrungen in Innovation Management*. Wiesbaden.
BCG (2007). *Innovation 2007: A BCG senior management survey*. Boston, MA: The Boston Consulting Group.
BCG (2008). *Innovation 2008: Is the tide turning? A BCG senior management survey*. Boston, MA: The Boston Consulting Group.
BCG (2009a). *Innovation 2009: Making hard decisions in the downtown*. Boston, MA: The Boston Consulting Group.

---

[19] On the list of top international design firms, we find companies such as Miele, Hilti, Hansgrohe, Festo and Kärcher, all of which follow integrated innovation and design strategies.

BCG (2009b). *Measuring innovation: The need for action*. Boston, MA: The Boston Consulting Group.
BERR (2008). *The 2008 R&D scoreboard*. London: Department for Business, Enterprise and Regulatory Reform (BERR).
Cooper, R. G. (2009). How companies are reinventing their idea-to-launch methodologies. *Research Technology Management, 52*(2), March–April, 47–57.
Cotterman, R., Fusfeld, A., Henderson, P., Leder, J., Loweth, C., & Metoyer, A. (2009). Aligning marketing and technology to drive innovation. *Research Technology Management*, September–October, 14–20.
Eisenhardt, K. M., & Martin, J. A. (2000). Dynamic capabilities: What are they? *Strategic Management Journal, 21*, 1105–1121.
Gerybadze, A. (2004). *Technologie- und Innovationsmanagement. Strategie, Organisation und Implementierung*. München: Vahlen.
Jaruzelski, B., & Dehoff, K. (2007). The Customer Connection: The Global Innovation 1000, Strategy and Business, Winter 2008.
Jaruzelski, B., & Dehoff, K. (2008). Beyond Borders: The Booz & Company Global Innovation 1000, Strategy and Business.
Jennewein, K. (2005). *Intellectual property management. The role of technology-brands in the appropriation of technological innovation*. Heidelberg, New York: Physica.
Jennewein, K., Durand, T., & Gerybadze, A. (2007). Marier Technologies et Marques pour un Cycle de Vie: Le Cas de Routeurs de Cisco. *Revue Française de Gestion, 177*, 57–82.
Jonash, R. S., & Sommerlatte, T. (1999). *The innovation premium*. Perseus Books.
Knott, A. M. (2008). R&D/returns causality: Absorptive capacity or organizational IQ. *Management Science, 54*, 2054–2067.
Knott, A. M. (2009). New hope for measuring R&D effectiveness. *Research Technology Management*, September–October, 9–13.
Roberts, E. B. (2001). Benchmarking global strategic management of technology. *Research Technology Management*, March–April, 25–36.
Teece, D. J. (1986). Profiting from technological innovation. *Research Policy, 15/6*, 285–305.
Teece, D. J. (2007). Explicating dynamic capabilities: The nature and microfoundations of (sustainable) enterprise performance. *Strategic Management Journal, 28*, 1319–1350.

Technology Strategy for the Corporate Research Center of a Diversified Global Enterprise

Reinhold Achatz and Hans Jörg Heger

Contents

1. Introduction 31
2. The Integrated Technology Company 32
3. Technology Strategy 34
4. Corporate Research and Technologies 36
5. Technology Portfolio Management 38
6. Global Technology Fields 40
7. Technology 41
8. Customer Care 42
9. Global Setup 42
10. Partners 44
11. Organization 45
12. Best People 46

1 Introduction

Siemens[1] is a multinational electronics and electrical engineering company operating in the sectors industry, energy and healthcare. Among other things, its product portfolio is comprised of industry automation equipment, electrical drives, trains, lighting, industry solutions, fossil and renewable energy plants, power grid equipment (transmission and distribution), in-vivo and in-vitro diagnostics such as medical imaging (for example, x-ray, MR, ultrasound, etc.), and medical solutions.

At the time of Siemens' founding, Joseph A. Schumpeter (1939) had not yet created the word "innovation." Yet, Siemens was a truly innovative company right from the beginning, transforming technological inventions into business (such as the

---

R. Achatz (✉)
Siemens AG, Otto-Hahn-Ring 6, D-81739 Munich, Germany

[1] http://www.siemens.com

telegraph in 1847 or the dynamo in 1866). Recent innovation breakthroughs include reliable industrial wireless communication, the full-body MR scan, and the world's most efficient gas turbine.

Siemens was also a truly global company from the beginning. The construction of the Indo-European Telegraph Line from London to Calcutta (1870) gained Siemens an international reputation. In the 19th century, Siemens offices could be found all over the world. Today, Siemens is a global powerhouse with activities in 190 countries.

Both the innovation and global approaches are part of the Siemens DNA. They have formed the basis for Siemens culture and entrepreneurial success for more than 160 years.

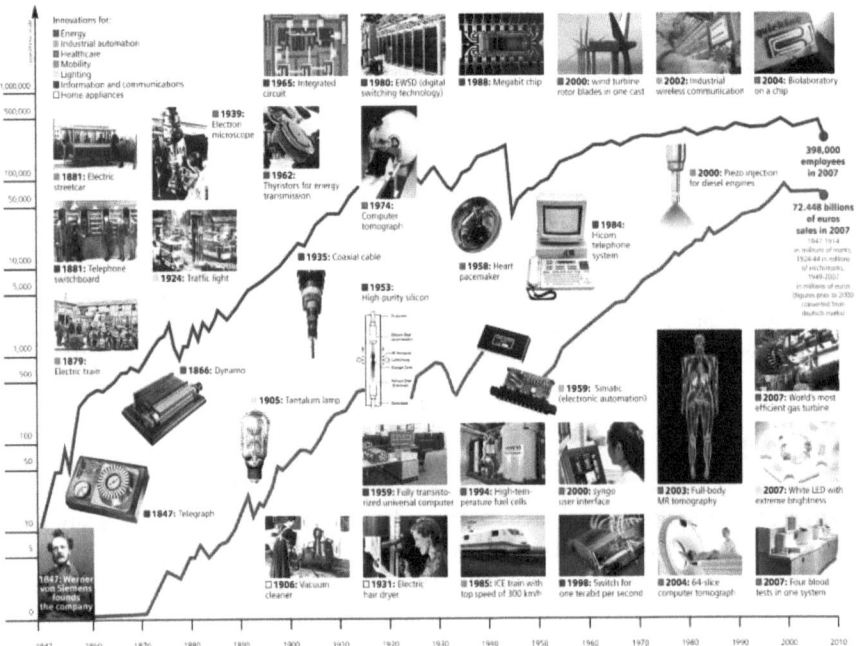

Fig. 1 150 years of innovation at Siemens

2 The Integrated Technology Company

The concept of Siemens is that of an integrated technology company. In this case, "integrated" means that the value of the company as a whole is greater than the sum of the parts ("conglomerate premium"). In contrast to a holding structure, that is limited to financial consolidation of independent groups, the stronger integration of the various groups leads to advantages in e.g. scale effects in supply chain management and also joint technology (pre-)developments and mutual intellectual property utilization. Thus, one important lever for achieving integration benefits is technology, which is a means of taking advantage of managing and reducing complexity and leveraging synergies.

The importance of technology has been underscored by the appointment of a Chief Technology Officer (CTO) to serve as both a member of the Siemens Managing Board and the head of Corporate Technology (CT) His task is to drive technology synergies between business units by orchestrating common technology development and platforms. Best practice sharing between the businesses is also a way of leveraging synergies. One example is the software initiative hosted by Corporate Technology. Siemens is not commonly recognized as a software company because the software is usually just embedded in Siemens products and solutions and not brought to market separately. Yet software is a key element of Siemens R&D, and roughly half of the Siemens R&D budget is invested in software development. The Siemens software initiative facilitates the sharing of knowledge about software development methods and techniques. It also aids in sharing knowledge about organization and workflow, for example, for global multi-site software development.

Corporate Technology (CT) also leverages synergies through activities on the corporate level. In addition to the CTO's office and a corporate intellectual property department, the main parts of CT are the separate units for "R" and "D" – the corporate research center (Corporate Research and Technologies) and the corporate development center.

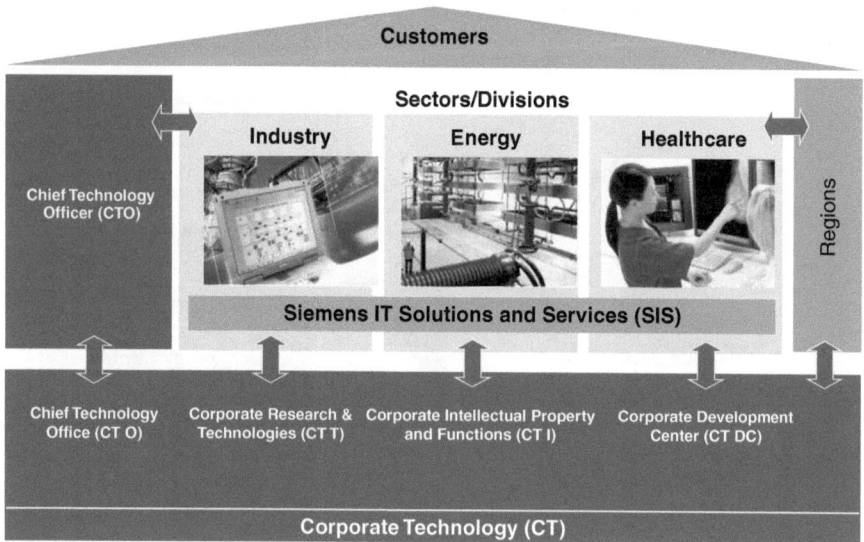

Fig. 2 Corporate technology: Innovation network powers an integrated technology company

Commonly, R&D is combined in one unit, particularly if "R" (research) is below the critical mass to build an own unit. However, looking closer at the topic it turns out that there is much difference between "R" and "D". Research consists of scientific activities that are still considered risky and exploratory. Research is always future-oriented, and its goal is to create prototypes and demonstrators, not fully developed products. This unit primarily employs scientists. Development, in contrast, is always linked to and determined by orders from the business units. Its goal is to help create concrete products, their manufacture and assembly. Development is focused on implementation of specified features, which is why this unit primarily employs "implementers" (e.g. software programmers). Thus, driving "R" in a research focused unit and "D" in a development focused unit allows to optimize the respective activities in all parameters like financing model, processes, global setup, human resources and so on.

In the pages that follow, this article will focus on the "R" unit of CT, Corporate Research and Technologies (CT T). CT T employs 2.250 employees worldwide and roughly 7.5% of the 3.8 billion Euro Siemens invests into R&D are invested in Corporate Research and Technologies.

3 Technology Strategy

The technology strategy for any corporate R&D center needs to be set in the context of the enterprise. While fundamental research seeks to extend the limits of feasibility, industrial research focuses on the topics with the highest business impact.

In this sense, the corporate research organization is the link between science and business, joining the R&D side ("money to knowledge") with the innovation side

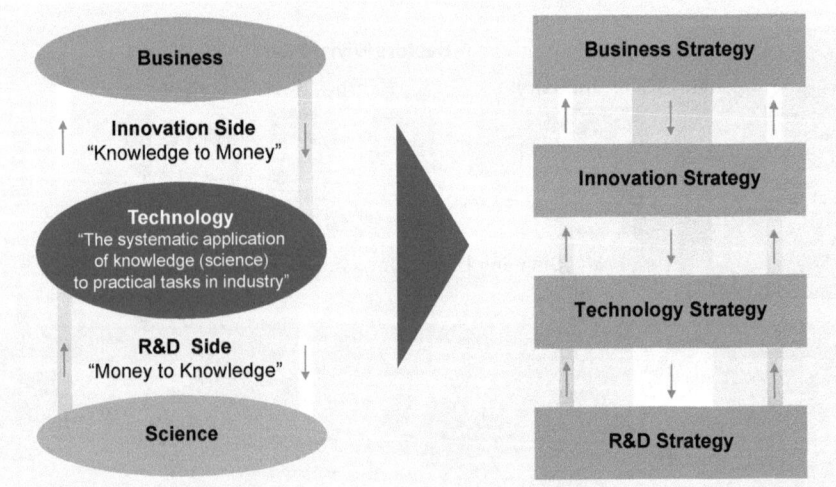

Fig. 3 Innovation strategy is part of Siemens' business strategy

("knowledge to money") through the systematic application of knowledge to practical tasks in industry. Thus, successful technology strategy development considers the innovation strategy of the company, which in turn is derived from its business strategy.

We distinguish three main innovation strategies: "first mover," "trendsetter," and "fast follower." Generally speaking, there is not necessarily a "best" innovation strategy, and no strategy is superior by definition to the others. However, each of these strategies requires distinct strengths. That means that the enterprise must be structured according to the innovation strategy:

- First movers address the early adopters. They need idea leadership and a creative R&D. Since they go beyond limits, they do not necessarily rely on industrial standards. Typical first movers are specialized, small to medium sized companies or even startups.
- Trendsetters combine market and technology power. They need an effective R&D, a broad and strong patent portfolio, and they control the (de facto) standardization processes. Typical trendsetters are technology-driven, medium to large companies.
- Fast followers are cost leaders who need efficient R&D. They license patents and rely on existing standards. Typical fast followers are companies in low-cost countries focused on manufacturing and distribution.

Although a typical implementation of the innovation strategies is given, the decision on the best innovation strategy is not trivial and has to consider the individual boundary conditions for each business. According to the above definitions, the vast majority of Siemens' businesses fall into the category of trendsetter.

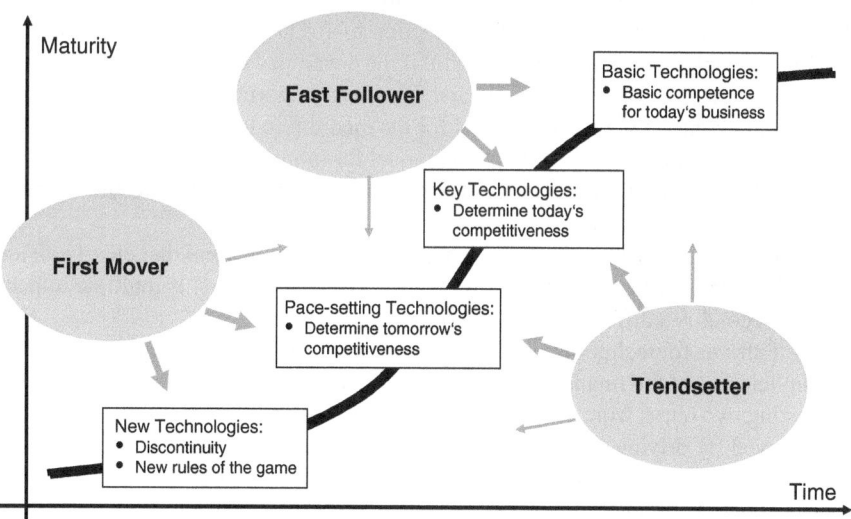

Fig. 4 Three types of innovation strategies and their positioning along the technology lifecycle

The trendsetter innovation strategy is based on market and technology power. Achieving market power is perfectly in line with the Siemens business strategy of being number one or two in all served markets. Technology power translates into a focus on key, pace-setting technologies – and that is the task of the business units' R&D. They hold the majority share of Siemens' R&D investments (>90%). However, remaining in a strong position in pacesetting technologies requires sustainable predevelopment and a research focus that is preparing for the next generation of pacesetting technologies. This is the task of the Siemens corporate R&D center.

4 Corporate Research and Technologies

The corporate research center is dedicated to linking fundamental research in universities and external research institutes to product development in the Siemens business units. Corporate Technology's own research activities focus on those technology fields of strategic importance to the Siemens businesses. The financing model of the research center ensures the link to the businesses, as roughly only one-third of the total budget is corporate funding. Corporate Research and Technologies invests this budget in the development of new technologies that will be important for tomorrow's business and for building up expertise for future topics. This expertise building is further supported by targeted applications of external funding (such as public funding) in all countries where CT T is active. The average share of public funding is 5% of the total R&D budget. The majority of the budget (roughly 60%) is generated by contract research and know-how transfer to the Siemens divisions and business units.

In fact, this financing model is not only a business rule, it is also the main key performance indicator (KPI). Corporate funding invested today must ensure contract research business for tomorrow. The selection of R&D topics driven by corporate funding can only be considered successful if the developed offerings pay off through contract research in the following years. Thus it is important to retain the 1/3 corporate – 2/3 contracted research model for the medium to long term. CT T has been successfully working with this financing model for more than a decade – clear proof that CT T's technology portfolio and R&D services are attractive to the Siemens divisions and business units.

Apart from the financing model, there are two key parameters that clearly distinguish the corporate R&D from the division R&D units. The first is multiple impact, and the second is a different focus in horizons.

CT T strives for technology development that results in multiple impacts – this is an implicit part of the financing model. Every invested corporate euro must result in harvesting two euros from contract R&D at least. Multiple impact technologies can be achieved by driving research topics that are needed by more than one Siemens business. Examples of multiple impact technologies include fundamental technologies in software development, information and knowledge management as well as

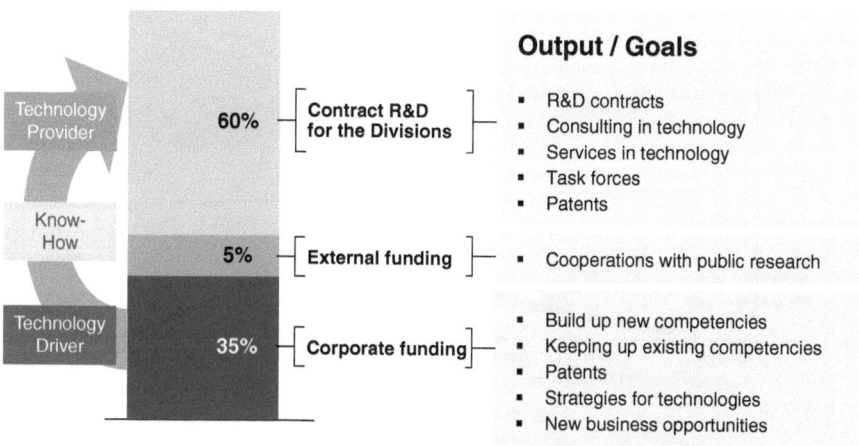

Fig. 5 Financing model for corporate research and technologies

materials science. It also includes more focused topics such as energy management, which has spawned a number of common developments benefiting applications in industry, energy and health. In this context, customer needs are the driving force for technology portfolio management. Another example of a multiple impact technology is a decision support system, which is applied in all three Siemens Sectors. Industry Automation applies self-learning systems for the control of complex plant processes. In the energy domain, power prognosis and forecasting for energy pricing relies on neural networks. And healthcare uses intelligent algorithms for mining large volumes of medical imaging data. From the viewpoint of the integrated technology company, the multiple impacts of CT T technologies are part of the payoff of technology synergies.

While businesses are rather hesitant to spend R&D financial resources on predevelopment topics (which are by nature risky and poorly defined), the corporate R&D unit can take that risk within its corporate funding. This leads to a different outlook and ensures the expansion of knowledge that's required for future product development.

R&D efforts can be roughly divided into three horizons: horizon 1 is product development for today's products; horizon 2 is development and predevelopment for future products; and horizon 3 is predevelopment and research that cannot yet be assigned to concrete products and product lines. The R&D units of the businesses focus on horizons 1 and 2, while the corporate R&D unit focuses on horizons 2 and 3. Consequently, while every individual business would hardly be able to drive topics in horizon 3, a corporate R&D can do this for strategic multiple impact topics. This in turn ensures that the company remains an experienced developer in pacesetting technologies, and at the same time guarantees the success of the "trendsetter" innovation strategy.

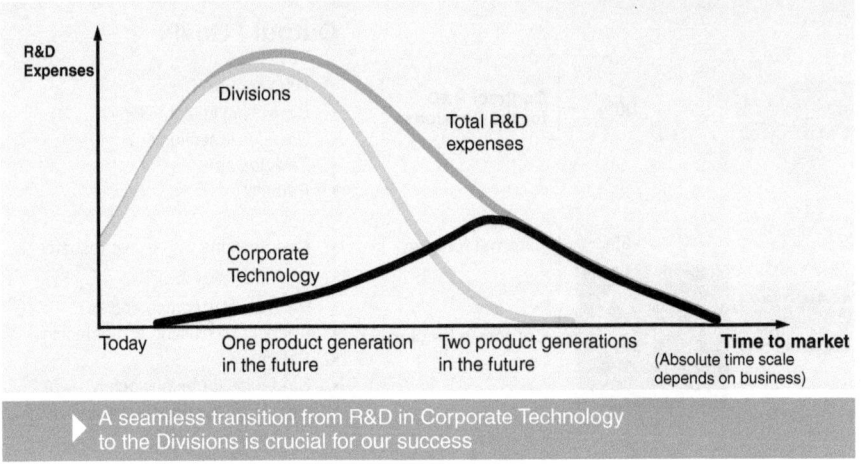

Fig. 6 Time horizon of R&D activities for divisions and corporate technology

5 Technology Portfolio Management

Successful technology portfolio management always balances two approaches: a *top-down approach*, which derives the future technology needs from trends and scenarios, and a *bottom-up approach*, which leverages the knowledge, ideas and the network of the laboratory researchers.

The top-down approach begins with the big picture, evaluates the influence of global megatrends on the served markets and the Siemens businesses, and subsequently derives the respective development requirements. To systematically drive this process, Siemens Corporate Technology has developed a method called "Pictures of the Future."

An extrapolation from current business trends and scenarios using roadmaps is a common method for technology development. However, this approach tends to focus on known fields, which leads to already-known streets. To identify additional, yet unknown fields, the Pictures of the Future method adds scenario development and performs a "retropolation" from the scenarios. Using both the extrapolation and retropolation techniques leads to a comprehensive view of the technology portfolio, and allows identification of white spots in the current technology portfolio. In this sense, the top-down approach ensures that the technology portfolio is customer-oriented and complete.

Just to give one example, in the picture of the future "energy" made in the beginning of this decade, the requirements for future energy grids were derived with the extrapolation-retropolation method. While large off-shore windparks were still a vision, the pictures of the future were already dealing with the consequences of a high share of fluctuating energy sources. A "dynamic load control" based on "sensors and automatic decision-making" ensures "safe operation of high voltage

Technology Strategy for the Corporate Research Center

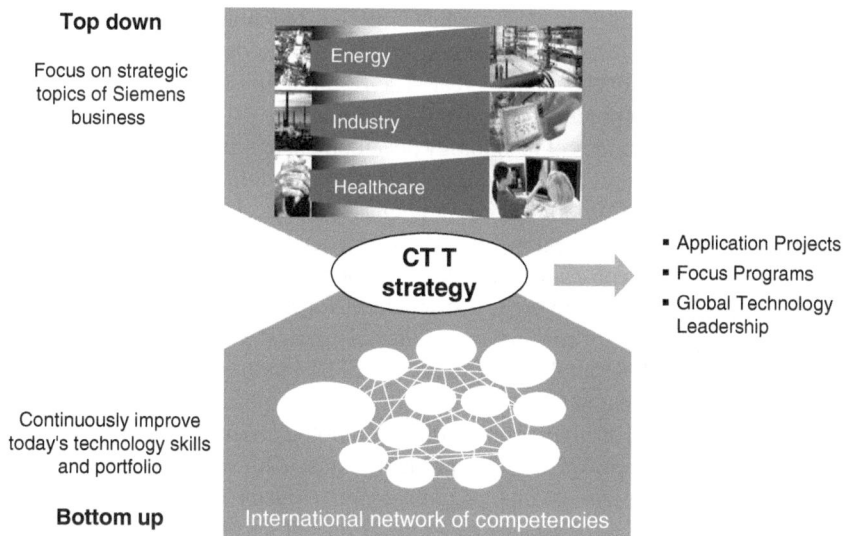

Fig. 7 The strategy process for corporate research and technologies

networks at the physical limit". Today, this sounds quite familiar and is known as the "Smart Grid" – of course, the pictures of the future did not anticipate this term, but much of its meaning. As a consequence, Corporate Research and Technologies is driving this topic in a big "lighthouse project" Smart Grid today.

In the bottom-up approach, a hierarchical, or "fractal" process ensures the local collection and evaluation of ideas on the technology field and research center levels,

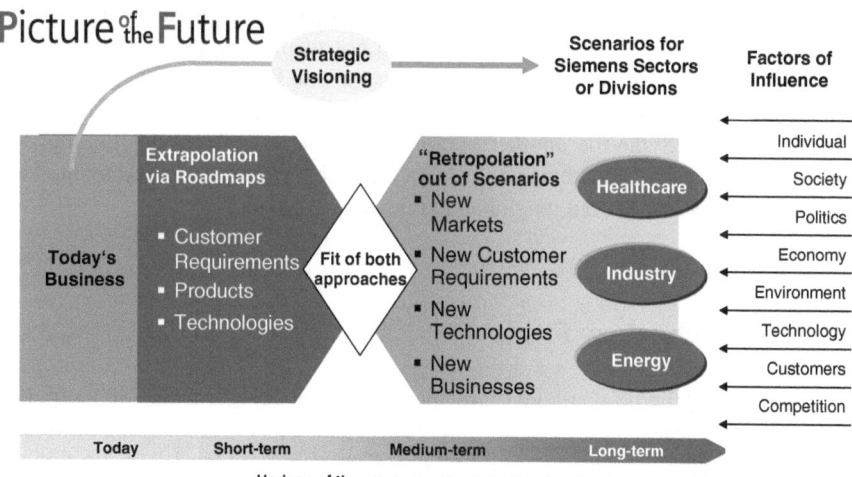

Fig. 8 Strategic planning: The combination of extrapolation and retropolation leads to the picture of the future

the main organizational units of Siemens Corporate Technology. Finally, top management of the research center meets once a year for a two-day retreat to discuss the technology portfolio for the coming year. The result is the set of technology fields currently driven by corporate research and technologies.

The increasing dynamics of technology development as well as business portfolio changes require measures to change the technology portfolio fundamentally, flexibly and fast. Siemens Corporate Technology has two "accelerator" units to achieve the necessary portfolio dynamics. First, there is a need to screen new technologies developed by start-up companies outside of Siemens. There is also the need to support the technologies that are a good strategic fit for Siemens, while leaving open the option to spin in the ideas and teams into Siemens R&D at a later stage. This accelerator is called "Technology to Business (TTB)" and it is situated in Berkeley in the U.S. and Shanghai in China.

An example for a spin-in transaction is the technology basis for Siemens Industry Automation's "Scalance W" WLAN product series. Although cableless, WLAN based communication has obvious advantages for industry automation, it has not been applied because of the lack of real-time ability of the standard WLAN protocol and equipment. With the technology-to-business unit a startup in California was supported in developing the idea of reserved bandwidth for time-critical communications. The "spin-in" of the successful developments then led to a market leading Siemens Industry WLAN product family.

Second, a complementary accelerator handles any idea that is developed within Corporate Technology but which turns out to have an insufficient strategic fit with the Siemens businesses. One of the risks of research is that some topics may end up failing completely. Others might become valuable in certain business environments, but not for the Siemens businesses themselves. To leverage the value of these topics, they are evaluated in terms of a potential spin-off, which Siemens may choose to further support. More than a dozen companies have already arisen from these kind of Siemens R&D developments that did not fit into the product portfolio of any Siemens unit. Recent spin-off transactions include Pyreos, which develops and produces simple, cost-effective infrared sensors based on a highly innovative thin film technology. Symeo is another spinoff that focuses on local positioning systems based on radar measurements. This accelerator is called "Siemens Technology Accelerator" and carries out its global activities from Munich.

6 Global Technology Fields

The "Global Technology Fields" are the main portfolio elements of Corporate Research and Technologies. The corporate R&D unit evolved from being renowned research institute to adding a strong link to the businesses in the 1990s through the introduction of the current financing model. From 2000 on, there was a focus on the Regions and an "internationalization initiative" that further leveraged their advantages by widening their scope. While there is a long tradition of partnering, and so-called "open innovation" has naturally been part of our R&D approach, we have recently put even more emphasis on the topic. The partner portfolio was added as the fourth key management task. That change ensures strategic partner selection and leverages the full partner potential. In sum, the Global Technology Fields today are responsible for technology planning and road-mapping. They are the customer interface, they drive their topics globally (with a single person responsible worldwide) and they orchestrate the partner network in their respective fields.

An example for a global technology field that is active in all regions is "software architecture and platforms". The global technology field is segmented into local programs. This local presence ensures the vicinity to the "customers", here: the Siemens business units with software development in the respective region. Furthermore, local technology partnerships with universities and research institutes can be exploited. The global technology leader ensures global coordination, he allocates the budget and sets the targets for the programs in each region.

7 Technology

The scope of the Global Technology Fields ranges from materials to software, including power components and sensors, information management, imaging and security – in short, all technologies that ensure Siemens leadership when it comes to tomorrow's products, solutions or services.

Generally, one of the challenges in an applied research organization is to link the relevant technologies to applications. To help meet this challenge, the offerings of the Global Technology Fields are specialized to different degrees in terms of technology versus application. For example, the GTF "Microsystems" is more technology-oriented, driving a number of topics based on common technology developments. In contrast, the GTF "Oil & Gas Field Development Technologies" has a clearly defined application focus. As a result, its offerings are more tailored to specific application needs. Basic R&D activities, such as the "Microsystems" example described above, however, are preferably done in cooperation with other GTFs that are already active in the respective fields. This kind of mutually beneficial GTF relationship is an important aspect of CT internal value chains; the collaboration of GTFs is essential to leveraging the full value of a comprehensive corporate R&D center.

It's also important to note that the portfolio also includes processes. Innovation is usually associated with "new features." Today, however, we know that there is significant potential in process and business model innovations. Siemens Corporate Technology covers the complete spectrum of innovation, including the PLM (product lifecycle management) process. This interrelationship between process consultants and technology consultants leads to efficient solutions for a wide range of challenges in the Siemens businesses. As an example, PLM process consultants from CT T are requested to support in a design-to-cost action in a Siemens business unit. During this consulting it turns out that the most significant cost driver is one of the used materials. In the next step, material experts from CT T are involved to investigate substitutional materials and to leverage the identified cost reduction potential.

8 Customer Care

When it comes to customer-oriented portfolio management, choosing the right technology fields is the strategic basis for a strong link to the businesses. Furthermore, the customer care of Corporate Research and Technologies ensures the transition of predevelopments into contract research for the Siemens businesses. From the point of view of corporate research, Siemens business units are their customers.

Customer care is one of the Global Technology Leaders' important activities, globally driving customer relations in their respective fields. To sell R&D capabilities, the "salesperson" has to be an expert in the field – someone who is trusted to solve a given problem. That's why the technology field management itself is responsible for pursuing customers.

In addition, account managers care for the internal customers of CT T. They devote themselves to their customers, spending the majority of their time with them, while accepted as trusted advisors to the business units. They know the main problems and action fields of the customers' R&D units and transform these into tailored CT T offers. Working together, the Global Technology Leaders and Account Managers are the successful customer interface of Corporate Research & Technologies.

9 Global Setup

In the past years, globalization more and more affected also Research & Development. The main drivers for internationalization of Corporate Research and Technologies are proximity to the R&D and production of the Siemens business units and networking with the scientific community.

The proximity to R&D and production is important because 60% of the budget of Corporate Research and Technologies has to be funded from the Siemens businesses. The R&D units of Siemens businesses in e.g. China would hardly use R&D support from Germany. Thus, a regional unit in Beijing was established to provide R&D services in China.

The networking with the scientific community gives access to the best brains worldwide and is the second main driver for R&D internationalization. Next to cooperation in projects, this is also an important factor in recruiting a new generation of researchers for the central R&D lab. Further drivers for internationalization are corporate citizenship, cultural diversity, flexibility and costs. However, it is important to emphasize that costs are not an important driver for R&D internationalization – particularly because excellent, state-of-the-art researchers do have global employment opportunities and thus are not "low cost" anywhere.

The enormous progress in international build up is reflected in comparison of the employee distribution in 2003, where 87% of the employees of Corporate Research and Technologies were in Germany, with the share of 2008, where for the first time more than half of the employees (52%) were outside of Germany. On the first sight, this looks like a threat for the German labs, but looking at the absolute numbers it turns out, that the number of researchers in Germany remained on the same level - but the growth took place internationally. This clearly shows, that not driving internationalization would mean to miss an opportunity to generate value for the company.

At Siemens, all Regions have equal status. In contrast, many R&D organizations have globalized in line with a strong central headquarter mindset, which means that headquarters always takes the technology lead. In contrast, from the beginning of its globalization process, Siemens Corporate Technology has intentionally given the responsibility for technology to the Regions. This in turn, has made the Siemens R&D laboratories attractive to ambitious researchers in their respective locations.

Currently there are global technology leaders in all Regions. German researchers, for example, are coordinated by a Global Technology Field leader from India and vice versa. Corporate Research and Technologies has locations in nine countries around the world. While global coordination and responsibility is important, it's also important that a local manager be as independent as possible in order to

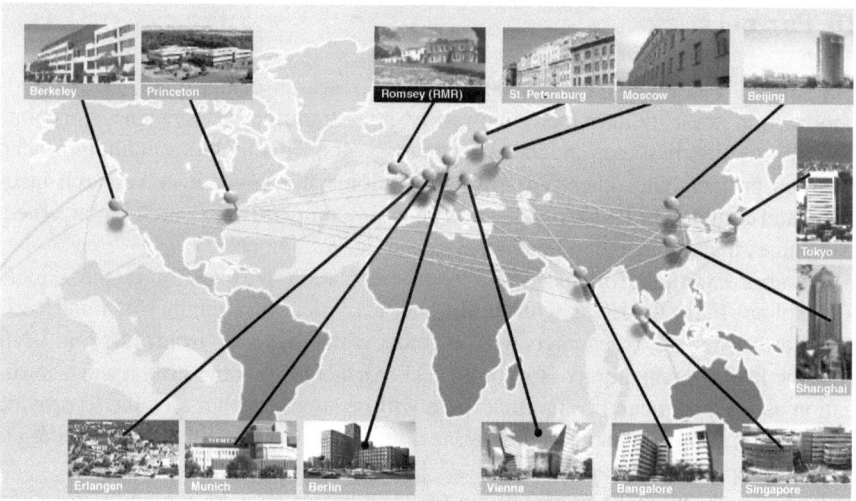

Fig. 9 Corporate research and technologies: Present in all leading markets and technology hot spots

take advantage of local opportunities. This is particularly true for emerging market locations.

R&D for emerging markets presents different challenges from premium product development. Successful products for emerging markets are not just reduced versions of high-cost products with stripped feature sets. Rather, they are specifically designed to be "good enough" at the lowest possible costs. Within Corporate Technology's Global Technology Fields, the R&D challenge for emerging markets is described by the term "SMART," which stands for "Simple, Maintenance-friendly, Affordable, Reliable, and Timely to Market." Searching for SMART solutions means searching for new simple, inexpensive and adequate ways to do the things. And these new ways offer new business potential.

An example for SMART developments is the "Rural Healthcare" project driven in CT China. In rural hospitals, there is not only the question of available and affordable equipment; there is also a question of trained personnel. Thus, simply downgrading a premium segment solution would not help. A more radical SMART solution will focus on the most relevant use cases, develop a simple solution to address only these use cases and add intelligent assistants to allow personnel with basic medical knowledge to operate the equipment. Thus, the SMART solution is a complete different approach from the beginning.

Once these SMART solutions have been successfully implemented in an emerging market, they might graduate to implementation in products in the premium segment. This could potentially disrupt existing premium businesses. Being present and active in emerging markets is thus not only a question of addressing additional markets, but of being a sustainable trendsetter in the global market.

10 Partners

The Global Technology Fields coordinate the research center's network of partners within their field. Open innovation is a tradition at Siemens Corporate Technology that has been practiced for many years. This includes publicly funded research projects and other types of cooperation with universities, research institutes, and companies. It also includes, as far as precompetitive research is concerned, companies that are active in fields that are similar to Siemens'.

Another example for successful open innovation partnership is the Global Technology Field Oil&Gas with its Russian activities. In a number of joint R&D projects the Siemens researchers work together with colleagues from Gazprom VNII Gas, the R&D organization of Gazprom. This partnership is obviously a win-win situation as Gazprom can be sure that there will be suppliers that are able to provide the required solutions and Siemens gains domain know-how and insight in oil&gas field development at Gazprom.

Today, web 2.0 technologies allow for even closer cooperation and partnering, internally as well as externally, and create new opportunities by leveraging the partnership network.

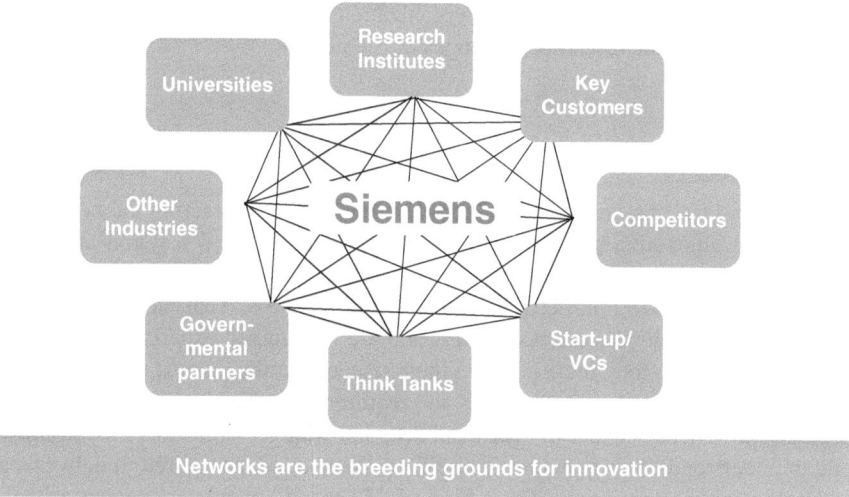

Fig. 10 The partner portfolio: Create global innovation networks

11 Organization

Progress in research is reached and measured in projects which transform virtual roadmaps into tangible deliverables. Because research is risky work with unpredictable results, project milestones are set that can either prove the project's success or suggest that that the project needs to be closed. Both results are valuable. Accepting these dynamics is important to providing researchers with the motivation they need to risk going down a challenging path that nobody has tried before.

To fully leverage the advantages of dynamic projects, Corporate Research and Technologies works with a project organization within a matrix of Global Technology Fields and Departments. That keeps the technology and project responsibility in the Global Technology Fields. Meanwhile, disciplinary leadership takes an arm's-length approach towards departments in the researchers respective regions. Thus, the Global Technology Fields source their knowledge and infrastructure needs from departments, the heads of which manage the people and their expertise.

Every global technology field is working in this setup, just as an example the comparably young Global Technology Field (GTF) "Renewable Energy" is described in more detail. The Global Technology Leader for this GTF is in the Indian branch of the research center. A number of the researchers working in this GTF are from departments in India, but there are competencies required, that already exist in the global corporate research network and there is no sense in duplicating these competencies in India. Thus, the Global Technology Leader asks for resources from a department in Germany. As the Global Technology Leader has the budget responsibility for his field, he/she is free to spend this budget for the best resources worldwide. In this way, the project organization ensures to have the best global teams in place for the research challenges of Siemens.

12 Best People

Last but not least, research is about people. The goal of "attracting the best brains in the most innovative regions of the world" is part of CT T's vision and is the guideline for its recruiting activities.

Siemens CT is the link between research and business. That means that on the one hand, the employees are researchers and are recognized experts in their fields. On the other hand, they must perform their research in the context of the Siemens businesses, understanding its domains and requirements. The ideal target employee profile for Siemens is a mix of broad and deep knowledge. Broad knowledge allows one to understand the big picture, while deep knowledge is required to deliver real technological breakthroughs. These kind of "T-shaped" people are needed when it comes to working at the intersection of research and business.

Our employees' other capabilities include technological leadership, business and industry expertise, international experience, and networking and communication skills. Given these highly skilled people, corporate technology also serves as a source for technology management candidates for the Siemens businesses. In addition to this close link between Corporate Technology and the businesses, there is also an expectation that careers will cross unit borders. These practices lead to the most efficient way of creating a business impact from research results. Since people are the repositories of knowledge, the most efficient way to transfer knowledge is to transfer people. The transfer of employees into Siemens business units is also an important measure of the success of a research center, since it ensures close cooperation and networking with the business.

Apart confidentiality and IPR issues (which cannot always be fully ensured by external partners), this link with businesses is what differentiates Corporate Technology from other external research institutes. An external research services provider aims to keep his best people, since they guarantee his business for the following year. In contrast, Corporate Technology has the unique goal of transferring the best researchers into the businesses.

R&D Internationalization in Multinational Corporations: Some Recent Trends

Heike Belitz

Contents

1 Introduction 47
2 Internationalization of R&D Driven by M&As 49
3 Increased Significance of Asset-Seeking Motives 52
4 Similar R&D Behavior of Domestic and Foreign Firms in Germany 53
5 Expansion of German Companies' R&D Abroad in the "Fifth M&A Wave" 54
6 Foreign R&D of German Companies Mirrors Technological Strengths at Home . . 56
7 R&D Abroad is Concentrated on Western Europe and the United States 61
8 Summary 62
9 Implications for German Innovation Policy 63
References 64

1 Introduction

The internationalization of Research and Development (R&D) in multinational corporations poses a challenge not only to the management of these companies but also to the national technology policy. A particularly controversial topic in the political arena concerns the global competitiveness of locations for research.

What are the driving forces of the internationalization of R&D in multinational companies? What are the motivations of multinationals in expanding R&D and other knowledge-based activities in their affiliates abroad? The question has prompted much debate in recent years.

There are two main reasons why companies internationalize their R&D activities:

H. Belitz (✉)
German Institute for Economic Research Berlin, Mohrenstrasse 58, D-10117 Berlin, Germany

- To penetrate foreign markets by adapting their products and processes to local conditions; and/or
- To take advantage of foreign expertise.[1]

To penetrate new markets companies must adapt their products to regional needs or even develop special products, as preferences in demand vary from country to country. The need to adapt products to address the special demands of international customers often leads companies to invest in R&D abroad. Some foreign customers even expect their suppliers to conduct development activities in close geographic proximity, to ensure a quick reaction to new product requirements.

A percentage of research and, to a greater extent, development activities are undertaken in order to upgrade production processes in foreign plants and to tool production lines for new products. In the case of such market-driven R&D activities, knowledge is primarily transferred from the company's home country abroad. However, in this internationalization scenario, the R&D conducted in a company's home country remains the most important source of innovation. Additional foreign research benefits the company at home to the extent that it serves the purpose of expanding in foreign markets.

Companies are better able to acquire new technical expertise from research institutions and universities when they are in close proximity geographically to one another. In order to absorb existing knowledge in foreign countries, companies must be embedded in local research networks with their own research departments. The opportunity to tap the know-how of scientific and technical experts in foreign countries is an important motivation for conducting R&D activities abroad.

An internationalization strategy based primarily on the acquisition of knowledge from foreign countries carries latent risks for domestic research, as such a strategy may result in the reduction of domestic research capacities, which in turn weakens a country's ability to absorb new knowledge from abroad.

This chapter begins with a short overview of recent trends of R&D internationalization in developed countries for which aggregate data are available. The driving forces of the internationalization of R&D and the motivations of multinationals in expanding R&D activities in their affiliates abroad are interpreted in the context of internationalization of production in multinational companies in general. Foreign direct investment is concentrated on developed countries and mainly driven by Merger and Acquisitions (M&A). In these countries the horizontal model of international division of labor is expected to dominate, in which companies conduct similar activities and produce similar products in different locations with the same factor endowment. Therefore our main hypothesis on the internationalization of R&D is the convergence of the R&D behavior of foreign and domestic firms in developed countries.

---

[1] See, for example, Les Bas and Sierra (2002) and Patel and Vega (1999), Kuemmerle (1997) and Dunning and Narula (1995).

We present some results of our studies of the R&D behavior of multinational companies for Germany – both inward and outward – based on aggregate data of the German R&D survey and on patent data for German multinationals and their most important foreign competitors.

After a summary of our results we close with some conclusions for national technology policy.

2 Internationalization of R&D Driven by M&As

The R&D data show that in important home countries of multinational companies, private sector R&D is to a large extent undertaken by foreign firms. In countries for which data on both directions are available (the United States, Germany, Japan, and Sweden), the R&D expenditure of foreign-owned multinationals and of their own multinationals abroad is balanced (Table 1).

The degree of internationalization of a production and research location, both externally and internally, is influenced decisively by a country's size, as well as by the dynamism and openness of its goods and capital markets. Small countries such as Switzerland and Sweden are, as a rule, more intensely internationalized than larger countries, such as the United States. MNEs in smaller countries have to locate more of their production and R&D abroad if they are to make use of the

Table 1 Internationalization of R&D and production in selected industrialized countries 2005/2006

Country	Business R&D expenditure			Level of foreign direct investment	
	Total	Share of affiliates abroad	Share of foreign firms	of foreign countries (inward)	in foreign countries (outward)
	2006	2005	2006	2005	2005
	In million. PPP $	% share	% share	% of GDP	
United States	247669	11.5	13.8	13.0	16.4
Japan	107078	2.9	5.4	2.2	8.5
Germany	46630	23.7	26.4[1]	18.0	34.6
France	26186	–	20.8	28.5	40.5
United Kingdom	21943	–	37.8	37.1	56.5
Canada	12754	–	33.5[2]	31.6	35.3
Sweden	8847	43.1[3]	42.3[1]	47.8	56.5
Switzerland	5515[2]	99.4[2]	–	46.9	107.4
Finland	4239	–	17.0	27.3	38.5

[1]2005; [2]2004; [3]Manufacturing only.
Sources: OECD, UNCTAD, U.S. DoC, SV-Wissenschaftsstatistik; own calculations.

advantages of scale and the many stimuli to innovation in international markets. Another factor affecting the higher degree of internationalization of multinationals from smaller countries is the need to master an increasing range of potentially useful technologies (Granstrand, Patel, & Pavitt, 1997).

Studies into the relationship between trade and foreign direct investment in industrialized countries confirm a strong positive correlation and mutually reinforcing interrelationships between the two forms of internationalization. The economic significance of foreign companies in a given country can be measured by foreign direct investment as a share of gross domestic product (GDP). Countries with a high weighting of foreign direct investment and direct investment abroad often also demonstrate a high share of R&D expenditure on foreign multinational firms and on affiliates abroad (Table 1). In the case of Germany, measured against the weighting of foreign direct investment, the share of research activities by foreign companies is in an internationally typical relation.

In Germany, UK and Sweden and to a lesser extent also in France, the share of affiliates under foreign control in total business sector R&D expenditure substantially increased between 1995 and 2001. By contrast, this share remained static in the United States, Canada and Finland (Fig. 1). After 2003, this is more or less true for all developed countries for which data are available, and can be seen as a temporary break of the internationalization of R&D in multinational companies.

Between 1995 and 2001, the increase in R&D expenditure of foreign companies in Germany, the United States and UK was higher than that of total business R&D. In this period, the internationalization of R&D was accelerated. After 2001, the growth rates of business R&D in these countries were reduced. As a result, the

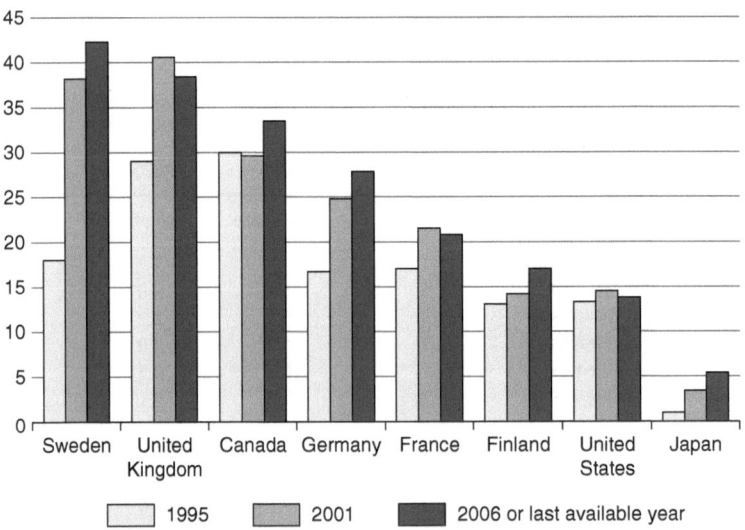

Fig. 1 Share of foreign companies in total business R&D of selected countries, 1995–2006. Sources: OECD, U.S. Department of Commerce, ONS UK, SV-Wissenschaftsstatistik; own calculations

Table 2 Total R&D expenditure by all and foreign companies in Germany, the United States and UK 1995–2005

	1995	2001	2005	2001	2005
	million PPP $			1995 = 100	2001 = 100
Germany					
All companies	26138	38041	43303	146	114
Foreign companies	4365	9434	12038	216	128
United States					
All companies	129830	202017	226159	156	112
Foreign companies	17240	29247	31099	170	106
United Kingdom					
All companies	14623	19121	20512	131	107
Foreign companies	4387	7759	8025	177	103

Sources: OECD, U.S. Department of Commerce, ONS UK, SV-Wissenschaftsstatistik; own calculations.

growth rates of R&D of foreign and of indigenous companies were now roughly the same (Table 2). All this provides further indications of an internationalization moratorium in business R&D. In contrast, a survey of the largest investors in R&D, undertaken at the beginning of 2005, suggests that the pace of internationalization in R&D is accelerating (UNCTAD, 2005).

Although there is no information in detail, the development of cross-border M&As suggests that the growth of R&D of foreign firms at the end of the century was mainly due to the acquisition of firms in research-intensive branches and only in a very few cases due to the establishment of new R&D sites of foreign firms (Greenfield investments).

The internationalization of companies in industrialized countries mostly takes the form of shareholdings in companies by foreign investors, and M&As. Cross-border M&As are the main driving force behind the surge in FDI, recently accounting for almost 90% of total FDI inflows in developed countries (Barba-Navaretti & Venables, 2004).

There is an important stylized fact as to the development of M&A activity over time: They come in waves. It is common to distinguish between five merger waves during the 20th century. The fifth and last wave started around 1995 and ended in 2000 with the collapse of the "New Economy" (Brakman, Garretsen, & van Marrewijk, 2006).

In Sweden, UK, Germany, and France the share of affiliates under foreign control in total business sector R&D expenditure strongly increased between 1995 and 2001 (Fig. 1). Not only in these countries but also in the United States and Canada the annual R&D expenditure of foreign companies has much more increased during the fifth M&A wave between 1995 and 2001 than in the following period (Table 3). This should provide an indication that cross-border M&As are the main driving force behind the internationalization of R&D in multinational firms. In most cases, the starting point of an internationalization strategy for R&D in a company is the acquisition of a firm abroad with its own R&D activities.

Table 3 Value of cross-border M&A in relation to business expenditure on R&D of selected seller countries

	Value of cross-border M&As, Cumulative 1995–2000	Business expenditure on R&D (BERD) 2001	Relation cum M&A/BERD
	In billion US $	In billion PPP $	
Germany	392	38	10.3
United Kingdom	620	19	32.4
United States	1090	202	5.4
Sweden	134	8	16.8
Japan	81	77	1.1
France	247	23	10.9
Canada	149	12	12.7
Finland	25	3	7.8
Selected countries	2738	381	7.2

Sources: UNCTAD, OECD; own calculations.

3 Increased Significance of Asset-Seeking Motives

More recent theoretical and empirical studies on the internationalization of multinational firms in industrialized countries assume the dominance of the horizontal model of international division of labor, in which companies conduct similar activities and produce similar products in different locations with the same factor endowment.[2] For horizontal firms, broadly defined trade costs constitute a location advantage, encouraging production abroad. Horizontal motivations for foreign activities are the need to place production close to customers and avoid trade costs. Skill differences between the international locations are relatively small. The dominance of the horizontal mode is consistent with the above mentioned facts of large volumes of cross FDI among the rich countries of the world and that the bulk of that FDI takes place through cross-border M&As.

In the literature, a "dichotomous set of motives" for the internationalization of R&D can be found, namely, that firms invest in R&D abroad either to exploit their existing stock of knowledge in foreign environments or to augment their knowledge base by gaining access to foreign centers of excellence. In "asset-exploiting" strategies they adapt products and processes to host markets whereas the need to acquire new knowledge assets is the essential motive of so-called "asset-seeking" strategies.

What can we expect to be the dominant motivation for conducting R&D abroad in a world of horizontal multinationals? These companies produce similar goods and services in multiple locations with similar factor endowments and therefore should to a growing extent demonstrate similar R&D behavior in their R&D locations in both home and host countries. In the horizontal model, market size and demand in the host country provide strong incentives for multinational activity. Accordingly,

---

[2] See, for example Markusen (2001/2002) and the literature cited there, Bloningen, Davies and Head (2002).

the innovation impulses generated by the market in particular should determine companies' R&D. This holds true for both domestically owned and foreign-owned companies. Thus, one consequence for R&D is that it may be dispersed globally as well and done near to different production sites if local market adaptations are needed, because of different local preferences or local regulations.

However, if knowledge generation is very much regionally concentrated or if there are locations with clear competitive advantages as to knowledge generation, R&D – and especially the research component – even in the horizontal mode may be done at one or a few selected locations (e.g., at headquarters or at the sites with the best knowledge assets) and then transferred to the global production sites. Such an "asset-seeking" strategy is becoming more and more important in those areas that are very knowledge-intensive and where a global specialization needs to be exploited. Empirical evidence suggests that this type of R&D is gaining importance in recent years (e.g. Kuemmerle, 1997; von Zedtwitz and Gassmann, 2002; Ambos, 2005; Hegde and Hicks, 2008).

Nevertheless, business R&D is dominated by the *development* of new and improved goods, services, and processes. Development applies scientific knowledge to the creation of specific marketable products. Despite the growing extent of "asset- or technology-seeking" strategies in foreign R&D locations, the market-oriented "capability augmenting" strategy is expected to stay the main motive for R&D abroad.

4 Similar R&D Behavior of Domestic and Foreign Firms in Germany

Aggregates related to R&D activities of foreign affiliates presented here cannot make a distinction between the two main strategies for R&D abroad. However, the assumption on the convergence of R&D behavior in domestic and foreign-owned firms in Germany can be supported by aggregate R&D data attributed to industries.

Since 2001, every fourth € invested in R&D in Germany has been spent by foreign firms, and one-quarter of those employed in R&D were working in these companies. With the expansion of R&D by foreign companies in Germany at the end of the 1990s, the sector-specific structures of R&D expenditure in domestic and foreign firms became more closely aligned (Table 4). The concentration of German companies' R&D expenditures in the three major R&D-performing sectors (chemistry, computer and electrical engineering, and transport equipment) closely matched the concentration of R&D of foreign businesses.

Transport equipment accounts for by far the biggest amount of R&D expenditures by foreign firms at more than € 4.9 billion, and the R&D capacities of this sector in Germany have been particularly expanded in recent years. The sector attracts nearly 40% of total R&D expenditures by both foreign and German firms. This is mainly due to the fact that some segments of the German market for cars and supplies are lead markets. A relatively large number of R&D activities are also to be found in the chemical industry, with foreign firms prioritizing the pharmaceutical industry. Computer, electrical, electronic and precision engineering attracts one

Table 4 R&D Expenditure by domestic and foreign companies in Germany 2005

	German firms	Foreign firms	German firms	Foreign firms
	million €		% share	
Chemical industry	5361	2526	15.2	20.0
Of which:				
Pharmaceutical industry	2724	1855	7.7	14.7
Mechanical engineering	3518	930	10.0	7.4
Computer, electrical, electronic and precision engineering	7098	2537	20.1	20.1
Transport equipment	13691	4911	38.7	38.9
Of which:				
Motor vehicles	13386	2365	37.9	18.7
Other industries	5679	1714	16.1	13.6
Total	35347	12618	100.0	100.0

Sources: OECD, U.S. Department of Commerce, ONS UK, SV-Wissenschaftsstatistik; own calculations.

fifth of the total R&D expenditures by both foreign and German companies in Germany.

Similar sector-specific structures of the R&D activities of foreign firms and their German competitors are in line with the theoretical expectation in the horizontal model of internationalization, namely that competitors in the same location demonstrate similar R&D traits. Accordingly, the innovation impulses generated by the market in particular determine companies' R&D.

The average research intensity per sector – measured by the share of R&D personnel in the total number employed – for foreign firms in Germany that engage in R&D is comparable to the figures for the German-owned companies (Fig. 2). This supports the thesis that companies competing in the same market also invest in R&D to a similar extent.

To sum up, foreign firms operating in Germany increased their R&D activities in Germany in the second half of the 1990s more than German companies. At the same time, their R&D behavior was similar to that of German firms. Hence both the European and German markets for new products and services and the competitive local R&D environment seem to determine the attractiveness of the R&D location for MNEs.

5 Expansion of German Companies' R&D Abroad in the "Fifth M&A Wave"

R&D expenditures by German subsidiaries abroad are estimated at about € 11.3 for 2005, roughly the same amount that foreign companies invested in R&D in

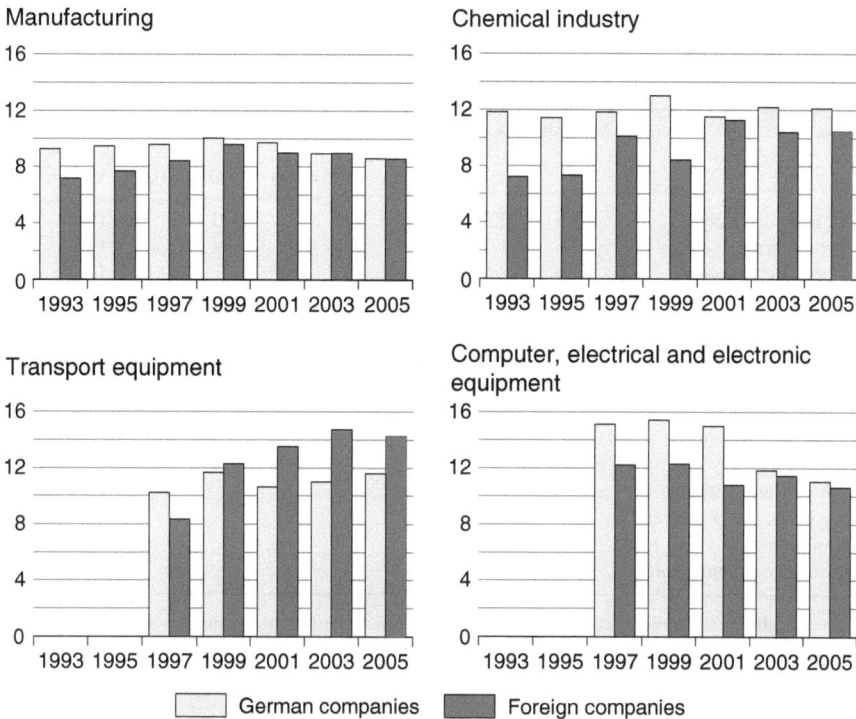

Fig. 2 R&D Personnel intensity in the German manufacturing sector 1993–2005, by Ownership (%). Sources: SV-Wissenschaftsstatistik; own calculations

Germany. In 1995 they were presumably € 5.1 billion (Table 5). R&D expenditure of German companies abroad rose by a good 120% in nominal terms from 1995 to 2005; it grew significantly more rapidly than total R&D expenditure in Germany (62%; Table 5). This increase in foreign R&D involvement is very probably due primarily to M&As rather than to an expansion in R&D in existing German companies abroad or to Greenfield investment in new R&D laboratories. Between 1995 and 2001 there was a strong worldwide increase in M&As, which fell sharply after 2000. In this M&A wave, Germany was the fourth-largest investor in cross-border M&As worldwide after the United Kingdom, the United States and France (UNCTAD, 2008). Among the motivations for firms to merge with or acquire an existing company, quest for strategic assets appears to be very relevant. This includes acquiring R&D or technical know-how, patents, brand names and so on (ECB, 2005).

Based on an empirical survey of laboratory sites established abroad by 49 German multinational companies Ambos (2005) finds an acceleration of the capability augmenting type at the end of the 1990ies. This does not seem to be simply a substitution of capability exploiting units but appears to be largely carried by additional establishment of capability augmenting labs.

German companies spent on average 30% of their overall R&D expenditure abroad. Motor vehicle manufacturers now have the highest amount of R&D

Table 5 R&D expenditure by companies in Germany and German companies abroad 1995–2005

Type	1995	2001[1]	2005	2001	2005	2005
	in billion euros			Change		
				1995=100	2001=100	1995=100
Firms in Germany	29.5	43.2	47.8	146	111	162
German firms	24.6	31.7	35.2	129	111	143
with R&D abroad	17.0	22.5	26.8	132	119	158
without R&D abroad	7.6	9.2	8.4	121	91	111
Foreign firms	4.9	11.5	12.6	235	110	257
German companies abroad	5.1	11.9	11.3	233	95	222
				As %		
Firms in Germany	100	100	100			
German firms	83	73	74	–	–	–.
with R&D abroad	58	52	56	–	–	–.
without R&D abroad	26	21	18	–	–	–
Foreign firms	17	27	26	–	–	–.
German companies abroad	17	28	24			

[1]Comparability to former years is restricted due to a new method of estimation.
Sources: SV-Wissenschaftsstatistik; own calculations and estimates.

expenditures abroad. The pioneers of internationalization, and not only in Germany, are the chemical and pharmaceutical firms. On average they spend nearly one half of their total R&D expenditures abroad (Table 6).

6 Foreign R&D of German Companies Mirrors Technological Strengths at Home

Publicly available information from patent applications provides insight into the foreign R&D activities of German multinational companies about where research is conducted and in which fields. For example, patent applications contain information about the company who has filed the application, about the location of the inventor – which usually corresponds to the location where research activities were conducted – as well as information as to which field of technology the patent belongs.

One shortcoming of using patents as indicators for R&D activities abroad is that the propensity to patent varies considerably between industries and nations – and changes over time due to changing patent strategies. Not every innovating activity leads to a patent. Often foreign R&D is devoted to adapting existing products or processes to local demand or to exploratory research of "listening posts" with no patentable outcome. Patents only refer to R&D activities leading to patentable outcomes. Another disadvantage of using patent data to describe the internationalization of R&D is the time lag between the R&D activities and the date of the

Table 6 Total R&D expenditure of German companies abroad by sector, 1995–2005

	1995	2001[1]	2003	2005
	in billion euros			
Manufacturing	4.9	11.6	10.2	11.3
Chemical industry	2.5	3.6	3.3	3.3
Of which: Pharmaceutical industry	.	.	1.7	2.1
Mechanical engineering	.	0.4	0.6	0.7
Computer, electrical, electronic and precision engineering	.	2.8	2.5	2.3
Transport equipment	.	4.6	3.6	4.8
Of which: Motor vehicles	.	.	3.5	4.8
Other industries	.	0.4	0.7	0.2
Total	5.1	11.9	10.9	11.4
	As % of total R&D worldwide			
Manufacturing	23.1	36.4	30.0	30.7
Chemical industry	35.6	48.0	41.2	40.4
Of which: Pharmaceutical industry	.	.	50.1	51.8
Mechanical engineering	.	39.5	32.2	27.2
Computer, electrical, electronic and precision engineering	.	37.7	36.5	31.6
Transport equipment	.	30.1	21.5	26.3
Of which: Motor vehicles	.	.	21.3	26.5
Other industries	.	13.7	30.8	10.1
Total	23.1	34.7	30.0	29.9

[1]Comparability to former years is restricted due to a new method of estimation.
Sources: SV-Wissenschaftsstatistik; own calculations and estimates.

application for a patent. On the other hand, one advantage of using patents is that they indicate the output of corporate research activities with a demonstrated market potential ("applied").

An analysis of patent applications filed with the European Patent Office of German companies and their main competitors between 1990 and 2005 was used to reveal in which technological fields they conduct R&D at home and abroad as well as their main foreign R&D locations. (Belitz, Schmidt-Ehmcke, & Zloczysti, 2008).

The data pool for this study consisted of transnational patent applications (including applications under the Patent Cooperation Treaty) filed with the European Patent Office between 1990 and 2005 by nearly 4,000 international companies.

By referencing the Derwent Patent Assignee Code, it was possible to identify common multinational corporate groups that have filed patents through various subsidiaries. The European Patent Office's Worldwide Patent Statistical Database

(PATSTAT, accessed in 1/2008) served as the source of the registration names and all additional information.

The geographic base of research for each company (i.e., home country) has been defined here as the country from which the largest percentage of patent applications originate at a given time. In this way, companies are not assigned to countries based on the ultimate beneficial owner of the patent, but rather based on the country with the most important research centers maintained by the company group. All previous studies have shown that multinational companies continue to focus their research in their respective home countries. The home country for each company (i.e., the location of the company's research base) was determined using the described method for two periods of time (1990–1993 and 2002–2005), as the assigned country could change over time due to M&As. The patent type, which is assigned by the patent office in accordance with the International Patent Classification (IPC), was used to sort the patents into 30 different technological fields.

The technological strength of German multinational companies relative to their competitors can be measured by using a specialization coefficient. This coefficient places the proportion of patents filed by the companies of one country in a given field of technology in relation to the proportion of filings by all companies in the same field. The coefficient tells us if the patent activities of a nation's companies in a field of technology are above or below average internationally.

German companies have specific competitive advantages (as indicated by a specialization coefficient greater than 1) in the fields of electrical engineering, drive technology, transportation and heat engineering, mechanics, consumer goods, and control technology. These are all fields experiencing rapid global growth. However, in some fields undergoing rapid growth, German companies did not exhibit any competitive advantage through specialization, including telecommunications, information technology, medical technology, and pharmaceuticals (Table 7).

The areas in which German companies specialize in their foreign research show that, on the one hand, they research in fields in which they already have a strong competitive advantage. In this way, the specialization coefficient is over 1 for foreign research in mechanical components, consumer goods and equipment, and transport and thermal processes and apparatus, indicating a strong concentration of foreign R&D activity in these fields (Table 7). In addition, foreign patent activity grew in these areas in near proportion to that within Germany. This can be interpreted as an expansion strategy in which foreign research activities primarily serve the effort to adapt products to local market conditions. This also accounts for the relatively low levels of foreign research in these areas.

By contrast, in the high-growth fields of pharmaceuticals, medical technology, telecommunications, and information technology, German companies cannot act from a position of relative technological strength. In these areas various patterns of R&D internationalization are observed. The number of patents filed by German companies abroad in the telecommunications and pharmaceutical branches is proportionally higher than that registered domestically (see Fig. 3). In these areas German companies are only specialized in foreign research. German companies attempt to close gaps in domestic research through knowledge acquisition abroad.

Foreign-based researchers, for example, play a role in 33% of the patents filed by German companies in the pharmaceutical branch, and 25% of patents in the telecommunications sector (Table 7). The increased research undertaken by German companies in areas of relative weakness – such as those above – is indicative of an attempt to compensate for disadvantages in domestic research by acquiring cutting-edge expertise abroad.

The proportion of research undertaken by German companies abroad in the medical technology and information technology sectors – which have witnessed particularly dynamic growth since 1990 – is also comparatively high. Yet German companies are not specialized in foreign or domestic research in either of these areas. Foreign patent activity has grown in proportion to domestic activity. This means that a greater proportion of foreign research does not automatically indicate that Germany has disadvantages as a location for research.

Table 7 Internationalization patterns of German companies in selected technological fields, 2002–2005

	Specialization coefficient		R&D abroad	Type of specialization
	Total	Abroad	%	
Mechanical components	+	+	12.3	Specialized at home and abroad
Engines, pumps, turbines	+	1	7.5	Specialized at home
Transport	+	+	9.2	Specialized at home and abroad
Thermal processes and apparatus	+	+	9.4	Specialized at home and abroad
Consumer goods and equipment	+	+	14.4	Specialized at home and abroad
Analysis, measurement and control technology	+	+	14.4	Specialized at home and abroad
Electrical devices, electrical engineering, electrical energy	+	–	8.9	Specialized at home
Civil engineering, building, mining	+	1	11.1	Specialized at home
Pharmaceuticals, cosmetics	–	+	32.7	Specialized abroad
Telecommunications	–	+	24.8	Specialized abroad
Medical technology	–	–	21.8	Not specialized
Information technology	–	–	17.9	Not specialized
Audio-visual technology.	–	–	18.6	Not specialized
Biotechnology	–	+	36.7	Specialized abroad
Organic fine chemistry	–	+	31.8	Specialized abroad

1) Above 1: +; below 1: –.

Source: PATSTAT; own calculations and estimates.

In the fields of biotechnology and organic chemistry – two areas which, in contrast to popular perceptions, are not among the fastest growing technology sectors worldwide – German companies have technological disadvantages, yet they concentrate their R&D activities abroad. In the field of biotechnology, nearly 37% of all patents filed by German companies – the highest in any sector – are the product of foreign research activities. In the field of organic fine chemistry this figure is nearly 32%. The proportion of foreign research undertaken in both of these areas provides an indication that Germany suffers from disadvantages as a location for research.

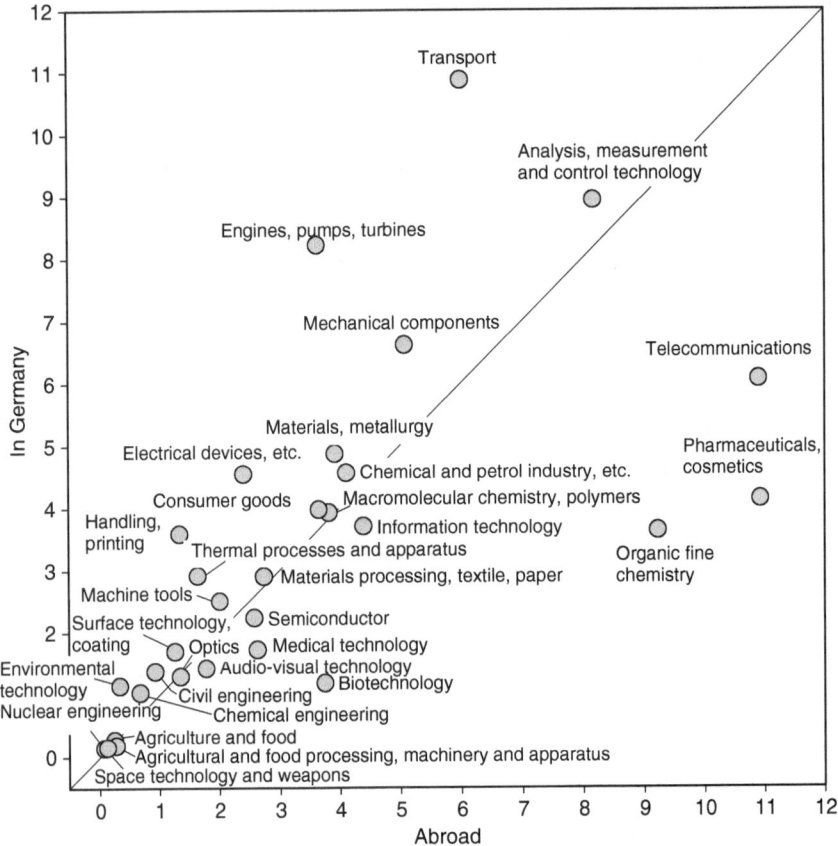

Fig. 3 Share of patent applications of German companies in Germany and abroad by technological sector, 2002–2005 (%). Source: PATSTAT; own calculations and estimates

These results are in line with other findings on the motivations for international R&D. It is, for example, confirmed by Les Bas and Sierra (2002) in a study of 345 multinationals with the biggest patenting activity in Europe (patent applications registered in the EPO between 1988–1990 and 1994–1996). Their study shows that companies primarily engage in foreign R&D in areas in which intensive research is also conducted at home. This focus on specific fields of research constitutes the basis for penetrating new markets (i.e., a "home-base-exploiting" strategy) or for acquiring new knowledge abroad (i.e., a "home-base-augmenting" strategy). 77% of patent applications could be ascribed to home-based internationalization strategies.

7 R&D Abroad is Concentrated on Western Europe and the United States

The patent-based analysis of foreign R&D activities of German firms reveals that the cross-border integration of companies' R&D locations and knowledge exchange occurs primarily within and between the knowledge-intensive regions of the United States and Western Europe.

The United States is the most important host country for research activities of German multinational companies, yet has lost somewhat of significance since the early 1990s (see Fig. 4). Their foreign research activities are concentrated geographically in Western Europe; the percentage of activities in Western Europe has experienced growth since the beginning of the 1990s.[3] Japan and other Asian countries continue to make up only a small percentage of foreign research. The most important European countries for German research are Switzerland (14% of all foreign activity), France (13%), and Austria (10%), followed by the UK (6%) and Italy (4%).

Between 2002 and 2005, the most important countries for German R&D in high-tech fields for which German companies had a domestic weakness and a high percentage of patent applications abroad were:

- In telecommunications: Austria,[4] the United States, and France.
- In pharmaceuticals, biotechnology, and organic chemistry: the United States, France, and Switzerland.

At the beginning of the 1990s, foreign research conducted by German companies was more concentrated in the United States and the industrial countries of Western Europe, which hosted 90% of all R&D activities. Since then, China, the new EU member countries, and the East Asian tiger economies – particularly South Korea – have captured a growing share of German foreign research. Nevertheless, their role remains relatively small – in 2002–2005, nearly 5% of foreign research by German companies took place in these countries, up from just over 1% in the early 1990s. Given the combined potential of their markets and skilled workforce, there is scope for expansion. The patent applications of German R&D units in the East Asian

---

[3] The position of Western Europe in relation to non-European research locations could be overestimated. Using patent applications filed with the European Patent Office we have to accept a bias to European applicants and European inventors.

[4] At that time Siemens AG Österreich ranked 2 amongst Austria's most diligent inventing companies. Nearly one half of the patent applications by Siemens Austria had their origin in its research and engineering division PSE with a strong focus on the telecommunications sector.

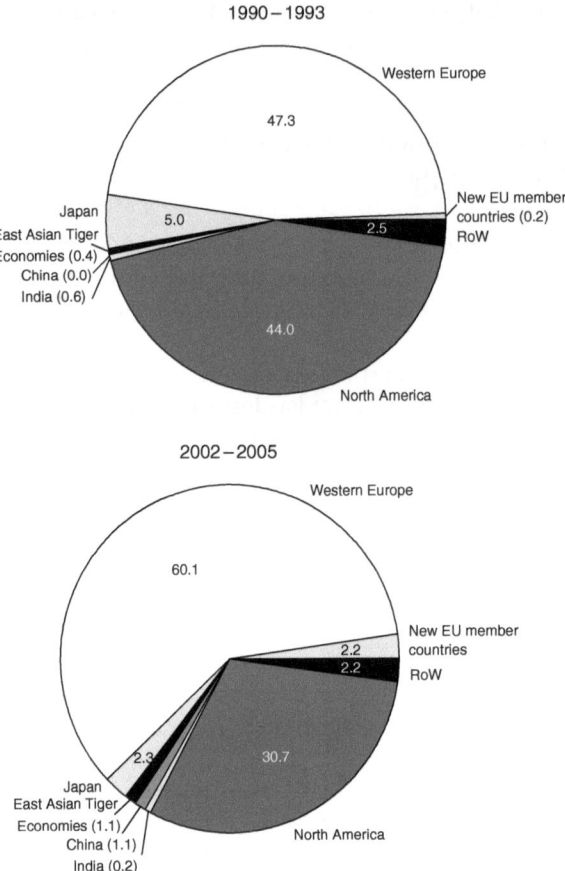

Fig. 4 R&D locations of German companies abroad, share of patent applications by inventor country (%). Sources: PATSTAT; own calculations and estimates

tiger economies and in China are concentrated on the technical fields electricity – electronics and chemistry – pharmaceuticals.

8 Summary

In Germany and several other developed countries (UK, Sweden and France) the share of affiliates under foreign control in total business sector R&D expenditure substantially increased between 1995 and 2001. By contrast, this share remained static in the United States, Canada and Finland. After 2003 this is more or less true for all developed countries designating an internationalization moratorium in business R&D. The cross-border M&A wave was the main driving force behind the push of R&D internationalization in multinational firms at the end of the last century.

Compared with other large industrialized countries, the internationalization of R&D in MNEs in Germany progressed considerably in both directions at the end of the last century. Germany became one of the leading home and host countries for R&D activities by MNEs. While German firms have increased their international R&D activities, this cannot be interpreted as a loss of attractiveness of Germany as a research location. Rather, internationalization is a two-way street for Germany, and the activity of foreign companies in the country has also increased. Foreign companies have contributed to the evident expansion of the R&D and innovation potential of the economy in Germany in recent years. They are involved in R&D – just as their domestic competitors are – particularly in those business areas that they consider provide new market opportunities in the medium term, based on the competitive advantages still prevalent in their home countries. At the same time, German companies need to go abroad with their own R&D activities, both to adjust to these markets and to exploit specialized forefront knowledge.

To summarize, our analysis provides evidence for more and more similar R&D behavior of multinational companies at home and in developed countries abroad. This is in line with the assumption of the dominance of the horizontal motivations for foreign activities.

The cross-border activities of R&D by companies and the exchange of knowledge are mainly within and between the knowledge-intensive regions in the United States and Western Europe. R&D undertaken by German companies abroad is concentrated in Western Europe and the US, where more than 90% of their foreign patent applications originate. Germany thus remains predominant alongside some of its Western European neighbors and the United States in the competition to attract companies with its research environment.

An analysis of patent applications filed with the European Patent Office reveals that German companies primarily expand their research activities abroad in high-tech sectors in which they already conduct intensive domestic R&D. In its core technological competencies, Germany remains an attractive location for corporate research, despite the increasing internationalization of R&D. Yet German companies have increased foreign R&D activities in several high-tech fields in which they have domestic research deficits. This can be indicative of a "catch-up" strategy. German companies conduct an above-average amount of foreign research in telecommunications and pharmaceuticals – two particularly high-growth fields worldwide – and in biotechnology and organic chemistry – which are experiencing somewhat lower growth. In these fields, the high rates of foreign activity are likely related in part to deficits in the domestic research environment.

9 Implications for German Innovation Policy

Altogether, the German system of innovation has largely been adequate to meet the demands of the internationalization of R&D by multinational companies. The internationalization of business R&D is part of the broader process of internationalization of innovation. It is the challenge of national innovation systems, and thus also of innovation and research policy, to constantly improve the conditions under which German companies and the German innovation system as such can take most advantage of international knowledge production and cross-national knowledge transfer.

All actors of the national innovation system need to adjust to the growing presence of international actors conducting R&D and strive to exploit cooperation and transfer opportunities. Policy must react to the demands made on the national innovation system through the progressive internationalization of knowledge generation and innovation. This requires policy approaches,

1. that will make a country an attractive location for doing business and to support attractive demand conditions for new products in the domestic market
2. to strengthen the national research and innovation system and
3. to remove barriers to cross-frontier R&D and innovation activities in both directions.

Measures to remove such barriers and to strengthen the international linkages of the national innovation system should i.e.,

- increase competence in languages and make occupational qualifications internationally comparable
- promote mobility in skilled personnel (work and residence permits, regulation of immigration)
- help to shape and implement international technical standards and norms
- give foreign firms equal access to national research promotion and pre-competitive research associations
- prepare publicly funded research facilities for joint research ventures with multinationals and for international competition between suppliers of research
- ensure internationally compatible protection of intellectual property.

References

Ambos, B. (2005). Foreign direct investment in industrial research and development: A study of German MNCs. *Research Policy, 34*, 395–410.

Barba-Navaretti, G., & Venables, A. J. (2004). *Multinational firms in the world economy.* Princeton, NJ: Princeton University Press.

Belitz, H., Schmidt-Ehmcke, J., & Zloczysti, P. (2008). *Technological and regional patterns in R&D internationalization by German companies*, Weekly Report, Vol. 4, No. 15/2008, 94–101.

Bloningen, B. A., Davies, R. B., & Head, K. (2002). *Estimating the knowledge-capital model of the multinational enterprise: Comment*, NBER Working Paper No. 8929, May 2002.

Brakman, S., Garretsen, H., & van Marrewijk, C. (2006). *Comparative advantage, cross-border mergers and merger waves: International economics meets industrial organization*, CESifo Forum 1/2006, 22–26.

Dunning, H., & Narula, R. (1995), The R&D activities of foreign firms in the United States. *International Studies of Management & Organization, 25*(1–2), 39–73.

ECB (2005). *Competitiveness and the Export Performance of the Euro Area. Taskforce of the Monetary Policy Committee of the European System of Central Banks*, Occasional Paper Series of the ECB, No. 30, June 2005.

Granstrand, O., Patel, P., & Pavitt, K. (1997). Multi-technology corporations: Why they have 'Distributed' rather than 'Distinctive' core competencies. *California Management Review, 39*(4): 8–25.

Hegde, D., & Hicks, D. (2008). The maturation of global corporate R&D: Evidence from the activity of U.S. foreign subsidiaries. *Research Policy, 37*, 390–406.

Kuemmerle, W. (1997). *Building effective R&D capabilities abroad*. Harvard Business Review March–April, 61–70.

Les Bas, C., & Sierra, C. (2002). Location versus home country advantages in R&D activities: Some further results on multinationals locational strategies. *Research Policy, 31*, 589–609.

Markusen, J. R. (2001/2002). *Integrating multinational firms into international economics*, NBER Reporter, Winter 2001/ 2002, 5–7.

OECD (2008). *The internationalization of business R&D: Evidence, impacts and implications*. Paris, 2008.

Patel, P., & Vega, M. (1999). Patterns of internationalization of corporate technology: Location versus home country advantages. *Research Policy, 28*, 145–155.

UNCTAD (2005). UNCTAD *Survey on the Internationalization of R&D, Current Patterns and Prospects on the Internationalization of R&D*, Occasional Note, United Nations, New York and Geneva, December 2005.

UNCTAD (2008). *World Investment Report 2008*. New York and Geneva.

Von Zedtwitz, M., & Gassmann, O. (2002). Market versus technology drive in R&D internationalization: Four different patterns of managing research and development. *Research Policy, 31*, 569–588.

The Chemical Industry Drives Innovation also in Times of Crisis: BASF as an Example

Andreas Kreimeyer

Contents

1 Introduction 67
2 Crisis as Innovation Engine 68
    2.1 Past Experiences: Innovations from the Chemical Industry in Times of Crisis . . 69
3 Long-Term Future-Oriented Corporate Strategy at BASF: The Beacon in the Crisis . 71
    3.1 Generating Knowledge in the Know-How Verbund 71
    3.2 Orientation to Megatrends 72
    3.3 Demographic Change 73
    3.4 Urbanization, Housing and Mobility 74
    3.5 Climate-Friendly Energy Supply 74
4 International and Interdisciplinary Networks: Important Success Factors 75
    4.1 Interdisciplinary Cooperations 75
    4.2 Importance of Internationalization 76
5 Innovation Sites in International Competition 77
6 Conclusions 79
References 79

1 Introduction

Scientific, technological and institutional innovations drive organic growth and thus are core elements of the strategy pursued by most companies. The chemical industry, as a supplier of innovative materials, drives innovation for many sectors of industry. The goal of a research oriented chemical company is to generate product and process innovations, something which often takes several years. Key elements of innovative activity are therefore continuity and long-term commitment – especially in times of crisis.

A. Kreimeyer (✉)
BASF S.E., D-67056 Ludwigshafen, Germany

The chemical industry bases its activities not only on topics of current interest to various industry sectors and consumers and on technological trends, but also increasingly on long-term megatrends. We are combining our spirit of innovation, entrepreneurship and knowledge to find answers to urgent challenges of the future, such as

- feeding the growing world population and securing supplies of the scarcest resource: clean water,
- meeting rising energy demand without harming the environment,
- satisfying increasing requirements for transportation and individual mobility.

Currently, we are potentially experiencing the most severe crisis of the past 100 years. Nevertheless, the chemical industry is demonstrating great dedication and commitment to research at this time. But we also depend on political support. Our innovative strength and international competitiveness must be sustainably supported by suitable framework conditions.

2 Crisis as Innovation Engine

Innovations from the chemical industry provide the vital impetus needed to develop innovative products in many industrial sectors. This has just been illustrated again by a study recently conducted by the Center for European Economic Research (ZEW), Mannheim, Germany and the Lower-Saxony Institute of Economic Research in Hannover, Germany.[1] This study shows that the chemical industry is one of the most important suppliers of innovative materials and is therefore an indispensable part of the innovation system. As a technology provider, it is on par with the mechanical engineering industry and the electrical industry.[2] For example, it gives the automotive industry essential innovative impulses by supplying new engineering materials, process innovations or competitive intermediates. With the highest innovation rate after pharmaceutical companies, four out of five chemical companies successfully launch at least one new product or process every three years.[3] Clearly, if the chemical industry calls a halt to projects in economically difficult times, this affects the entire innovation chain in many sectors of industry.[4]

Globalization is a key stimulus in the competition for innovations and was the factor that allowed Germany to establish itself as a successful, industrially based exporting country.[5] This is also reflected in the chemical industry: about 11 percent

---

[1] Rammer, Sofka, Legler, Gerke, and Krawczyk (2009).
[2] Rammer et al. (2009), 31.
[3] Rammer, Schmiele, Legler, Krawczyk, and Sofka (2007), 5.
[4] Davis (2009).
[5] Hüther (2009).

of total global investment in research and development in the chemical industry are spent here in Germany; this is substantially more than Germany's share in global chemical production of just over seven percent.

Over the last 100 years, we in the chemical industry have weathered many crises. But each time we emerged stronger than before, and one of the reasons for this is that we remained faithful to innovation as a success factor.

2.1 Past Experiences: Innovations from the Chemical Industry in Times of Crisis

During an economic downturn, companies in all the affected industries see themselves forced to cut costs. This can also impact investment in research and development, although innovations are and will remain the principal sources of long-term growth. The economic and financial crisis we have been experiencing as a global phenomenon since 2008 has not reduced the need for innovations in areas such as nanotechnology, green gene technology, material sciences and clean energy.

Crises and technological progress are not mutually exclusive. On the contrary, experience shows that crises actually generate innovations. The great world economic crisis which started in 1929 and overshadowed the 1930s is a good example: research by industrial historians has revealed that from 1929 onwards, chemical companies developed more than 50 important basic products within a single decade, more than ever before.[6] Second only to the electronics industry, the chemical industry was the most research intensive sector at that time. In fact, the period from 1929 to 1941 is even regarded as the most technologically progressive periods in U.S. economic history,[7] and chemistry was one of the key industries of science-driven U.S. society.[8]

In the 1930s, the aim of companies striving for innovations was to reduce costs and simultaneously develop new and better products instead of merely returning to the status quo of the pre-crisis period, as American economic historian Michael Bernstein writes. Moreover, the strained economic situation acted as a stimulus to improve the utilization of reaction by-products and waste materials.[9]

The German chemical industry also achieved major breakthroughs in that period, although cost cutting was still the order of the day in research. In 1929, at the outset of the world economic crisis and in the Depression years that followed, the German chemical industry suffered a severe setback.[10] When the Depression came to a head in 1930–1931, research and development expenditures were drastically cut back:

---

[6] Bernstein (1989).
[7] Field (2003).
[8] Bruland and Mowery (2005).
[9] ICIS (2008).
[10] Abelshauser (2003), 254.

laboratory personnel were laid off, a hiring freeze was imposed and the research budget for new processes was slashed to less than a tenth. Despite everything, the crisis of 1929 also resulted in great innovations within Germany (see Box 1).[11]

Box 1: Examples of Innovations in Times of Crisis

Despite huge budget cuts, the 1930s were marked by spectacular research achievements, including some in promising new fields of activity for the BASF Group, which in 1925 was merged into I.G. Farbenindustrie. The resulting products entered markets that experienced a rapid upturn at the end of the Second World War.

Kaurit One example from the world of BASF is the invention of the industrial glue Kaurit in the 1930s. In 1922 and 1925, BASF had initiated the industrial scale synthesis of urea and formaldehyde and in 1929 received the patent for the gluing of wood. Following its market launch in 1931, Kaurit glue influenced all areas of the wood processing industry. From then onwards, it was possible to permanently glue wooden elements together. Moreover, Kaurit glue also paved the way for chipboard, a revolutionary innovation in wood technology. Without Kaurit, there would have been no chipboard, and without chipboard there would have been no optimal and environment friendly utilization of the material wood. Kaurit therefore indirectly created new markets for cost effective furniture made from chipboard and it is still being developed today.

Magnetic Tapes In 1934, production started on the first magnetic tapes for sound recording that were to provide the basis for whole generations of magnetic storage media. The background was a cooperative development project with AEG commenced in 1932: while AEG worked on the recording machines, researchers in Ludwigshafen developed the actual tapes. The cellulose acetate tapes coated with carbonyl iron powder were manufactured using a process in which pigments could be applied in an extremely fine dispersion in solvents. The coating material had already been available since 1927 when an iron powder plant was established in Ludwigshafen. The resulting magnetophone was the sensation of the Radio Exhibition in 1935.

Styrene Synthesis The era of plastics was ushered in by the first technical styrene synthesis in 1929. Styrene synthesis provided the basis for starting production of Styropor in 1951 and laid the foundation for today's innovative materials such as Neopor.

Reppe Chemistry The renowned "Reppe chemistry" also has its roots in the depression era: during the emergence of polymer chemistry in the late

---

[11] Abelshauser (2003), 255; Gallecker and Hesse (2009).

1920s, Reppe had made it his goal to produce the monomers needed for plastics manufacture from simple building blocks. He therefore performed catalytic reactions with acetylene under pressure, such as vinylation to produce vinyl ether. In 1939, the vinyl ether factory was inaugurated. This was followed four years later by the butanediol plant for acetylene-based buna synthesis. Butanediol is still one of BASF's important intermediates and is used, for example, to produce elastic Spandex fibers used in sportswear and swimwear. It also served as the source of engineering materials used in high-frequency technology. Reppe chemistry was one of BASF's great strengths for many years: the six reactors constructed by Reppe so long ago to produce butanediol are still working today.

3 Long-Term Future-Oriented Corporate Strategy at BASF: The Beacon in the Crisis

3.1 Generating Knowledge in the Know-How Verbund

Innovations are based on processes of knowledge generation and dissemination. Satisfying the growing need for knowledge and increasing the contribution of knowledge to create value are among the most important challenges facing the industrialized societies, both in good and bad times.

Competitive advantages are gained by the company which produces an innovation with limited resources in a short time; in other words achieving market success with new solutions to existing problems. This applies especially to countries, such as Germany, having few raw material resources. Transforming knowledge into market successes and thus into added value allows companies to differentiate themselves from each other. Many years ago, we established a process at BASF which now allows us to distinguish ourselves from our competitors through our innovations. One of the elements of this was to consistently place research and development on an international footing while simultaneously reducing complexity. Innovation is a key component of our long-term corporate strategy.

Today, BASF's global R&D organization is based on four central technology platforms embedded within a global network of customers, universities, research institutes, high-tech joint ventures and industry partners. This Know-how Verbund is a concentration of the highly specialized expertise of some 8900 employees in research and development and that of the external partners.

We have established a learning organization. This allows us to identify tomorrow's trends and the challenges of the future in good time and to address them consistently and continuously – independently of crises of the kind we are currently experiencing.

3.2 Orientation to Megatrends

The need to project developments into the future is not an invention of the 21st century. More than 200 years ago, the political economist Thomas Malthus was analyzing the worsening living conditions in the England of his time.[12] Malthus assumed that humanity was increasing exponentially and would double within one generation. Since food production was only increasing linearly, however, he saw mankind threatened by famine. In methodological terms, the scenario described by Malthus in 1798 was a precursor of what we now call megatrends – our assumptions about the future. Megatrends stand for regional and global development drivers which are constantly present: they are the major social developments that will influence all areas of life, state, market and civil society for long periods.[13] Megatrends describe the possible future and therefore represent an important starting point for our innovation strategy.

Global megatrends drive our innovations because the scenarios we deduce from the trends are an indispensable component of entrepreneurial activity – and also extend beyond fluctuations in the economic cycle. In the absence of warning signs of approaching economic crises, social critics see themselves confirmed in their belief that statements about the future are condemned in advance to failure.[14] Economic crises such as those in 1929 or 2008, however, cannot fundamentally change the long-term trends. The emerging global developments retain their validity as long as the basis for these scenarios remains unchanged: a constantly growing world population which is also increasingly aspiring to prosperity.

The importance of these long-term trends is now being recognized not only in traditional industrialized societies, but globally. There is a general appreciation of the fact that in town planning, for example, new solutions are needed to deal with the environmental impact of urbanization. Countries for which energy conservation was not highest priority have now started planning environmentally friendly cities in the open countryside. One of these projects is Masdar City in Abu Dhabi (Arabic for "the source"), which was planned taking many megatrends into consideration (see Box 2).

Box 2: Example Ecotopia – The Future Belongs to the Energy-Efficient City

In August 2009, Masdar and BASF entered a strategic partnership for the construction of Masdar City as the world's first zero-carbon and zero-waste city. As the preferred supplier of construction materials and system solutions,

---

[12]Malthus (1798).
[13]Larsen (2006).
[14]Taleb (2007).

> BASF with its innovative solutions for sustainable and energy efficient construction will be making a key contribution to reducing the demand for energy.
>
> The zero-carbon, zero-waste city will be automobile-free: a city in which driverless "PodCars" running on invisible tracks will convey their passengers to the desired destination, while outside the city hybrid automobiles driven by lithium ion batteries will be available for long distance travel; the energy supply will be secured in the wind farms and on the vast areas dotted with solar collectors and even garbage will become a source of energy.
>
> Many environmental technologies incorporate chemical innovations which, for example, increase the energy efficiency of buildings. With modern facade systems and innovative construction and insulating materials, building owners can operate buildings efficiently and drastically reduce the costs of air conditioning systems. In Masdar, traditional energy saving systems are combined with alternative energy systems. The classical Arab architecture with its high walls and narrow, shadowed lanes keeps the hot desert wind at bay and guides cool breezes into the city. Competition now no longer focuses exclusively on size, but is all about sustainability, both material and conceptual: the heart of Masdar will be the Masdar Institute of Science and Technology which as a center of energy and environmental research, will attract scientists from around the world and which aspires to set new standards of environment friendly city planning. The desire for sustainable living and housing space is something that unites people worldwide. It is not by accident that numerous ecocities such as Masdar are currently springing up in regions that are *still* rich in raw materials and in emerging countries. Creativity needs a fertile substrate. "Innovation needs freedom" is the watchword of the Arab principals: "We are testing and creating green and efficient living units in our society. With their innovations and technologies, our partners are contributing to our success".

As an industrial company we have learned to address the needs and challenges of people worldwide. We orient our activities towards long-lasting trends to direct our innovative strength in the right direction – towards the most important issues. Chief among these are demographic change and globalization with all the resulting effects such as urbanization, growing demand for food, water and energy and climate change. Not to be overlooked, however, is the fact that globalization not only opens up markets but also brings political, social, religious and social systems closer together.

3.3 Demographic Change

The world's population is growing and is also becoming steadily older. According to current estimates, up to 9 billion people will be living on our planet in 2050. Every year 50 million people are added, especially in emerging economies like India and

China. This has created and will continue to create numerous challenges for the chemical industry: how can we keep a growing world population supplied with the most precious raw material – clean water? How can we feed so many more people? Can we increase agricultural yields without causing environmental harm? What role can/must green biotechnology play in these endeavors? With growing life expectancy, expectations are also growing for quality of life in older years, especially in the aging industrialized societies of Europe and Japan: compared to their parents' generation, the "young" elderly are healthier, more active and more mobile, better educated and more consumption oriented. Inevitably, the demand is growing in the health care sector for life science, wellness disciplines, health vacations, anti-aging products or trend sports. Since we are becoming older and older, how can we assure the medical care of an aging society with constantly growing aspirations?

3.4 Urbanization, Housing and Mobility

Urbanization is the second trend we can observe all around the world. The cities, especially the metropolises of the emerging and developing countries with their increasingly difficult to control structures, act as gigantic magnets attracting inhabitants from the surrounding regions. At the beginning of the century, every second person lived in a city – in 2025 it will be two thirds of the population. City dwellers are demanding an increasingly high quality of life: they want to be mobile and live comfortably in a limited space. What can chemistry contribute to this scenario? We are thinking about how we can make urban living resource conserving and are searching for solutions for the sustainable city of the future. Fast, simple and cost-effective construction is what is needed. As in our example Masdar, this includes the use of energy conserving construction materials but also new means of transportation and infrastructure concepts, an efficient and sustainable energy and water supply and finding solutions to disposal problems.

3.5 Climate-Friendly Energy Supply

The growing world population and the rapid pace of industrialization in emerging economies such as India and China has led to a an additional increase in energy demand in recent years. Based on existing concepts, we expect the demand for primary energy to increase by 50 percent by 2030. Especially the increased consumption in emerging markets with their high rate of development will increase. As a chemical company we have long been concerned with providing intelligent energy saving concepts and have developed products that reduce greenhouse gas emissions – at our own and our customers' sites. What is needed is higher energy efficiency in construction and housing and in the automotive sector. Permanent solutions to such problems as energy storage and alternative energy production are being sought.

The equitable distribution of global energy reserves remains one of the foremost geopolitical challenges of this century. All industries are searching for the ideal energy mix to deploy and distribute this vast quantity of energy cost effectively, reliably and in an environmentally friendly manner. Developing innovative solutions and future oriented projects in this area is the wellspring of our research.

4 International and Interdisciplinary Networks: Important Success Factors

4.1 Interdisciplinary Cooperations

Achieving peak innovative research performance over the long term means we have to lead the field in the competition for the best minds. After all, both as an industry and as a company we are competing not only in the sales and procurement market but also in the talent market. Cooperation in science and economy is an important success factor for this. We therefore purposely encourage partnerships within the chemical industry and promote interaction and networking.

This joint search for bright minds is only one aspect of networking. One other task that is at least as important: research topics are now so complex that innovations can only arise from cooperation within a research community. Creativity arises when different perspectives, diverse competences and distributed knowledge are brought together. Einstein's 130th birthday in 2009 reminds us of how much our research model has changed. The age of the universal scholar is coming to an end and is being succeeded by flexible social organizations which are increasingly determining the pace of research progress: networks.[15] Modern projects demonstrate how much working methods in science have changed. Today, innovations arise from alliances and networks of companies and institutions of all sizes.

This networking concept is also reflected by the European Union, for example by the metropolitan regions which have existed at the German Federal level since 1995. These regions have acquired great international importance and play a key role in the social, societal and economic development not only for one country, but for Europe as a whole (see Box 3).

Box 3: Research in the Rhine Neckar Metropolitan Region

Since the end of April 2005, the Rhine Neckar region has been a "European Metropolitan Region". The region not only houses Europe's largest cluster

---

[15] Osterhammel and Petersson (2003), 22.

of chemical industries, but is also among the leading life sciences locations and has thus become a dynamic knowledge region. The Rhine Neckar Triangle, encompassing territory from the three German Federal States Baden Württemberg, Hesse and Rhineland Palatinate, has become a driving force of the high-tech strategy of the German Federal government and a center of attraction for companies and scientists alike – not only in the Rhine Neckar region itself, but internationally.

As one of eleven metropolitan regions in Germany, the Rhine Neckar region is an important driver for growth and innovation and symbolizes the multi-institutional, international cooperation often involving numerous partners as a basis for innovation. Initiatives of this kind are also an important source of innovation for the chemical industry. For example, BASF also participates with numerous projects.

Innovative processes in the chemical industry need active cooperation with external partners to transform shared knowledge into market success. Between 2002 and 2004, almost 30 percent of all research driven chemical companies established innovation partnerships with other companies and research institutions – compared to 17 percent in other industries.[16] The exchange carried on in "open innovation networks" and the shared dedication to seeking the best possible practical solutions in development has now become a routine approach in cooperation with important corporate partners, customers, startup ventures and academia.[17]

4.2 *Importance of Internationalization*

In research and development it is repeatedly necessary to overcome not only institutional but also national boundaries. The challenges of the future can only be solved through international cooperation pursued with a global outlook. This is especially true for the chemical industry. For example, within two decades BASF has almost doubled the number of employees in research and development outside Europe and has set up regional research platforms. This internationalization has also been driven forward by partnerships and the establishment of new laboratories around the world. BASF is pursuing important and successful cooperative projects with, for example, Harvard University, ISIS, the Institut de Science et d`Ingénierie Supramoléculaires in Strasbourg, and the National University of Singapore.

Asia, especially, is gaining importance as a research location. Asia's growing economic strength is simultaneously increasing the region's innovative power. Countries like China and India have made enormous efforts in recent years and

---

[16]Rammer et al. (2007), 16. Every second innovative chemical company involves science as important partners in innovation projects (Rammer et al. 2009, 20).

[17]Schneider and Wysocki (2009).

have set up productive research centers at their universities. This is illustrated, for example, by their growing number of scientific publications. The contribution now being made by the emerging countries of Asia to international scientific chemistry publications has reached almost 30 percent.

During the process of globalization, many chemical companies have therefore not only established production centers in Asia but are now increasingly pursuing research in this region. The companies are working on special close-to-customer applications in direct and close cooperation with local and international customers. For many years now, we have not considered it self-evident to spend our research budgets only in Germany. We are increasingly utilizing the resources of the Asian knowledge and research network. For example, an average 30 percent of the overall research budget of German chemical companies is now allocated to research and development projects abroad.

5 Innovation Sites in International Competition

Even before the economic crisis of 2008, it was already clear how much tougher the international competition is becoming. In the 1980s and 1990s the traditional industrialized countries had already lost their monopoly on research intensive production to the East Asian "small tiger states" of Taiwan, South Korea, Singapore and Hong Kong. In the new century countries like China, India and Brazil as well as Russia and the Middle and East European countries are now setting the pace.[18] Emerging economies have now become "catching-up countries" which are no longer just extended work benches and raw material suppliers, but are expanding with great force and vitality and offering technology-intensive products. China, for example, has increased its research expenditure sevenfold since the mid 1990s.[19]

The traditional industrialized countries like Germany must be sure to move with the times. As a company with its roots in Germany, this is of course particularly relevant for BASF. Nevertheless, we see that Germany's chemical industry is continuously losing shares in world trade (1995: 15.2 percent; 2007: 12.4 percent). Moreover, research expenditure for "pure" chemistry in Germany is stagnant, whereas research networks abroad are being established. To avoid "missing the boat", we must tap into new areas of activity such as nanotechnology or green biotechnology. This requires an innovation-friendly environment. Instead, in the public sponsorship of nanotechnology, Germany has dropped from third place after the USA and Japan to fifth place – after China and Russia.[20]

So how does Germany shape up with regards to the technology acceptance and innovation friendliness we need to remain competitive in the long term? For the classical chemical companies to step up the pace of research again in Germany, we

---

[18] Rammer et al. (2009), 47.
[19] Duerand (2009), 47.
[20] Werner, Grabbe, and Oden (2009).

need the right political framework conditions. Especially now, social investments in research and education are more necessary than ever: the Institute of the German Economy in Cologne estimates that in ten years industry will be lacking about 230,000 engineers, scientists and technologists.[21] This will naturally also impact company startups: the Center for European Economic Research has found that many fewer research intensive high-tech companies are being founded.[22]

An industrialized society like Germany and an industrial region like Europe cannot do without a high educational standard if they want to remain competitive. At the same time, we have long been aware from demographic surveys that, for example, the lack of specialized personnel will become even more critical. This segment of the population is shrinking. Apart from their adverse influence on producing a new generation of scientists, educational deficits also have another negative impact: ignorance in addressing new topics leads to diffuse fears and skepticism that have no basis in fact. Public dialogue is therefore urgently needed to improve knowledge of new technologies and dispel anxieties. This is also the task of politics – by sponsoring science education, improving the basic conditions for traineeships in chemistry, chemical research at the university level and by fostering the new generations of university teachers.

A focus on our core competences – research-intensive goods and knowledge-intensive services – is long overdue. In Europe, it is often the risks that are seen first rather than the opportunities presented by the new directions. Europe must quickly and substantially improve its own background conditions for research, development and production and must remain attractive and open to the world's brightest minds, including China. Knowledge creates the future. And by guiding our knowledge with a strong vision, we shall jointly be shaping this future. But a society which through reluctance to take risks in everything new and first demands proof that any element of risk is ruled out, loses time, opportunities, markets, jobs, tax revenues, growth and dynamics. The consequence is the migration of industries and research topics abroad.

The development of lead markets, improved cooperation between science and the economy and faster implementation of research results require suitable sponsorship. Germany needs an instrument to reward the innovative efforts of industry. As a supplement to existing project sponsorship models, the instrument of fiscal promotion of research offers research driven companies greater opportunities for innovation. Research-led companies should be able to deduct at least 10 percent of their entire self-financed investments in research and development from their tax burden. Tax credits of 8–20 percent have long been customary in the major industrialized countries. Two thirds of the OECD countries have introduced fiscal promotion of research. Fiscal promotion of research for large and small companies is essential if Germany is to remain a top address for our research and development departments.

---

[21] Ibid.
[22] Metzger et al. (2009).

6 Conclusions

In our global world, the chemical industry delivers decisive stimuli for innovations across all industries. Especially in times of crisis, this innovation driver must be a smooth running engine. To ensure this, we must maintain the continuity of research and development. Innovations are indispensable for solving the problems of the future which transcend national concerns, such as supplying a growing world population or safeguarding energy supplies.

The economy must demonstrate the will to show a long-term continuous commitment to research and development. At the same time it must remain open to interdisciplinary cooperation with partners in both economy and science, because the increasingly complex research projects can only be handled through a joint effort. This cooperation has long transcended all international boundaries both in the form of international cooperations and through the internationalization of research and development within the companies. Only with a global perspective can the global challenges we want to tackle be solved.

Technological progress alone, however, is not enough unless accompanied by institutional and societal change. The fear of the new must not lead to general taboos being imposed either on thinking and research, or on decision making and implementation. Our task is therefore to see innovations more within their sociopolitical context than we did in the past.

Only if politics and economy stand shoulder to shoulder can a climate favorable for innovation arise. For the chemical industry to sustainably prevail in international competition, we need optimal operating conditions. We need the best chemists, the best physicists, the best engineers. We need patience, resolve and the will to pursue projects with a long-term perspective. Because both in good and bad times, innovations shape the future.

References

Abelshauser, W. (Eds.). (2003). *Die BASF. Eine Unternehmensgeschichte*. München: C.H. Beck.
BASF (2009). *BASF presents sustainable construction concepts at the World Future Energy Summit in Abu Dhabi*.
Bernstein, M. A., & Bernstein, M. A. (Eds.). (1989). *The great depression: Delayed recovery and economic change in America, 1929–1939, Studies in economic history and policy*. Cambridge, UK: Cambridge University Press.
Bruland, K., & Mowery, D. (2005). *Innovation through time, The Oxford Handbook of Innovation*. Oxford: Oxford University Press.
Davis, N. (2009). INSIGHT: R&D will suffer despite stimulus packages. *ICIS Chemical Business*, 10.06.2009.
Duerand, D. (2009). Bilanz mit Makel. *Wirtschaftswoche, 34*, 19.08.09.
Field, A. (2003). The most technologically progressive decade of the century. *American Economic Review, 4*, 1399–1413.
Gallecker, G., & Hesse, K. (2009). Crisis management at BASF. Interview with Dr. Hans-Ulrich Engel. *BASF-Online Reporter, 15*, June 2009.
Hüther, M. (2009). Gute Diagnose, schlechte Therapie. Steinmeiers "Deutschlandplan" bleibt trotz einzelner guter Ansätze insgesamt enttäuschend. *Handelsblatt*, 14.08.2009.

ICIS (2008). History recommends innovating toward recovery. *ICIS Chemical Business*, 29.12.2008.
Larsen, G. (2006). Why megatrends matter. *Futureorientation, 5*, 8–13.
Malthus, T. R. (1798). *Essay on the principle of population*. London: J. Johnson.
Metzger, G., Heger, D., Höwer, D., Licht, G., & Sofka, W. (2009). *High-Tech-Gründungen in Deutschland – Optimismus trotz Krise*. Mannheim: Zentrum für Europäische Wirtschaftsforschung (ZEW).
Osterhammel, J., & Petersson, N. P. (2003). *Geschichte der Globalisierung: Dimensionen, Prozesse, Epochen*. München: C.H. Beck.
Rammer, C., Schmiele, A., Legler, H., Krawczyk, O., & Sofka, W. (2007). *Innovationsmotor Chemie 2007: Die deutsche Chemieindustrie im globalen Wettbewerb*. Studie im Auftrag des Verbands der Chemischen Industrie e. V., Technical Report. Mannheim: Zentrum für Europäische Wirtschaftsforschung (ZEW).
Rammer, C., Sofka, W., Legler, H., Gerke, B., & Krawczyk, O. (2009). *Innovationsmotor Chemie 2009: FuE-Potenziale und Standortwettbewerb*, Verband der Chemischen Industrie e.V. Mannheim: Zentrum für Europäische Wirtschaftsforschung (ZEW).
Schneider, K., & Wysocki, M. V. (2009). Unternehmen und Unis Hand in Hand. *Handelsblatt, 19*, 20.04.2009.
Taleb, N. N. (2007). *The black Swan: The impact of the highly improbable*. New York: Random House.
Werner, K., Grabbe, H., & Oden, M. (2009). Kleine Zukunft für kleine Teilchen. *Financial Times Deutschland (FTD)*, 26.08.09.

The Innovation Premium: Managing for High Return on Innovation

Tom Sommerlatte

Contents

1 Introduction 81
2 Innovation Strategy 83
3 Innovation Process 85
4 Innovation-Friendly Organizational Solutions 87
5 Open Innovation: Leveraging External Resources 88
6 Innovation Culture 89
7 Conclusions 89
References 90

1 Introduction

At Arthur D. Little, Fortune Magazine's (FM) annual ranking of "America's Most Admired Companies" (Fortune Magazine, 2008) triggered the question which of the ranking criteria used[1] correlate more or less strongly with stock performance, after all the most relevant evaluation aspect from an investor's point of view.

We found that none of the FM ranking criteria shows a significant correlation with stock performance as best expressed by total shareholder return, TSR (Total shareholder return, TSR: increase in share value plus dividends). However, by taking the average of each of the companies' TSR over the last 5 years and plotting it against their respective innovativeness[2], the correlation turns out to be pretty

---

T. Sommerlatte (✉)
Arthur D. Little GmbH, D-65185 Wiesbaden, Germany

[1] Fortune magazine ranking criteria: Innovation, People Management, Use of Corporate Assets, Social Responsibility, Quality of Management, Financial Soundness, Long-term Investment, Product/Service Quality.

[2] Innovativeness: Percentage share of new and substantially improved products/services introduced over the last 5 years in total sales or total contribution over innovation expenditures as percentage of total sales or total contribution.

good (Sommerlatte, 2004). Why is that? Apparently, the TSR average over a longer period eliminates one-time effects of rationalization measures, restructurings and other changes affecting performance in a given year but shows the inherent value creation of a company based on its innovativeness, its return on innovation.

We identified the companies positioned in the quadrant of high innovativeness and high average TSR, as opposed to those positioned in the quadrant of low innovativeness and low average TSR, and investigated the particularities of their innovation management. We found that innovative companies with a sustainably high TSR achieve their "innovation premium" thanks to a strategic innovation management which goes beyond just effective R&D management and technology leadership (Jonash & Sommerlatte, 1999).

"Premium companies" such as Canon, Pfizer, Boston Scientific, Millenium Pharmaceuticals, Sun Microsystems, Cisco, Lucent Technologies, Alcoa, Eveready Battery Company, S.C. Johnson & Son or Gilette in the United States and, in Europe, Beiersdorf, Bosch, BP, Hilti, Nokia, Porsche, SAP, Sartorius or Sixt have indeed demonstrated successful innovation management and a clearly above average TSR performance over an extended period of time. They have been able to maintain their innovation premium throughout the turmoil of stock market ups and downs and are likely to also do so during the current global financial and economic crisis.

Many other companies commit the classical mistake of reacting to market shrinkage by not only cutting costs but also reducing their development expenditures. In terms of their innovation management, they behave pro-cyclical instead of counter-cyclical and jeopardize their ability to respond to new market opportunities once they reappear.

The reason is that they are not convinced of and therefore do not manage the correlation between innovativeness and company value. Their skepticism stems from the fact that they too narrowly interpret "innovation" as "technological innovation". Investment in technology development, however, without ensuring that clear customer benefits are achieved does not qualify as innovation nor does it lead to market success. Premium companies know that only products, services and ways of doing business can be innovative and that technologies are just one of the means for achieving added customer benefit.

Our investigations show that premium companies distinguish themselves from other participants in their industry in that they direct their innovation efforts much more explicitly and rigorously to creating customer benefit, market success and company value. They do this by having implemented a value-based innovation management going far beyond technology development. Their innovation management has five interrelated thrusts:

- An overriding strategic objective to conquer market share and achieve premium margins with innovative products and/or services based on a competence platform enabling them to respond swiftly to customer needs,
- top priority attached to the innovation process, over and above all other business processes, which they understand to extend from systematically generating and assessing innovation ideas, to selecting and pursuing the most promising ones

The Innovation Premium

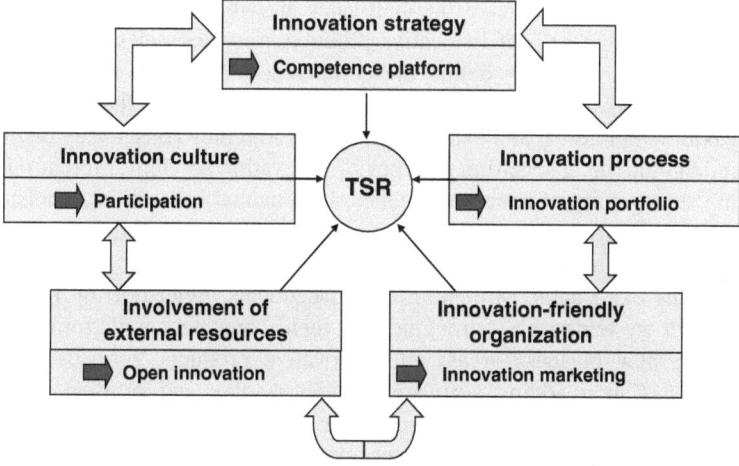

Fig. 1 Components of the best-practice model of value-based innovation management

to culminate in the successful market introduction of innovative products and services,
- the creation of internal structures favoring the constructive interaction of the various functional areas throughout the innovation process, and differentiating clearly between the efficiency requirements of their current operations, on the one hand, and the creative, explorative nature of new product and service development, on the other,
- the proactive involvement of a broad range of external resources such as customers, business partners and research establishments, in idea generation as well as in adding to their competence platform and exploring market needs, and
- fostering a corporate culture in which all of their employees align themselves with the company's innovation strategy and actively support the innovation process.

None of the premium companies has fully implemented these five thrusts – in fact, most of them excel in only one or two of them – but from the collective sample of premium companies studied we were able to derive a complete best-practice model of value-based innovation management (see Fig. 1). The components of this best-practice model of value-based innovation management are described in the following paragraphs and illustrated by selected company examples.

2 Innovation Strategy

Premium companies realize that international competition is increasingly based on innovation and therefore deploy their innovation capabilities more strategically than others. They set themselves explicit sales growth and profitability targets to be achieved with the help of innovative products, services, sales channels and/or value chain solutions. They define the competences that they need in order to enhance their innovativeness: their competence platform which they build systematically both with internal and external resources. And they consciously distinguish between, and pursue, a spectrum of types of innovation serving different purposes but being strategically of equal importance: incremental innovation to bolster current sales, breakthrough innovations to open up new markets, and radical business model innovations to create new businesses.

Companies like Canon, Millenium Pharmaceuticals, Beiersdorf or Tognum are examples for systematically strengthening their competence platform in order to penetrate new business areas with innovative products, services and business models: Canon adding software, sensor, display and printing know-how for office automation systems and computer periphery products (copiers, printers, telefax and integrated copier-printer-telefax machines) to their optical, opto-electronic and mechatronics know-how of their more traditional business lines (cameras, wafer steppers etc.), Millenium Pharmaceuticals combining robotics and genomics competences to build a contract research business in drug discovery and development, Beiersdorf moving from base competences of skin-care into a broad range of innovative female and male dermatological cosmetics, and Tognum adding to its heavy diesel engine technology for ships, tanks and construction vehicles a growing competence in combined heat and power generation.

In all these cases, top management knew that the success of their innovation efforts was not only a question of their own technical competences but required access to external ones plus a thorough understanding of unsatisfied customer needs in new application areas as well as completely new marketing/sales competences.

To make sure that these capabilities are available and are maintained at a superior level, they nurture a competence platform from which their people can generate a sustainable flow of innovative products and services. Thus they are able to out-innovate their competitors who may have similar technical competences but do not sufficiently understand latent customer needs, can therefore not tailor new products and services specifically enough to creating outstanding customer benefit, and typically underestimate the sales and application know-how required for ensuring market penetration.

German companies pursuing a platform-based innovation strategy are, for example, Linde Material Handling GmbH and Dräger Medical AG & Co. KG. Linde Material Handling have combined unique competences of hydrostatic drive systems with those of highly ergonomic cockpit design to offer superior operating productivity of their lift trucks which customers are willing to pay a premium price for. Dräger Medical's competence platform includes the intimate knowledge of the transport chain of critically ill patients in emergency and perioperative units of hospitals, enabling the company to offer a critical care system that enhances the survival chances of patients and the effectiveness of medical staff. Competitors typically think only in terms of individual technical apparatus.

3 Innovation Process

Premium companies do not confound R&D management with innovation management. Instead, they incorporate R&D management in their more comprehensive innovation process. This process begins with the thorough exploration of deficits that customers consciously or unconsciously experience in the application area at stake. Market research is only one of the instruments used; much more important for premium companies is that they ask and train their sales, marketing and development people to put themselves into the shoes of the customers and to discover deficits these customers suffer from, even though they may not (be able to) articulate them. The experience of several of the premium companies shows that recognizing and precisely formulating a customer problem is at least as important a step as developing ideas for solving it.

Therefore, companies like Hilti AG and Nokia separate the search for and characterization of problem ideas from those of solution ideas, both of which they collect from a broad spectrum of sources. They systematically tap, in addition to their own R&D, sales and marketing people, customers, suppliers, business partners, external research establishments, consultants etc.

To make collecting innovation ideas effective, most premium companies have designated a responsible unit or person whose task it is to promptly evaluate the ideas submitted applying clearly defined criteria, to decide on how to deal with them and to communicate the decision to the owners of the ideas. Whenever ideas or combinations thereof (e.g., a problem idea and one or several solution ideas) pass the first screening, they are further substantiated, usually by the owner of the idea(s) or the potential project leader together with a team of relevant technical, marketing and regional experts plus a representative from controlling.

The objective is to submit a well-founded project proposal to the innovation steering committee. This committee (called Innovation Steering Committee, Technology and Innovation Council, Innovation Review Board or so, depending on the company) is typically composed of a top management representative and responsible managers from the various functional areas involved in the innovation process. It evaluates the project proposals on the basis of their economic and strategic attractiveness and technical and economic risks. If deemed necessary, it requests further verification and information. Once satisfied, it asks the potential project leader and his team to elaborate a fully fledged project plan and budget.

On this basis, the proposed project is included in the overall innovation portfolio where it has to compete for available resources with all other on-going and proposed projects. The innovation steering committee selects the projects to be pursued at defined review points (milestones) as a function of the overall expectation value and pipeline profile of the innovation portfolio (Sommerlatte & Krautter, 2005). At these review points, the innovation steering committee reexamines the attractiveness and risk status of all projects and aligns their time to completion with the strategic timing requirements of the company. This can lead to discontinuing certain projects and shifting resources to other ones which help to improve the expectation value and pipeline profile of the innovation portfolio as a whole. Innovation portfolio management has proven to be a highly effective way of enhancing innovation performance in companies like Canon, Nokia, SAP, Metabowerke, Dräger Medical or Linde Material Handling which make sure that their innnovation steering committee is powerful and interdisciplinary.

Premium companies also apply a special marketing approach to their innovation projects which differs distinctly from marketing for established products and services. Already in the early development stages they begin to convince their own people of the significance of the innovative products and services under development for the future of the company, thus soliciting their people's commitment (internal marketing). This enables their people to competently dialogue with existing and potential customers and to discern which features of the products and services under development resonate with the market. On this basis, ongoing development work is, if necessary, modified in order to better zero in on customer needs (premarketing). At the same time, this interaction activates customer awareness of the deficits of existing products or services and arouses their interest in innovative solutions. Particularly in the area of investment goods, premium companies are working with so-called lead users with whom they ventilate already in the early development stages how their innovative product or service concepts can be improved to offer the highest possible benefit. Lead-users in turn are generally interested in cooperating because they expect to gain a competitive advantage by benefiting early-on from innovative products or services.

Another important element of innovation marketing as practiced by several premium companies is their intellectual property business which serves to broadly market their technical and application know-how as such. They aim at more rapidly amortizing their investment in know-how development and/or at licensing it out in exchange of external know-how needed. To do this, they create a close link between their IP management, R&D management and innovation process management.

The example of Nokia shows the key elements of the innovation process. Nokia is permanently looking for unmet needs in its various customer segments. People at Nokia hold intensive discussions with business users of mobile telephones and observe their communication profile. Thus they realized early on the problem of many managers having to deal with a growing number of communication forms: messages via fixed line and mobile telephone, per SMS, e-mail, fax and physical mail. From this they derived the vision of a miniature office integrating all communication forms and fitting into a pocket. This vision was translated into rough specifications and a competence platform which would be required to build and market such a mobile digital assistant. After having elaborated a technically feasible concept and assessed the benefits and attractiveness from a user's point of view, top management launched a highly ambitious development project for which the entire Nokia organization was won: the "Communicator" project. Missing technological know-how was quickly sourced from partners. Throughout the development process, Nokia had its people carry out acceptance tests with lead users from which the Nokia sales and marketing people derived their user benefit arguments. While still finishing development work on the first product generation, Nokia had a more advanced concept of a second generation ready and started to use the know-how acquired for other product categories such as mobile responders for logistics organizations. The market introduction of the "Communicator" (today Nokia E90 Communicator) was the beginning of a completely new category of mobile gear for business users which is by now a must in travel-intensive sectors such as consulting, the insurance business and banking. It contributed substantially to the remarkable growth of Nokia's TSR.

Among the few German companies having implemented a comprehensive innovation process are, for example, Dräger Medical and Lufthansa. Dräger Medical supports its innovation strategy characterized earlier by continuously interacting with hospitals and medical doctors positioning the company as a systems designer and complete solutions provider in the area of life-critical patient care. Lufthansa supports its strictly customer-oriented brand values and its design identity by accompanying all its innovation projects from the initial idea through to implementation, i.e., through the entire innovation process, by a feedback process with test users.

4 Innovation-Friendly Organizational Solutions

Premium companies have gone beyond the classical project organization which one finds in most R&D departments. Their project approach is no longer function-internal within the R&D organization but company-wide. Leadership of innovation projects has become a corporate mission which is entrusted not only to R&D staff but to qualified managers from other functional areas as well. Project leaders are elevated out of their organizational unit while they work on a project plan and, in case the project is approved by the innovation steering committee, for the entire project life-time during which they report directly to the innovation steering committee. One of their initial tasks is to select the most qualified team from the various functional areas and to develop the project plan with the team. They jointly submit the project plan and apply for the required budget and resources from the innovation steering committee. They position the project following defined criteria of economic and strategic attractiveness, technical and economic risks and in terms of estimated lead time to market introduction. On this basis, the innovation steering committee can compare the proposed project with all other projects in the innovation portfolio. If approved, possibly after some modifications, the project leader and his team are authorized to implement the capacities and means as planned. In the premium companies investigated, project leaders are not, as in many other companies, pro-forma project leaders without authority vis-à-vis the functional directors who "own" the resources but have the say over their team and budget, regardless of whether the team members are fully or only partly allocated to the project. In the event of conflict, it is the innovation steering committee (in which top management and the functional responsibles from R&D, sales/marketing, production, finance and controlling are represented) that decides. At defined review points, the project teams report on their interim results and an updated assessment of the attractiveness, risks and estimated time to completion of their projects, allowing the innovation steering committee to decide on the continuation or discontinuation of individual projects in the context of the overall innovation portfolio. Projects are considered completed only once the new products or services have been introduced into the market so that business success in the introductory phase becomes an important part of project evaluation.

The example of BP demonstrates the dynamizing effect of an organizational change to this type of project approach. Since the creation of BP's Peer-Assist-Program (PAP) at top management level, project leaders from any organizational unit can submit proposals for innovation and improvement projects to this committee and recommend a project team from other organizational units.

If approved by the Peer-Assist-Program, the project teams (called Performance Improvement Teams) are extracted from the three-dimensional matrix organization of BP (business areas, process units or regions) and report directly to the Peer-Assist-Program. This change led to a major increase of innovation initiatives and of the effectiveness of innovation and improvement projects and contributed to the increase of BP's TSR from initially low levels.

5 Open Innovation: Leveraging External Resources

Premium companies are good examples for open innovation: They aim at reducing the cost and duration of their innovation projects by engaging in various forms of cooperation and partnering to gain access to know-how and innovation ideas (Chesbrough, Vanhaverbeke, & West, 2006). Partners are typically suppliers with whom they share major development projects, contract research or engineering firms and specialized companies in neighboring sectors.

The success of open innovation arrangements depends greatly on the partners' readiness to equitably share the benefits of their cooperation. Premium companies willingly engage in contractual agreements and cooperative modes to safeguard the success of joint innovation projects.

The example of SUN Microsystems shows the stability of well-managed innovation partnerships. While SUN Microsystems carries out systems design und development of its specialized work station systems for medical, architectural and layout applications, the company forms so-called "single teams" with suppliers such as Zytec, Seagate and Solectron, develops Internet software in close cooperation with Novell and coordinates the development of monitors with Sony and of keyboards with KBM. The "single team" partners are ready to finance their share of the development effort because they secure an attractive share of the procurement volume of SUN Microsystems. In fact, they have located manufacturing facilities near SUN Microsystems in Palo Alto, California. Thanks to this cooperation, SUN Microsystems is able to maintain its innovative edge and to offer unusually short delivery lead time in spite of low warehouse levels. German car manufacturers have learnt to share their innovation effort with their systems suppliers and thus to shorten their development lead time significantly. This openness and co-innovation approach pays off particularly in tough times as currently experienced where new developments (e.g. the electric car) have to be fastened up.

6 Innovation Culture

Premium companies enjoy a corporate culture which supports innovative thinking, team behavior and flexibility of their people. This is what can be called an innovation culture. They have brought it about by getting their people to understand and support the objectives of the company's innovation strategy, by having them participate in the decisions for innovation projects and assume responsibility for the success of these projects, by providing the framework for internal cooperation and effective coordination, by creating transparency of the innovation process and by incentivizing commitment. Top management of premium companies is seen to communicate openly and to behave in a cooperative and participative manner.

The example of Metabowerke GmbH shows how a change at top management level can favor the development of an innovation culture. The new CEO who was appointed in 2004 by the family shareholders when the company was stagnating came with the vision of making Metabo the innovation leader in the electric handtool business. He convinced the company's leadership group to go for innovation as a major pillar of Metabo's corporate strategy. Business took off, and in 2008, Metabo won the "Top Innovator" prize for exemplary innovation management.

The company changed from being dominated by technical priorities to understanding and satisfying customer needs in the first place. Members of the company's leadership group are themselves assuming the patronage for major innovation projects and involve middle management in determining common innovation targets. These are then broken down into "innovation scorecards" for the various organizational units defining their contribution to the innovation process. An "innovation roadmap" serves to show the direction of the company's innovation efforts over a perspective of several years, ready to be adapted, when necessary, to new insights into market and technology developments. The top leadership group meets monthly with the company's innovation steering group, thus demonstrating the priority it gives to innovative products and business approaches. Creativity workshops, design contests and prizes for successfully executed innovation projects have stimulated a dynamism which motivates the sales and marketing staff to convincingly position Metabo as a provider of premium products with exceptional customer benefit.

7 Conclusions

The best-practice model of value-based innovation management characterized above represents a challenge for most companies, including the "premium companies" cited as examples for one or the other aspect of the model. But even having only partly implemented a value-based innovation management so far, the "premium companies" investigated have been able to achieve a clear innovation premium in terms of above average TSR.

It is therefore amazing how few of the companies claiming to be innovative have actually recognized the importance of managing their competence platform and their innovation process, of creating the organizational conditions for effective innovation project management and of developing an innovation culture.

In many cases, companies consider their level of R&D spending and of investment advanced technology development as a proof of innovativeness. Although there is no doubt that technological competence is an important ingredient, it does not suffice. Premium companies manage a much more comprehensive competence platform including the ability to focus on customer needs and tailoring products, services and business approaches to offering clear customer benefits, including outstanding ergonomic design and systematic branding. They have organized the innovation process from the early stage of "ideation" through innovation portfolio management to innovation marketing with the aim of achieving a high return on their innovation investment. Systematic innovation portfolio management helps them to optimize the expectation value and pipeline profile of their development program. Their project managers enjoy entrepreneurial responsibility and authority, different from the pro-forma role project leaders play in many other companies. Premium companies are actively influencing their culture to favor initiatives of their people towards innovative ideas, interdisciplinary cooperation and openness to external competences. Their top management is actively encouraging and expecting innovation initiatives.

Clearly, an imbalance between technology mindedness and non-technological competences, one way or the other, hinders companies from achieving an innovation premium. In that sense, the notion of "technological innovation" should be recognized as severely misleading. A high "return on innovation" is earned with innovative products, services and business approaches in the market place, not with technologies alone. Striving for the innovation premium needs to be the objective, not for technological leadership only.

References

Chesbrough, H., Vanhaverbeke, W., & West, J. (Eds.) (2006). *Open innovation: Researching a new paradigm*. Oxford: Oxford University Press.

Fortune Magazine. (2008). America's Most Admired Companies, http://money.cnn.com/magazines/fortune/mostadmired/2008/index.html

Jonash, R. S., & Sommerlatte, T. (1999). *The innovation premium*. London: Random House Business Books.

Sommerlatte, T. (2004). Capital market orientation in innovation management. *International Journal of Product Development, 1*(1), 1–11.

Sommerlatte, T., & Krautter, J. (2005). Innovationsportfolio-Management. In: Albers, S., & Gassmann, O. (Eds.), *Handbuch Technologie- und Innovationsmanagement*. Wiesbaden: Gabler.

Innovation in the Interplay of Organization and Culture: The TRUMPF Story

Christian Koerber, Gabriela Buchfink, and Harald Völker

Contents

1 TRUMPF: Leading High-Tech Company 91
   1.1 Company Portrait 91
   1.2 Innovation as a Corporate Strategy 92
2 Corporate Culture as the Key to Innovation 94
   2.1 Introduction 94
   2.2 TRUMPF Corporate Culture 95
3 Impact on the Company 97
   3.1 Strategic Impact 97
   3.2 Methodical Impact 102
   3.3 Organizational Impact 105
   3.4 New Function Areas 107
4 Conclusions 109

1 TRUMPF: Leading High-Tech Company

1.1 Company Portrait

Swabian model company, innovation leader, global player or financially independent family company – any of these can be used to describe TRUMPF. TRUMPF defines itself as a high-tech company whose goal is to lead the way in terms of both technology and organization in its fields of activity.

C. Koerber (✉)
TRUMPF GmbH + Co. KG, Johann-Maus-Straße 2, D-71254 Ditzingen, Germany

1.1.1 Fields of Activity and Business Areas

Today, a separate business area is allocated to each of the five fields of activity. The Machine Tools business area is the largest of the five and occupies the number one position for laser machining, punching, forming and bending machines worldwide, as well as for automated production solutions. The Electric Tools business area is the oldest of the five: the product portfolio includes handheld electric tools for cutting, joining and forming aimed at the professional market segment. The Laser Technology business area ranks as the world's best positioned supplier of laser technology for materials processing: it offers lasers, laser systems and process technology for welding, cutting, marking and for surface finishing at the macroscale, microscale and nanoscale. The Electronics business area manufactures high- and medium-frequency generators for the power supply process. Fields of application include induction heating, as well as plasma and laser excitation. The focus of the Medical Technology business area is on products for fitting out operating theaters, intensive care units and for hospital logistics.

1.1.2 Family Company

TRUMPF is and always has been a family company – even if it was not always owned entirely by the Leibinger family. Berthold Leibinger had already forged links to TRUMPF during his education and while writing his diploma thesis before he joined the company as Head of Design Engineering in 1961 after returning from a period in the United States, where he gained his first professional experience. He became a partner in 1964 and successively increased his stake in the years that followed. Between 1978 and 2005 he managed the company as Managing Partner and Chairman of the company's management board. In 2005 he handed over the chairmanship to his daughter Nicola Leibinger-Kammüller and took on a new role in the supervisory board, which he now chairs. The company's executive board also includes two other family members: Peter Leibinger, Deputy Chair of the company's executive board and Head of the Laser Technology and Electronics business units, and Mathias Kammüller, Head of the Machine Tools and Electric Tools business units.

1.2 Innovation as a Corporate Strategy

1.2.1 Innovation Means Converting New Ideas into Market Success

Innovation is "... the process of finding economic applications for inventions," wrote Austrian economist Josef A. Schumpeter in his 1912 book "The Theory of Economic Development." Today, financial success is a key focus, alongside the degree of innovation and the existence of meaningful applications. Accordingly, TRUMPF defines innovation as "converting new ideas into market success."

1.2.2 Strategy for Innovation Leadership

A company's innovative strength is a crucial factor of its success. Innovation leaders can position their products in the premium segment because of their unique selling points, penetrate the market as the first mover, and build up lasting competitive advantages. This paves the way for above-average results in terms of growth and profitability.

The TRUMPF business model is based on the realization of claims to technological and organizational leadership. TRUMPF has chalked up an average annual growth of 15% in turnover and operating profit since 1950. The values were frequently significantly higher than this, but TRUMPF did not escape the crisis that hit the German tool machining industry in the early 1990s unscathed. In the 2007/2008 fiscal year, around 8,000 employees generated a turnover of 2.14 billion euros. 57 subsidiary companies in 26 countries, 21 of those with production facilities, guarantee customer proximity throughout the world. TRUMPF operates profitably, with an operating margin of 14.1% before tax.

Continuous innovation is therefore vital for TRUMPF and is a firmly established business objective. Complementary measures include above-average investment in research and development, close cooperation with leading institutes and research establishments and measures to establish a working atmosphere that is conducive to innovation. In order to secure and build on the company's role as an innovation and technology leader, between 15 and 20% of the turnover is invested in research, development and expanding the infrastructure each year.

1.2.3 Technological Leadership Instead of Invention Leadership

Companies with a focus on innovation can distinguish themselves by being the first on the market (invention leader) or through technology (technological leader). Invention leaders address markets such as research establishments, for whom the early availability of a product is more important than its degree of technological maturity.

The industry, on the other hand, demands high-performance products with round-the-clock availability and sophisticated technology. New products for their own sake offer minimal added value for this sector. For manufacturers, this means additional development time and thus an extended time-to-market.

As a manufacturer of products for industrial production, TRUMPF pursues the goal of technological leadership. "We want to be first to place industrially viable products on the market." By the same token, the company accepts that in exceptional cases it may also be an "early follower".

2 Corporate Culture as the Key to Innovation

2.1 Introduction

2.1.1 Factors of Success in Innovation Management

The task faced by innovation management is establishing the necessary environment within the company and supporting and encouraging the relevant activities. Successful innovation management is based on a fine balance between the key variables of employee creativity, an ability to recognize successful business ideas, entrepreneurial implementation focus and the availability and organization of the necessary resources.

There is no such thing as an ideal combination of these aspects – each company faces the challenge of finding its own optimum combination. The same applies to the organization of the innovation management function and the degree of systematization: disruptive innovations necessitate the courage to call existing ideas into question and to come up with new and creative ones. This is opposed by the need for systematics and standards. It is crucial that dynamism is injected into the activities relevant to innovation and that the internal and external framework is continuously challenged and adjusted. The corporate culture provides the frame of reference for this.

2.1.2 Corporate Culture as a Defining Factor

The corporate culture encompasses all of the standards, ideals and ways of thinking that shape employee behavior and the company image. Part of this culture is readily accessible and documented in mission statements, guidelines and objectives. Other documents such as work instructions and forms reflect the corporate culture indirectly. Memories and anecdotes bring the corporate culture to life, and managers reinforce it by setting an example. The other part is not so obvious. It includes basic assumptions – also known as the unwritten laws. They are followed without being discussed and are normally not called into question.

The corporate culture plays a role in establishing uniform behavior within the organization and dictates the interaction between company and environment. It offers the company's employees behavioral guidelines and supports them in achieving the goals they have been set. An efficient corporate culture can save the need for formalization.

Like an individual's personality, each corporate culture differs in its characteristics and is unique, thereby giving the company a distinct character. As long as the entrepreneurial context is aligned to the values entrenched in the corporate culture, a company's strengths (and possibly also its weaknesses) will be maintained.

2.2 TRUMPF Corporate Culture

The specific characteristics of the TRUMPF corporate culture both encourage and demand a high degree of innovative strength. It has been shaped largely by the company's leaders, initially Christian Trumpf and then increasingly by Berthold Leibinger. The topic of innovation rates highly in the corporate culture.

2.2.1 Leadership Based on Expertise

An interest in technical innovation is a powerful driving force behind the company's actions. The aim here is not a series of incremental improvements, but rather radical innovation, also known as a "technology push", in order to safeguard distinguishing features for TRUMPF products in the long term.

This innovation strategy does not focus on determining customer requirements or on analyzing competitive products. Instead, future customer requirements need to be anticipated and met by innovations at an early stage. The prerequisite for this is profound knowledge of customer applications and an excellent understanding of the market.

2.2.2 Continuity

Knowledge of customer applications and understanding of the market can only be acquired through close contact and personal experience. On the one hand, this calls for an enduring and cooperative relationship between customer and supplier, and on the other hand continuity at the management level. TRUMPF managers have many years of experience in their specialist areas and the upper management posts are filled almost without exception by employees from within the company.

The principle of durability applies similarly with regard to the ties between the Leibinger family and its company – the family shapes the company and brings it to life. The second generation of family members is now actively leading and shaping the company on the basis of its expertise.

2.2.3 Courage and Willingness to Embrace Change

At the value and behavioral level, the claim to technological leadership is translated into expertise, enthusiasm for one's own area of work, creative will, using one's own initiative and a desire for continual improvement. This applies not only to development, but also to all areas of the company.

This is accompanied by a need for intellectual openness and a willingness to risk change and question established ideas. Instead of drawing up five-year plans, existing ideas are constantly challenged and adapted to meet changing requirements where necessary.

2.2.4 Creative Freedom

Innovation presupposes creative freedom, which is why unconventional ideas are also welcomed. Innovators have to take risks, so mistakes are unavoidable. We learn from our mistakes – but we shouldn't make the same mistake twice. Instead of subjecting all good ideas to a myriad of formal approval stages and stifling them in compromise, they are pushed through. This involves an open culture of debate and a willingness to make and receive fair criticism.

2.2.5 Focus on Implementation

Managers are responsible for innovation in their area of work and encourage their staff to improve existing products and to develop new ones. At the same time, managers are expected to demonstrate their entrepreneurial focus on implementation and be responsible for fostering new ideas and motivating their team to achieve a common goal. The involvement of internal and external bearers of knowledge and drivers increases an innovative idea's chance of success. This involvement should be kept within limits, however, so as to avoid unnecessary complexity.

2.2.6 Western Culture and Christian Ethics

TRUMPF's ideals are based on the values of the Christian West. The cultures and religions of other countries are treated with respect, however. Fairness, moderation, trust, respect and courteousness when dealing with others are a top priority. These values lay the foundations for constructive and open teamwork. Even if they do not always bring rapid and spectacular results, these ethics are not seen as a contradiction, but rather as an internationally valid basis for lasting success.

2.2.7 Family Company

The term "family company" applies not only to TRUMPF's ownership structure, but also more importantly to the Leibinger family's leadership. Three of the six executive board members are members of the Leibinger family and Berthold Leibinger is the Chair of the supervisory and administrative board. The family members' willingness to dedicate themselves fully to the company through the good and the bad is also expected of everyone employed by the company.

Social and cultural commitment is an important part of TRUMPF's family company culture. Its focus is on the company's location and employees, training and research, as well as medium-sized businesses and family companies.

A visible outward sign of the Leibinger family's focus on its employees and identification with the company is the corporate design reflected in the architecture of the buildings, communication and products.

These elements of the TRUMPF culture have shaped the company. Important strategic decisions on vertical integration or diversification, the need for internationalization, the pursuit of independence or the choice of distribution channel can largely be derived from this understanding. The same applies to organization – with regard to the leadership structure, lean management methods or the parameters for human resources management. The effects of the "corporate culture frame of reference" will be examined below.

3 Impact on the Company

3.1 Strategic Impact

3.1.1 New Business Development

Back at the start of the company's history, TRUMPF was a manufacturer of electric tools for sheet metal forming and was already known around the world for its high-quality products in the pre-war period. Its core competence was designing cutting heads and process know-how. Expansion of the business activities into the thick sheet sector was the obvious next step: the manufacturing methods used and customer requirements are initially the same. Technology, business model, expertise and market reputation can be credibly transferred. The tool design is new, however: thick sheet cannot be machined using hand tools. So in 1948 the decision was taken to construct a stationary machine for processing thicker sheets, known as a nibbler or cutting-out shears. This was the world's first nibbler.

Nibbling is an advanced punching process which uses partially overlapping strokes and can generate any contours in sheets with a single tool. This contour flexibility is also its Achilles' heel, however. The low repetition accuracy of manual sheet guidance limits its use in an industrial environment. To put it another way: to address a new market segment by means of series production, the stationary machines had to be developed from a tool into a machine tool, i.e. equipped with a control system. Berthold Leibinger's diploma thesis laid the foundations for the copy and coordinate nibbler introduced in 1958. The machine's innovative feature was its ability to scan a sample and transfer it to the workpiece guide using a type of pantograph – essentially a cam disk for accurately reproducing preset geometries. Thanks to this solution, TRUMPF evolved into an international company for machine tool building virtually overnight, and its turnover also rose dramatically. By the 1990s, over 13,000 machines of this type had been built.

Several years later, advances in electronics made the next technology leap possible. Berthold Leibinger observed the developments in the electronics sector from the very early stages and adapted the machine control systems developed for precision machining to the requirements of sheet metal forming. In 1967, TRUMPF introduced a numerically controlled punching and contour-nibbling machine. The use of NC technology to control a nibbler was a world first and gave TRUMPF a further boost in terms of turnover, internationalization and technology. Even today, the control system is one of the most important and enduring unique selling points of the company's machine tools.

The primary motivation behind the third major technology leap twelve years later was improving productivity and quality in the separation process. At that time, potential substitution technologies for nibbling were plasma cutting or laser beam cutting. Berthold Leibinger, who had meanwhile taken over as Managing Partner, opted for the laser as the tool for materials processing. He soon realized that the lasers available on the market did not meet TRUMPF's quality expectations, however. The decision to develop and build the company's own lasers guaranteed the availability of laser beams that were up to the job. The strategy of vertical integration and the use of systems in the company's own production facilities still offer an important competitive advantage even today: machines, lasers and processes are mastered individually and also as a complete system. Development and production from a single source offers customers not only a perfectly tuned system, but also a technically mature and proven solution for their production tasks.

It was this same strategy of vertical integration that led TRUMPF to take over Hüttinger Elektronik in the 1990s. Hüttinger supplied one of the key components of the TRUMPF CO_2 lasers: the generators for supplying the electricity. The acquisition guaranteed the availability of generators and a development roadmap geared towards application requirements. Other possible uses of the generators in the fields of semiconductor technology, large-scale coating of architectural glass or induction contributed to the desired horizontal diversification of the technology portfolio.

Laser technology was not only a substitution technology for mechanical nibbling, but at the same time a key technology that opened up the path to a multitude of new products, applications and markets for the first time. By consistently expanding its technology portfolio TRUMPF was pursuing the same strategy as it had done with machine tools. Its aim was to transfer the competency in macro applications that it had developed with the CO_2 laser to the precision and fine mechanics markets. At the time, these were the domain of solid-state lasers. By acquiring Haas Laser GmbH, which had its foundations in the clock and watch industry, TRUMPF took its first steps into the solid-state laser and microprocessing technology sector in 1993. The following years were spent developing and building on its role as a technological leader: continuous improvements were made to laser power, beam quality and controllability. A future-oriented technology platform was established by replacing conventional flash lamps used as pump sources with laser diodes (2001) and the laser rod with disk laser technology (2006). TRUMPF presented its first fiber laser in 2007.

Like with machine tools, TRUMPF's aim was once again vertical integration. Its acquisition of the world's second largest fiber laser manufacturer SPI in 2008 was particularly aimed at acquiring the expertise and equipment it required to manufacture its own fibers. TRUMPF is currently in the process of developing its own laser diodes which will be far superior to conventional diodes in terms of beam quality and power. These will be available on the market in 2009 and will not only pump the solid-state lasers (vertical integration), but also offer a range of new application possibilities as a direct radiator (new laser type). Today, TRUMPF is the world's leading provider of lasers, laser systems and process technology for welding, cutting, marking and for surface finishing at the macroscale, microscale and nanoscale.

Besides laser technology, TRUMPF has also expanded its business activities into conventional technologies. TRUMPF has focused its innovations on the sheet metal process chain, which essentially consists of three process stages: cutting – forming – joining. This focus on execution has resulted in products that can be used flexibly as standard tools or machines. The target markets are metal-working companies that manufacture small- and medium-sized batches of products that undergo frequent changes, such as on construction sites, in workshops or for job piece producers. For TRUMPF, the focus on standard products with manageable variance guarantees that they are capable of being manufactured in (small) series.

This focus becomes clear when one considers the company's first steps into the bending technology sector. In the 1990s, Linz-based Voest-Alpine disassociated itself from its bending activities. These included both flexible machines for bottom bending and highly automated hemming machines for batch production in the thin sheet sector. Leibinger decided to add forming to the existing technologies for cutting and joining sheets and to position TRUMPF as a supplier for the complete sheet metal process chain. He confined himself to bottom bending, however. This technology addressed the same target customers and was based on the same business model as the other TRUMPF technologies. Confining the innovation to the technology contained the risks associated with entry into the bending technology sector.

TRUMPF opted for a different route when entering the medical technology market by way of the Hüttinger acquisition. Here, the primary focus was on diversification by means of transferring the competency acquired in the machine tool construction and laser technology sectors to a new market with a different dynamic. Since Hüttinger was already active in HF surgery, thanks to its competency in HF, MF and DC generators, this field was initially addressed by using the laser for laser-induced thermotherapy (LITT). Due to better compatibility with the TRUMPF business model, the decision was then taken to focus on fitting out operating theaters

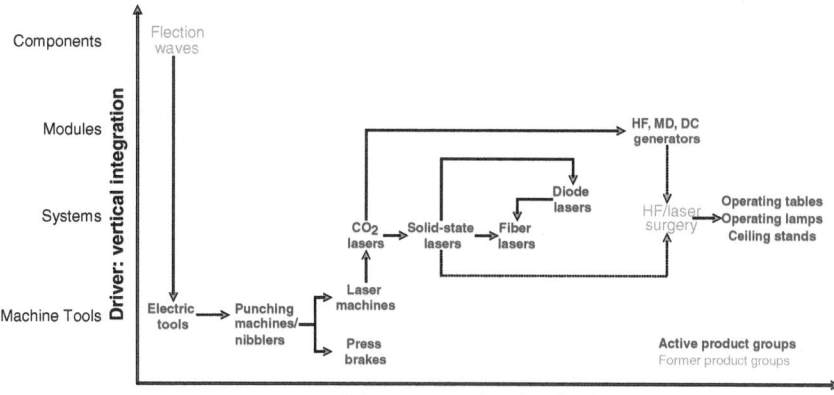

Fig. 1 TRUMPF innovation strategy

instead, however. Examples of this technology transfer are the operating tables, ceiling stands and LED-based operating lamps.

The development of new business areas described followed similar strategic patterns. The company's expansion into new technologies – by way of both internal development and acquisition – is motivated primarily by its ongoing quest to improve the quality and performance of existing products. Its focus on expertise and continuity is reflected both in the depth of added value (in-house production of key components) and in (external) growth. The prerequisite for expansion into new technologies is a profound knowledge of the relevant target market and its requirements – acquired either through integration into the company's own products or through market presence over many years and a close rapport with customers. This facilitates the application of technology, results in new fields of expertise being established and simultaneously increases the possible applications of the new technologies in other fields.

3.1.2 Growth and Internationalization

The innovation strategy described above, together with the leadership approach, had a direct effect on the company's development. Looking back, three major growth phases can be identified, each of which was shaped by one of the three technology leaps: copy nibbling, CNC control and laser technology. The following diagram exposes this interrelationship.

The "technology push" and the leadership approach led almost inevitably to conflict with other markets and made intense internationalization necessary at an early

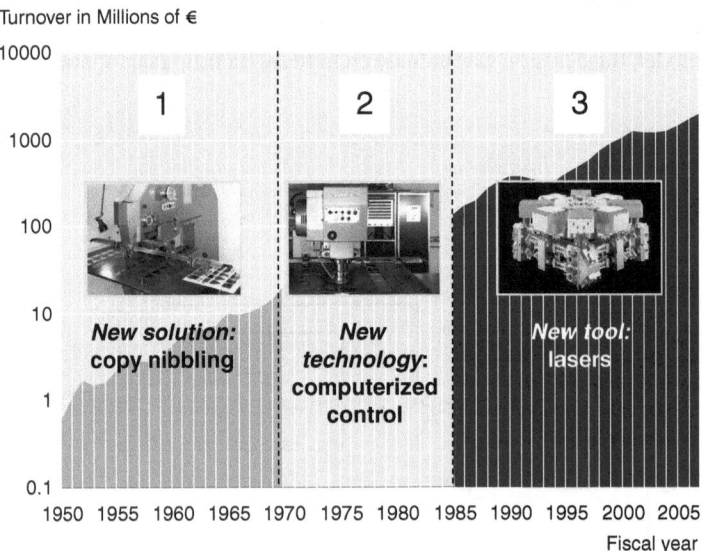

Fig. 2 Average innovation-based growth of 15% in turnover and operating profit

stage. An understanding of the processes and knowledge of customer requirements are prerequisites for leadership based on expertise. These can only be established and developed through competent, continuous and cooperative contact. Direct marketing therefore offered the ideal distribution channel. In the area of After Sales, the quick-response and professional service were not merely a point of contact to the customer, but much more a customer retention measure and additional mainstay of turnover in one. Direct marketing and rapid-response service necessitated a local presence.

New markets were initially handled by a representative and then by the company's own sales office, and finally by a service center. In key strategic markets, the creation of value was successively expanded to the application laboratory, production and procurement, and partly even to development. In many respects, local presence is vital for the TRUMPF business model: from a strategic perspective, it guarantees proximity to the customer and its regionally specific requirements. At the same time, it sends a clear signal to the customer that it is highly valued. From an operational point of view, shorter delivery and response times can be guaranteed. Furthermore, the sales department is able to demonstrate the machines in the company's own production environment. From a financial perspective, regional presence offers the possibility of controlling the cost structure in line with the market and using natural hedging to cushion the exchange rate risks associated with procurement and sales.

This focus on growth and globalization resulted in the establishment of subsidiary companies throughout the world. Here, TRUMPF led the way for other medium-sized businesses. A subsidiary company was founded in the United States as early on as the 1960s. At a time when other companies were still deliberating over involvement in neighboring European countries, TRUMPF had already recognized the opportunities in the Far East and was active in Japan. And this tradition continues to this day: when it purchased its own production plant in Japan in 2008, TRUMPF became the first non-Japanese company in the machine tool sector with its own development and production facilities in Japan. TRUMPF is currently present in 26 countries with 57 subsidiary companies. Production takes place at 21 sites in Western and Central Europe, in North and Central America and in the Far East.

3.1.3 Financial Independence

Financial independence is both a prerequisite and also a consequence of the TRUMPF corporate culture. Independence means being able to use the company's resources to finance growth, without borrowing large amounts or bringing in external shareholders. As a family company, TRUMPF is able to operate a conservative dividend policy. Most of the profits earned stay within the company and enable the company to finance its own growth from the cash flow.

This means the family is able to shape the company according to its own goals, to develop long-term strategies in the cyclic machine construction industry, and to implement these sustainably, that is to say independently of business partners.

3.2 Methodical Impact

The principle of continual improvement achieved by taking large and small leaps that has always existed in the technology sector was transferred to structural, process- and behavior-related topics in the mid-1990s. Examples of method innovations include the introduction of lean management methods and the inventor program, as well as the further development of quality management.

This process is founded on the belief that only companies which continuously improve in all areas by taking small and large steps will achieve long-term success. It is important to be able to react quickly to changes in the market and in the overall environment. Technical precision, engineering virtues and networked actions help overcome challenges. A primary goal for players on the world market is meeting the requirements of users in all regions. And that demands consistency and flexibility in equal measure: being large enough to develop high-technology in the first place, but at the same time staying flexible enough to apply the technology to various applications.

3.2.1 Project SYNCHRO

Following the example set by the Japanese car manufacturer Toyota, TRUMPF developed its own synchronous production system in 1998: SYNCHRO. It consists of a system of various methods for systematically optimizing production processes and simultaneously improving the quality of the products. SYNCHRO carries the idea of lean production and lean management over into machine construction. The central idea of lean management is the elimination of waste. This is based on the recognition that not everything that employees do during a typical workday adds value. SYNCHRO employs three principles to eliminate wastage:

- The largest source of waste is overproduction. For this reason, production takes places exclusively in response to orders placed.
- This is only possible if perfect quality is achieved at all times as a result of secure processes.
- The basis for any improvement is the knowledge and creativity of all employees, which is why they are involved at all stages.

To this day, the key advantages remain self-organization and self-optimization of semi-autonomous production units. Each staff member is able to recognize the direct outcome of his own actions. Competency and responsibility are present at the point of action. The interfaces between production units are diminished. Communication becomes simpler and more direct. The content of employees' work is enriched. Administrative and supervision expenses are reduced.

The principle of decentralization was the basis for introducing the production system. In the implementation stage, this meant establishing so-called production units, each with between 30 and 100 employees. They take on all direct and indirect production tasks required to manufacture a finished product or a clearly definable

Vision	World class in all processes Optimal fulfillment of market needs. Permanent avoidance of waste.					
Goals	Best quality		Least cost	Highest flexibility		Shortest throughput time
Principles	Quality and assured processes		TRUMPF culture Employee integration			Just-in-time production
Methods and Tools	Six Sigma	5A	Team work	MIT	Flow production	SMED
	Transport on wheels	Audits	I-Point	CIP	Supermarkets	Level smoothed production
	TPM	Standards	Management by Objectives	Qualification	KANBAN	JIT
	Footprint	Value Stream Mapping	Benchmark	KIS	One piece flow	Pull principles
	EKIB	Poka Yoke	Time Managemen	Problem solving methods	Part Set	Synchronized pull
Evaluation	World-class processes ⇒ Ranking Goal attainment ⇒ Key figures Method application ⇒ SYNCHRO audit					

Fig. 3 The "SYNCHRO House"

sub-assembly. The production units are responsible for their respective product and act as a sort of factory within the factory. As a result, central functions such as production planning and control, work scheduling and material planning are dispensed with. The production units determine their own requirements and schedule their parts in their own inventory-controlled warehouses. Within the corporate group, the relationship between the production units is that of customer and supplier. The three SYNCHRO principles were implemented from 1998 onwards on this basis.

The principle of "just-in-time production" is based on the approaches of synchronized continuous assembly, single piece flow, parts set production and control using kanban. "Quality and secure processes" are the prerequisite for being able to realize the quality demanded by customers on a permanent basis. A high level of standardization is aimed for at all times, in order to reduce diversification and variations in the process. Standards are defined using methods such as 5A, Six Sigma, poka-yoke or value stream mapping. The "involvement of employees" takes place as part of a continuous improvement program with regular SYNCHRO workshops, by applying the approach of systematic problem solving, control using key figures and cross-workplace qualification.

TRUMPF's modification of the methods to match its own requirements instead of blind adoption, as well as consistent implementation by the management, all of whom were convinced by the SYNCHRO principles in advance, were crucial to success. TRUMPF allows numerous employees to dedicate all of their working hours to reviewing and refining the changes: in 2007, 2.4% of the production staff were made available as SYNCHRO specialists. This results in an improvement mentality that is strengthened further by the regular publication of targets and figures. Conscious competition between departments and entire companies to perform best in audits is also very beneficial to the entrenchment of the SYNCHRO principles.

SYNCHRO is a prime example of how the corporate culture influences the company's success through organization and behavior. It is based on the pursuit of continual improvement. Prerequisites for its success are institutionalization of the principle as a process and the active participation and creativity of employees as bearers of knowledge. SYNCHRO lives from courage and desire for change, as well as direct implementation of optimization approaches. The success of SYNCHRO using the example of a punching laser machine speaks for itself: inventories and machining times were halved. Surface area productivity was almost doubled. Group-wide productivity has increased by more than 5% annually, thanks to the institutionalized improvement program.

True to the strategic success model for developing new business, a method for improving the organization of indirect functions ("BüroSYNCHRO") was derived based on the principles of success of the SYNCHRO production system.

3.2.2 Inventor Program

New structures and processes have been established not only for production and administration, but also for development. The aim of the inventor program, for instance, is to offer an incentive to register more inventions around the world and then turn them to account as patents. For this, the different legal requirements in the individual countries had to be taken into consideration. All inventors worldwide – if the reported invention shows promise – receive financial recognition for their inventive work relatively early. In countries where the law stipulates remuneration commensurate with the economical success of a patent (in Germany, for example), an initial agreement about the subsequent procedure is reached at this early stage already. The advantages that the system offers both parties are obvious: the inventor receives early financial recognition, regardless of the patent application's success and the utility of the invention, without foregoing any subsequent claims to compensation. On the other hand, TRUMPF's administrative expenses are reduced and it is able to gear inventions towards the continual review and further development of TRUMPF products, or their replacement by new TRUMPF products.

3.2.3 Quality Management

For TRUMPF, as for many other companies, process optimization and quality were important themes in the 1990s. In 1996, faced with the prospect of certification according to DIN ISO 9001 for the first time, the company was confronted with the task of firmly establishing the quality objective of "customer satisfaction achieved through first-rate products and services" in the consciousness of employees and providing them with tools to achieve this objective. Instead of introducing a formal quality policy, nine simple quality principles were developed and communicated using icons and catchwords.

These principles contain several traditional quality management values such as error avoidance, adherence to delivery dates and controlling. But they also include values typical of the TRUMPF corporate culture: order, cleanliness, simplicity, taking on responsibility and teamwork.

3.3 Organizational Impact

The company organization as an operational and organizational structure encompasses processes, leadership structures, competency profiles, target objectives and control. Important influencing factors are the purpose, culture and size of the company. The company's size, agents and, consequently, its interpretation of functions change over the course of time. An altered understanding of roles directly affects the leadership style and thus the corporate culture.

3.3.1 Organizational Development

The TRUMPF corporate culture has made a lasting mark on the leadership and the organizational structure, particularly in the growth stages. Leadership based on expertise, combined with a pronounced focus on implementation, has resulted in a leadership culture known as "management by delegation and by exception". Decisions were made quickly and locally; conversely the headquarters was quick to intervene in the event of deviations or problems. The exchange of information was very intense and communication direct, without "official channels" being adhered to compulsorily. This enabled targets to be pursued without the need for formalized strategy descriptions or target agreements.

This leadership principle was reflected in the organizational structure. The hierarchies were flat and the paths short. Key members of staff and localized distribution units were afforded a high degree of independence. Minimal formalization in the description of responsibilities offered them the opportunity to find their own place within the company and to make an impact.

The focus on functions that had emerged in the early stages did not change significantly during the growth phase. The company was managed by the CEO and the heads of development, production, sales and finance. The increasing technological diversity and internationalization did not alter the centralized leadership structure and top-down management.

The enormous growth and associated complexity due to increasing internationalization and a growing number of sites, technologies and business areas were a driving force for change in the organization. Analogous to the introduction of lean management methods for processes, a structure was sought that would enable the inherent strengths of a medium-sized company such as flexibility, transparency, market proximity and response speed to be retained.

A divisional organizational structure was selected as the proposed solution. Key measures include the introduction of a holding structure in 1999/2000 and the subsequent decentralization of research and development. The five technology groups, electronic hand tools, machine tools, laser technology, electronics and medical technology were transformed into business areas and consolidated into three business units. The heads of the business units are part of the Group executive board. Besides this, each of the six executive board members takes on Group-wide responsibility for individual business functions. Last but not least, the subsidiary companies are allocated to the Group executive board members according to a regional key. On the one hand, this structure reduces the complexity of the company into "manageable" areas, but on the other hand it emphasizes the inter-divisional overall responsibility of each of the Group executive board members.

The decentralization of research and development transformed a functionally structured central division into regional product centers each with responsibility for researching, developing and producing a product group such as punching, bending, laser machines, gas lasers, solid-state lasers, medical technology, automation systems, etc. Group-wide bodies made up of managers and experts, such as the Technology Development Team (TET), guarantee transparency and cross-divisional cooperation. Furthermore, an advanced technology development which fulfills two functions was outsourced: basic development and the transfer of the development of critical components and systems. The development resources can be better managed and lead times for development projects are reduced.

The transformation from a function-based corporate structure to one organized by division allows complexity to be mastered. At the same time, TRUMPF continues to be managed by a powerful headquarters and the leadership structure remains horizontal. However, operative managerial responsibility is delegated cross-functionally. It was clear that this restructuring would require the key players to re-interpret their own functions. No development manager had sufficient experience and expertise to command all technologies, for instance. Consequently, the interpretation of the function had to be altered from "first developer" who develops creative solutions, takes important decisions and leads his staff more or less as a vicarious agent, to competent manager of a development portfolio and the associated resources. Conversely, the middle management had to take on not only operative, but also strategic responsibility.

3.3.2 Human Resource Management

The influence of cultural identity on the corporate strategy also became evident in the recruitment of new employees and employee development. Regardless of function, the initial focus is on expertise. People whose only aim is to lead don't stand a chance. Social skills and compatibility with the corporate culture are even more important, however: "Attitude comes before competence." The background to this is the understanding of leadership and organization described above. Employees act on the basis of their cultural identity, their socialization to TRUMPF behavioral patterns and their tight network. This applies particularly to leadership positions. Managers are encouraged to form a tight network at business unit manager meetings, manager seminars, international conferences for executive managers, head of sales or head of production meetings, etc. One result of this approach which might seem bewildering to outsiders is that TRUMPF managers display a very similar strategic understanding and behavior, even though no vision or self-contained strategy has ever been formulated.

This innovation-conducive atmosphere is an essential prerequisite for creativity and preserving the potential for taking major steps forward in the future. Accordingly, the real motivation for TRUMPF employees lies in this openness, the interesting tasks they are set, and not least in the respect they are afforded by the management and company members. In contrast, it is TRUMPF's belief that monetary incentives would result in motivation by a devious route. This attitude correlates directly with the ethical values exemplified by the Leibinger family: diligence, dedication, modesty, groundedness, cultural openness and responsibility. Intellectual openness and ethical stance result from the family's roots in Korntal Pietism and also from the cultural openness of Berthold Leibinger's parental home. His father dealt in East Asian works of art.

3.4 New Function Areas

3.4.1 From Functions to Function Areas

In small companies, the entrepreneur is frequently also an inventor. The factors of creativity, a feel for success, entrepreneurial focus on implementation and assertiveness, both within the defined processes and transgressing these, are tied to human resources. The organization is aligned to these key players. Each innovation highlights the significance of these central figures, since the turnover generated has a significant impact on the company's growth, due to its relatively small size.

The company's growing size and internationalization altered the competitive environment and increased the competitive pressure. The continually expanding range of technologies and products increased the complexity of innovation. In order to continue to achieve growth rates on a par with those realized in the past, an array of innovations had to be accomplished simultaneously in difficult conditions and an increasingly broad technology field. The technological diversity and number of innovations required increased. To achieve the same growth rates as in the company's early days, the innovative strength needed to be improved dramatically. Innovation projects initiated and conducted by individuals were no longer sufficient.

The principle of delegation offered a possible solution by consistently tying in employee creativity and using additional external resources, where necessary. In order to nevertheless safeguard expertise, experience and the entrepreneurial feel for success, the activities relevant to innovation needed to be systematized and dynamized. Consequently, innovation-related tasks and functions that had previously been carried out by managers were transferred to their own function areas for the first time. Examples of such measures included the establishment of a strategic marketing department together with its customer and knowledge management tools, the establishment of a central department for mergers and acquisitions, the Group-wide concentration of SYNCHRO activities into an internal organization development department and the introduction of company-wide innovation management.

TRUMPF demonstrated courage and willingness to change on several different levels. Innovation was delegated. Methodical expertise was introduced alongside technical expertise. A balance needed to be struck between creative freedom and processes geared towards effectiveness and efficiency. The function areas were reconstructed as central service providers. The transformation is described below using the example of innovation management.

3.4.2 TRUMPF Innovation Management

Peter Leibinger, Deputy Chair of the company's executive board and Chair of the laser technology and electronics business unit is responsible for the topic of innovation at TRUMPF. TRUMPF innovation management faces the task of converting strategic goals into concrete decision-making templates for innovation ideas and subsequently initiating or realizing their implementation. Innovation ideas can be related to business units, products, technologies, business concepts or processes, for instance. Systematic ideas management ensures that ideas are generated and recorded, assessed, selected and implemented. New Business Development is responsible for the stage gate-based development, assessment and implementation of innovative business ideas and strategies which cannot be allocated to the existing units.

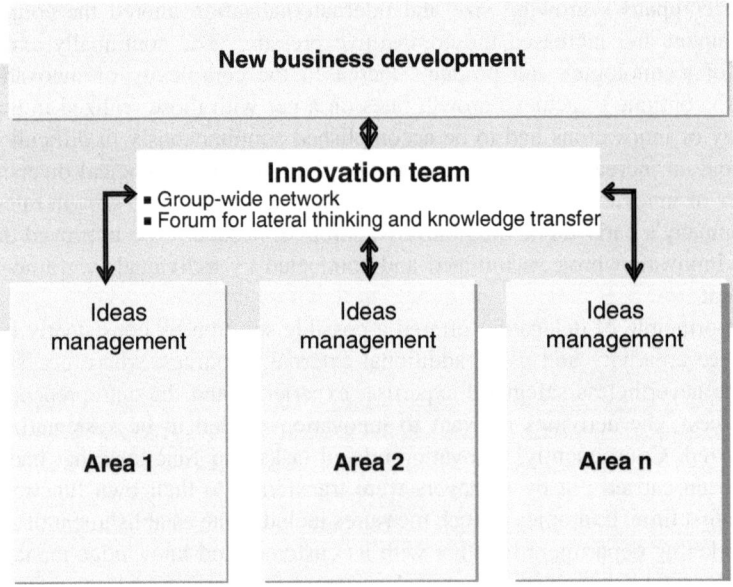

Fig. 4 Structure of TRUMPF innovation management (schematic)

Innovation management is organized based on the principle that actively involving all of the company's employees lastingly strengthens the capacity for innovation. Ideas management is decentralized for this reason. In the individual TRUMPF product centers, professional ideas managers initiate the innovation process. They are responsible for systematically gathering or generating as many raw ideas as possible, recognizing good ideas and converting them into concrete measures such as development projects.

TRUMPF's New Business Development division is attached to the holding company and has overall responsibility for the innovation process within the company. Part of its work involves the technical coordination of decentralized ideas management, as well as processing new themes. The New Business Development division and ideas management together make up the innovation team, a group-wide working team which links employees and knowledge across divisions, makes sure that experience is exchanged and offers continuing professional education and a forum for lateral thinking and knowledge transfer.

4 Conclusions

This analysis of the company history reveals that every great innovation is linked to an existing product or business area. Even when the innovation aspect concerned (position in the value creation chain, technology or market) involved a high degree of risk, the risk remained manageable thanks to this policy of linking. A further factor of success is proximity to the market. The profound understanding of customer processes enabled future requirements to be anticipated and new technologies and solutions to be developed for these. And last but not least: Without an openness to new ideas, a willingness to get to grips with other technologies and cultures and a desire for implementation, these opportunities could not have been taken advantage of.

The strategy of innovation leadership not only relates to products and services, however. The corporate culture influences the methods, processes and management organization in equal measure. For this strategy, the company as an innovative total package is shaped by four areas of innovation, the Four Ms: The first M stands for *machines (and components)*. When this concept is applied to other business areas, it also includes all other products and services, however. *Markets* not only relates to market penetration, but also to developing new markets with new and existing products. This applies to user groups as well as to regional markets. *Manpower* is not reduced to a production factor, but rather every employee contributes to the company's success through their dedication, richness of ideas and loyalty. TRUMPF therefore attaches great importance to employee dedication and a corporate culture that is conducive to this. The fourth area of innovation concerns *methods*: Anyone who wants to innovate successfully needs the correct tools, for instance to improve processes in direct and indirect areas.

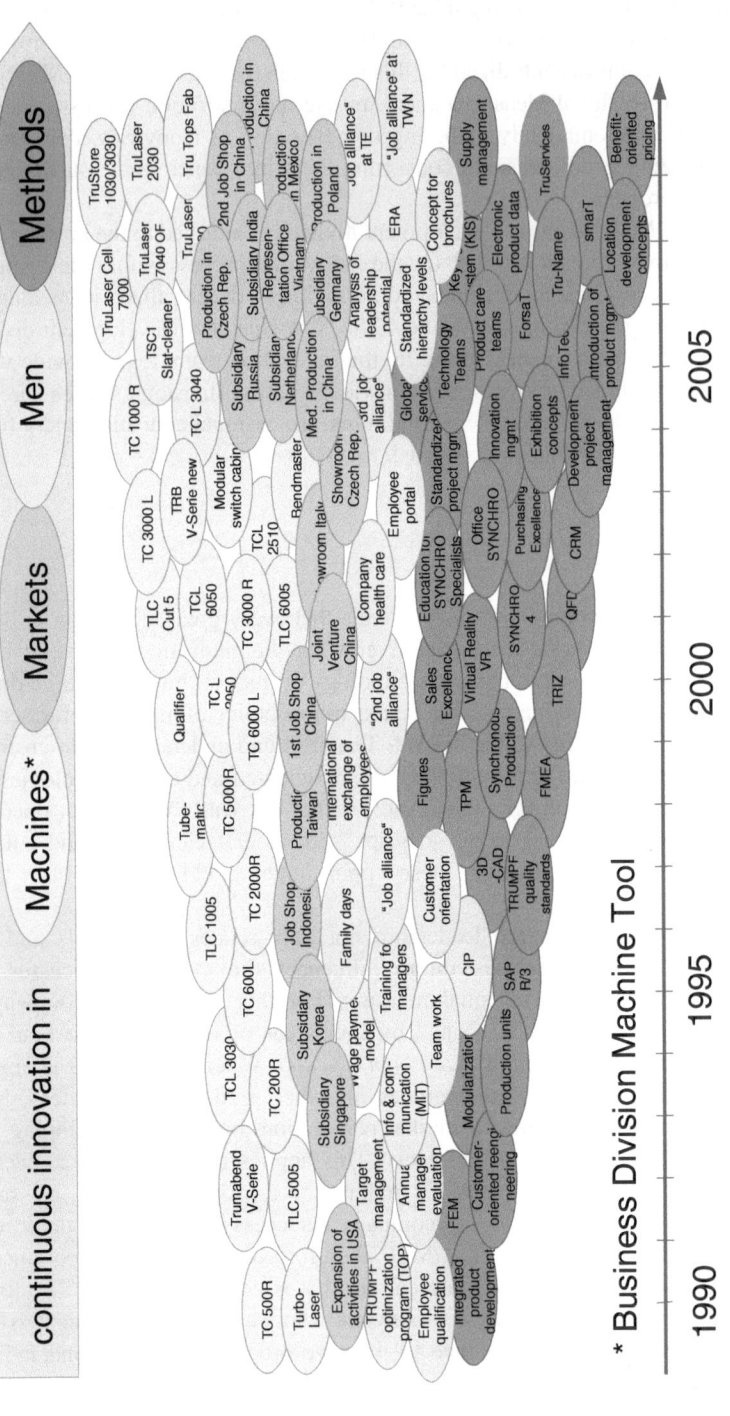

Fig. 5 The company as an innovative total package

So is the "TRUMPF story" a classic? Are the approaches to the solution a benchmark for other companies? Certainly not. The strategies and approaches to the solution are common knowledge. The success is the result of the company's own specific adaptation and combination, as well as the consistent application and implementation of these approaches. Ongoing improvement at all levels and the exemplification and implementation of clearly defined values are recurrent themes of the company's history. They allow the balance between constancy and change to be successfully preserved. Even as a global player, an organizational form was found which permitted the traditional strengths of flexibility, transparency, market proximity and response speed to be retained.

The TRUMPF company history reveals that innovations are the basis for financial success and growth. But innovations inevitably entail modifications to the organization. There can therefore be no such thing as a model organization and culture conducive to innovation. On the contrary: The key to success is injecting dynamism into activities relevant to innovation and preserving a breeding ground that is conducive to innovation by means of continual change. A company's ability to master this continual change thus determines its success.

A clearly defined and implemented corporate culture simplifies change processes by offering a point of reference. It does not necessarily require elaborate standards in organizations, however. Of course, the corporate culture should not be set in stone and threaten to become a dogma. Continuity is only necessary for the core values; the peripheral values can and should be altered along the way.

Growth and Internationalization: Renewable Energy and New Technology-Based Firms

Christian Schwens, Holger Steinmetz, and Rüdiger Kabst

Contents

1 Introduction 113
2 Theoretical Background 114
3 Data 116
4 Growth and Internationalization Patterns 117
    4.1 Timing and Degree of Internationalization 117
    4.2 Foreign Market Development and Entry Mode Choice 118
    4.3 Networks 119
    4.4 Learning 120
5 Summary and Implications for Practitioners 121
References 122

1 Introduction

Due to changing market conditions, a lowering of trade-barriers, and new technological developments, the competitive environment of firms has changed dramatically over the past decade. Consequently, even small firms face increasing international competition from the beginning of their organizational life. The increasing importance of trans-national and multi-national economic activities exposes the firm to a number of opportunities and challenges alike (Schwens, 2008). On the one hand, internationalization offers fundamental market opportunities that help to realize economies of scale and scope; on the other hand, firms experience liabilities of foreignness such as legal, cultural, linguistic or political barriers while internationalizing. Due to the increased international integration of markets, firms have to develop strategies that allow them to develop the domestic market as well as to venture into foreign markets in order to achieve a sustainable firm development.

C. Schwens (✉)
University of Giessen, Licher Str. 66, D-35394 Giessen, Germany

Against this background, the aim of this chapter is to gain a deeper understanding of the growth and internationalization patterns of firms in high-growth industries such as *Renewable Energies*, *Microsystems*, *Biotechnology* and *Nanotechnology*. These industries are well-known for their technological excellence and have been asserted as the very future industries of the German economy. Knowledge about innovations and path-breaking technologies in these sectors has been shared to a great extent in recent years. However, studies applying a management perspective to explore dominating growth and internationalization strategies of these firms are largely absent. There is reason to believe that firms from these industries apply different strategies in order to grow and survive. For instance, the corporate strategy of Renewable Energy firms is often dominated by strategic actions induced by legal regulations, support programs, or subsidies. In contrast, Technology firms (i.e., Microsystems, Nanotechnology, and Biotechnology firms) are highly knowledge intensive and therefore have a strategic orientation which is more towards niche markets.

We suggest that these industry-specific characteristics will be manifest in the growth strategies of the firms. In the present study, we explore the different growth strategies of the firms. In this regard, we focus on the timing and intensity of internationalization, foreign market development, and entry mode choice as well as the role of networks and learning in the course of internationalization activities. We base our arguments on two well established theories from the international business literature – the Process Theories of Internationalization (Johanson & Vahlne, 1977/1990) and the International New Venture Theory (Oviatt & McDougall, 1994). The next section introduces the Process Theories of Internationalization and the International New Venture Theory and identifies differences and commonalities among the two. Along with these commonalities and differences, we then analyze the internationalization patterns of German Renewable Energy and Technology Firms. We use data collected via a postal survey ($n = 335$).

2 Theoretical Background

Over the past years, a large part of the international management literature has intensively discussed the contrasting views provided by the Process Theories of Internationalization (PTI) (Johanson & Wiedersheim-Paul, 1975; Johanson & Vahlne, 1977/1990) and the International New Venture (INV) Theory (Oviatt & McDougall, 1994/1995/1997).

The basic idea of Process Theories of Internationalization (PTI) is that companies lack knowledge about foreign markets, which hampers foreign market entry and subsequent internationalization patterns. Furthermore, knowledge can only be acquired stepwise as "the model focuses on the gradual acquisition, integration and use of knowledge about foreign markets and operations, and on the incrementally increasing commitments in foreign markets" (Johanson & Vahlne, 1977: 23). Hence, the internationalization process is depicted as a learning process. The acquisition of knowledge over time is considered to be a firm's resource. One key characteristic of PTI is the *psychic distance* (Johanson & Wiedersheim-Paul, 1975) which is defined as "the sum of factors preventing the flow of information from and to the market" (Johanson & Vahlne, 1977: 24). Such hindrances of information flow result from differences in legal, political, linguistic, cultural or economic norms between country markets. Through gradual internationalization from psychically close to more psychically distant markets, the firm reduces the frictions resulting from psychic distance incrementally. Finally, the process takes place along the *establishment chain*. That is, foreign market treatment occurs in the steps "no regular export", "independent representative (agent)", "sales subsidiary", and "production" in the final stage (Johanson & Vahlne, 1977: 24). Firms begin to export as a response to requests received to sell products abroad (Aharoni, 1966). Thus, internationalization is perceived to be a reactive process.

International New Venture Theory (INV) (Oviatt & McDougall, 1994), on the other hand, contrasts traditional internationalization process theories by pointing to the young firm age at which companies start internationalization and by stressing that firms do not necessarily venture into foreign markets incrementally but internationalize on various steps of the establishment chain ("leap-frogging"). Hence, internationalization efforts may start by directly establishing an international joint venture instead of first penetrating the foreign market with export activities.

The main purpose of INV theory is to explain why firms internationalize right from or shortly after inception. The theory found widespread acceptance in the international entrepreneurship literature and has been regarded as having "started an important and influential research stream, whose contributions have been insightful, powerful, and varied" (Zahra, 2005: 27). INV has made valuable contributions to research by examining why young companies internationalize rapidly at various steps of the establishment chain.

Table 1 shows the core elements of PTI and INV and in how far these are complementary or contradictory. Both theories make assumptions about the timing and intensity of internationalization: PTI regards internationalization as a slow and reactive process based on an established domestic market, whereas the INV theory describes a rapid and proactive internationalization approach shortly after firm inception. Both theories emphasize the role of networks and learning in the course of internationalization. Whereas PTI argues that networks form a mechanism to acquire foreign market knowledge in the post-entry phase, the INV theory exemplifies the role of networks as a pre-entry mechanism allowing for early internationalization.

In the present chapter, we draw conclusions from INV and PTI theories to elaborate the internationalization behavior of German Renewable Energy and Technology firms (i.e., from the Nanotechnology, Biotechnology, and Microsystems sectors). Using survey data, we illustrate the timing and degree of internationalization, market development and entry mode, networks, and learning.

Table 1 A comparison between Process Theories of Internationalization (PTI) and the International New Venture Theory (INV)

Feature	Process Theories of Internationalization	International New Venture Theory
Timing and degree of internationalization (Age at entry, international intensity; number of foreign markets)	Slow internationalization activities based on a stable domestic market; Older firm age at timing of foreign market entry; limited intensity of internationalization; limited number of foreign markets; gradually increasing internationalization efforts	Rapid and proactive internationalization shortly after firm inception; high degree of internationalization in terms of intensity and number of foreign markets
Foreign market development and entry mode choice	Gradual resource commitment along the establishment chain mostly starting with export activities	Leap-frogging; internationalization not necessarily along the establishment chain
Networks	Networks play an important role in the (post-entry) internationalization process of the firm (Johanson & Vahlne, 2003)	Networks as an enabling resource allowing for an early foray into foreign markets (pre-entry)
Learning	Regulating role of learning once the internationalization process has been started (post-entry emphasis)	Learning as an enabling factor in the pre-entry phase of internationalization

3 Data

The sample used consisted of $n = 248$ firms with international activities and $n = 87$ firms with activities on the domestic market only. Data collection took place in 2007 via standardized questionnaires. The survey was accomplished in close cooperation with the Association of German Engineers (VDI/VDE-IT) and the German Energy Agency (dena). It targeted the total populations of all German Renewable Energy ($N = 821$), Nanotechnology ($N = 305$), Biotechnology ($N = 558$), and Microsystems ($N = 348$) firms. Questionnaires were sent to CEOs, export managers or owners of the firms as these persons are perceived to have the most profound knowledge of the firm's internationalization practices and strategic decisions. As we had sent out $N = 2032$ questionnaires, the response rate was about 16.5%.

4 Growth and Internationalization Patterns

Due to increasing globalization and competition in the international environment, German firms face stronger demands to orientate towards international markets. Particularly for firms operating in future-oriented technology areas, internationalization is of major importance as the potential of the domestic market is often limited or saturated. Internationalization is therefore a key element to guarantee sustainable growth and company survival. However, the question is when to enter, where to enter, and how to enter distant markets. We will elaborate on these questions below.

4.1 Timing and Degree of Internationalization

The decision to internationalize early in the firm's life cycle is often a trade-off between opportunities and risks of internationalization. On the one hand, internationalization exposes the firm to increased market potential and new customers, while, on the other hand, it implies fundamental risks caused by distinctive political, legal, and cultural features of the foreign environment. Whereas PTI assumes internationalization to take place after years of activities on the domestic market, the INV theory examines internationalization right from inception. The following section will show, however, that the internationalization behavior also depends on the industry of the firm.

The Renewable Energy industry is highly regulated by legal and institutional norms. Both national and international support programs prevail. These programs impact the strategic growth patterns of the firm to a high degree. A significant number of firms first focus on the domestic market because legal regulations secure a robust and calculable income. The Act on Granting Priority to Renewable Energy Sources for instance guarantees a fixed reimbursement rate for suppliers of renewable energies. Therefore, companies from that industry are likely to first focus on their domestic market and then to internationalize into markets which guarantee similar legal security and calculable incomes.

Technology firms, in contrast, experience less regulatory and legal protection. Companies operate across different industries and often apply niche strategies which have to be marketed around the globe. The domestic market is often too small and restricted for the technological solutions of these firms.

The internationalization data of our Renewable Energy and Technology firms support our arguments for the different importance of international markets between the two types of firms. Our data show that for Technology firms, the average age at internationalization is 2.7 years and around 36% of the firms internationalize in the first year after firm foundation. Renewable Energy firms, in contrast, start internationalizing 3.3 years on average after inception and only 20% internationalize within one year after inception. Although these results show that both firm types internationalize early in their life cycle, Technology firms show a higher proactivity as suggested by the INV theory: The Technology firms in our sample generated about 43% of sales from 11 different international markets on average. The Renewable Energy firms, in contrast, had a lower international exposure with about 29% of foreign sales to total sales and on average six international markets the firm has international activities in. Thus, as suggested by the INV theory, Technology firms pursue a more proactive and early international approach, whereas Renewable Energy firms more incrementally approach the foreign market with strong emphasis on a stable domestic market.

4.2 Foreign Market Development and Entry Mode Choice

As outlined before, internationalization in the Renewable Energy industry is strongly influenced by public support programs. Several European countries have established similar legal prerequisites as guaranteed by the German Act on Granting Priority to Renewable Energy Sources. Spain, for instance, supports the solar energy sector, whereas France offers wind energy support programs which has resulted in a leading position of French firms in that sector. In contrast, internationalization of Technology firms is more strongly influenced by customer demands for highly knowledge-intensive products than by legal norms. In particular, demand comes from highly industrialized areas such as West European, North American, and some Asian markets, which are major target regions for technologically sophisticated products.

Figure 1 shows the major international target markets of the firms in our sample. The figure shows that both Renewable Energy and Technology firms have their major international revenues from West European countries. On the one hand, this is because West European countries are highly industrialized and thus show a high demand for technology products. On the other hand, the results indicate that firms prefer to internationalize to countries with a smaller psychic distance in order to

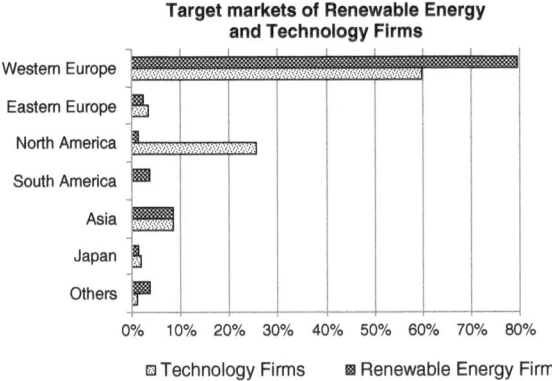

Fig. 1 Target markets of renewable energy and technology firms

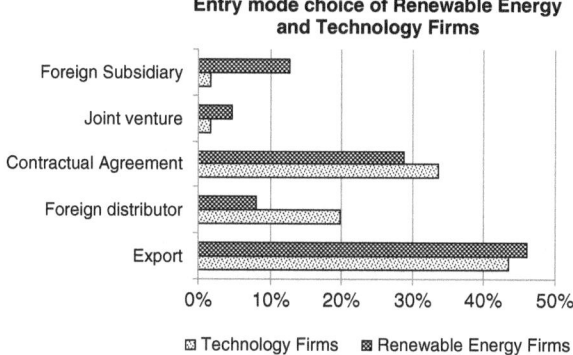

Fig. 2 Entry mode choice of renewable energy and technology firms

overcome the liabilities of foreignness that result from psychically distant markets. Apart from the focus on the European market, Fig. 1 shows that Technology firms have important activities in North American markets whereas Renewable Energy firms (in particular, solar companies) focus some activities on South America. It can be expected that this tendency will change due to some recently established Renewable Energy support programs by the U.S. legislation – particularly in California.

Figure 2 illustrates differences in the market entry mode of the two types of firms. The figure shows that export activities and contractual agreements prevail for both types of firms. However, Renewable Energy firms show a higher probability to utilize foreign direct investments like equity joint ventures and foreign subsidiaries, whereas Technology firms prefer to stick to non-equity modes of market penetration.

4.3 Networks

Using networks is an important mechanism for firms to get to know "the rules of the game" in the foreign market and thus to overcome the hurdles of being a "non-domestic" firm (so called liabilities of foreignness). Access to critical knowledge stored within the network reduces uncertainties concerning the foreign market. Network relationships may help to overcome resource constraints and to be fitted better directly to the demands of the foreign market (Grandinetti & Rullani, 1994). Our analyses show that both Renewable Energy and Technology firms internationalize quite early in their life cycle – besides limited resources and scarce international knowledge. Networks may help the firms at this early stage in two ways: first, networks can provide access to resources (such as sales channels or contacts) of others. Second, networks may help the firms to learn about foreign market particularities

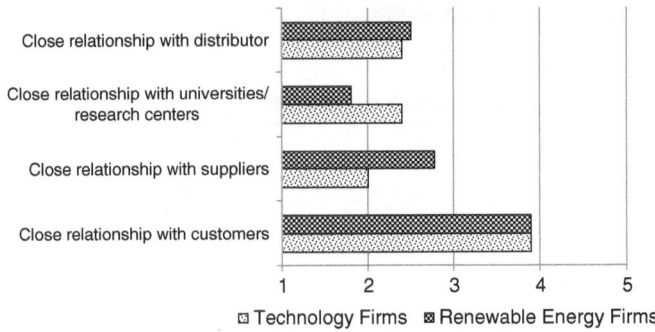

Fig. 3 Importance of network contacts for foreign market development of renewable energy and technology firms

and to gain knowledge, for instance, about customers' preferences or technological trends.

Figure 3 shows the importance (from 1 = *not important at all* to 5 = *very important*) of different network contacts for the foreign market development of the firm. The figure shows that both Renewable Energy and Technology firms use networks to develop their foreign market engagement. The access to foreign customers is most important for both types of firms. However, both differ in their use of suppliers and research centers. Renewable Energy firms strongly use their supplier contacts to overcome the hurdles of the foreign market, whereas the Technology firms primarily use contacts to universities and research centers. Thus, Renewable Energy firms profit from contacts with customers and suppliers, whereas Technology firms need to exchange with knowledge-intensive institutions to develop the know-how basis of the firm further.

Both the INV theory and recent developments of PTI emphasize the role of networks in the internationalization process. Our results support the two views in terms of the importance of networks in order to secure access to foreign markets and to develop foreign markets further.

4.4 Learning

Internationalization exposes the firm to new knowledge embedded in the particularities of the foreign market. According to Ghoshal, the differences in routines and norms in the foreign country market provides the firm with a superior knowledge base as it exposes the firm "to multiple stimuli, allows it to develop diverse capabilities, and provides it with a broader learning opportunity" (Ghoshal, 1987: 431).

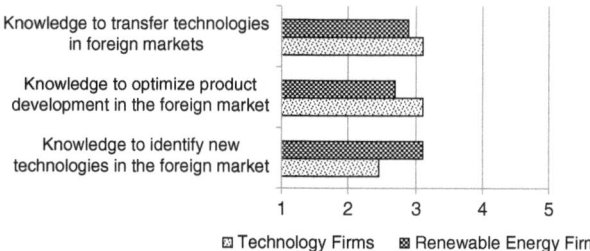

Fig. 4 Learning activities of renewable energy and technology firms in the foreign market

Some of these particularities to which a firm is exposed while doing business in foreign markets "are stored in the firm's routines and processes, thereby transforming the firm's current stock of knowledge" (Eriksson, Johanson, Majkgard, & Sharma, 2000: 28; cf. Nonaka, 1991). Firms are exposed to a higher degree of diversity in the foreign market, therefore, get a larger amount of information out of which some may be beneficial for the firm's knowledge stock in general. According to both PTI and INV, firms learn by operating on the international market. They gather information which is beneficial for the future development of the firm. It may, however, be expected that this knowledge acquisition and learning process differs between Renewable Energy and Technology firms. Renewable Energy firms offer mostly standardized products whereas Technology firms have more knowledge intensive products with a lower degree of standardization.

Figure 4 shows in how far Renewable Energy and Technology firms have gained new knowledge during their internationalization process in selected areas (from 1 = *limited new knowledge learned* to 5 = *fundamental new knowledge learned*). The figure shows that Renewable Energy firms prefer to develop their existing technology base further and to optimize product development. Technology firms, in contrast, primarily learn to transfer knowledge that was acquired in one particular market into other international markets. Further, Technology firms aim at developing new technologies in the foreign market in order to target other niche markets and to expand their product range. This is less likely for Renewable Energy firms capitalizing more on scale and scope and transferring existing know-how to a broader customer base.

5 Summary and Implications for Practitioners

Several key implications can be derived from our analyses. For firms – no matter from which industry – a thorough selection of the target market, a fit of entry mode choice, learning via networks, and the selection of business partners are essential. A prior market analysis is important to identify the right target market. Further, internationalization activities need to be adapted thoroughly to the prerequisites of the different customers in the different foreign markets. These may vary due to diverse environmental conditions in the target markets. Internationalization into psychically closer markets as is suggested by the PTI models is still the main practice. Most firms in our sample have conducted stepwise international development into close international markets, although their internationalization activities have taken place early in their life cycles. Concerning the entry mode in foreign markets, exports and contractual agreements remain the preferred choice. In respect to foreign direct investments, Renewable Energy firms utilize equity joint ventures and foreign subsidiaries to a higher degree than Technology firms that prefer to stick to non-equity modes of market penetration.

Another key issue for internationalization of the firm is knowledge acquisition and learning through network partners. Whereas Renewable Energy firms utilize contacts to the supplier as the dominant learning mechanism, Technology firms focus on close learning cooperation ventures with universities or other research institutes. Such cooperation provides for a promising mechanism to generate knowledge and to access the foreign market in a sustainable manner. Our analyses have shown that internationalization is not only an option to increase the customer base and develop demand further but is also a mechanism in order to learn new knowledge and to increase technological sophistication. This is a major benefit in particular for Technology firms that continuously have to adapt and develop their products further in order to achieve sustainable corporate growth and development.

References

Aharoni, Y. (1966). *The foreign investment decision process.* Graduate School of Business. Boston, MA: Harvard University.

Eriksson, K., Johanson, J., Majkgard, A., & Sharma, D. D. (2000). Time and experience in the internationalization process. *Zeitschrift für Betriebswirtschaft, 71,* 21–43.

Ghoshal, S. (1987). Global strategy: An organizing framework. *Strategic Management Journal, 8,* 425–440.

Grandinetti, R., & Rullani, E. (1994). Sunk Internationalisation: Small Firms and Global Knowledge, *Revue D' Economie Industrielle, 67,* 238–254.

Johanson, J., & Vahlne, J.-E. (1977). The internationalization process of the firm: A model of knowledge development and increasing foreign market commitments. *Journal of International Business Studies, 8*(1), 23–32.

Johanson, J., & Vahlne, J.-E. (1990). The mechanism of internationalisation. *International Marketing Review, 7*(4), 11–24.

Johanson, J., & Vahlne, J.-E. (2003). Business relationship and commitment in the internationalization process. *Journal of International Entrepreneurship, 1,* 83–101.

Johanson, J., & Wiedersheim-Paul, F. (1975). The internationalization of the firm: four Swedish cases. *Journal of Management Studies, 12*(3), 305–322.

Nonaka, I. (1991). The knowledge-creating company. *Harvard Business Review, 69*(6), 96–109.

Oviatt, B. M., & McDougall, P. P. (1994). Toward a theory of international new ventures. *Journal of International Business Studies, 25*(1), 45–64.

Oviatt, B. M., & McDougall, P. P. (1995). Global start-ups: Entrepreneurs on a worldwide stage. *Academy of Management Executive, 9*(2), 30–44.

Oviatt, B. M., & McDougall P. P. (1997). Challenges for internationalization process theory: The ease of international new ventures. *Management International Review, 37*(2), 85–99.

Schwens, C. (2008). Early Internationalizers: Specificity, Learning and Performance Implications, Dissertation. Hampp.

Zahra, S. A. (2005). A theory of international new ventures: A decade of research. *Journal of International Business Studies, 36*, 20–28.

Part II
Efficiency of Innovation Processes in International Enterprises

Part II
Efficiency of Innovation Processes in International Enterprises

Determinants for Failure and Success in Innovation Management

Dieter Thomaschewski and Alexander Tarlatt

Contents

1 Introduction 127
2 Fundamentals on Innovation 128
   2.1 Strategic Planning and Innovation Strategy 129
   2.2 Features of the Innovation Strategy 129
   2.3 Success and Failure Rates 130
3 Innovation Management 131
   3.1 Elements of Innovation Management 132
   3.2 The Innovation Process 133
   3.3 Roadblock to Success 134
   3.4 Key Questions for Innovation Management 135
4 Determinants for Success 136
   4.1 Performance Measurement 136
   4.2 Empirical Studies on Success Factors 137
   4.3 Perspectives of Success Factors 138
   4.4 Determinants for Success 139
   4.5 The "Human Side" of Innovation 146
5 The "Ten Commandments" of Innovation Management 147
References 148

1 Introduction

Continuous and discontinuous development processes in the business environment change the rules of the game for corporations. Changes in the framework conditions of entrepreneurial activity are taking place faster and faster. Growing competitive pressure is forcing companies to adjust to the changed market- and environmental conditions in ever shorter periods. The increasing speed of competitive

---

D. Thomaschewski (✉)
Ludwigshafen University of Applied Sciences, Ludwigshafen, Germany
e-mail: dieter.thomaschewski@fh-ludwigshafen.de

A. Gerybadze et al. (eds.), *Innovation and International Corporate Growth*,
DOI 10.1007/978-3-642-10823-5_9, © Springer-Verlag Berlin Heidelberg 2010

advantages that emerge and disappear again requires companies to constantly review potentials and opportunities to hold and improve their own positions.[1] Successful, long-term profitable growth requires not only operative excellence but also a market-driven strategic positioning and a permanent willingness to change and to innovate.[2] Innovation is one of the most important long-term success factors for prosperous entrepreneurship. The awareness that innovations are the key to the organization, development and growth of companies has grown sustainably and manifested itself inter alia as a result of growing globalization.

Innovations are the driving forces of growth. Innovations ensure competitiveness and offer opportunities for differentiation. Therefore "innovation" is to be related to all that has to do with permanent, substantial customer benefit and a perceivable competitive edge: the development of new successful business models. Hence, the entrepreneurial challenge lies in the successful management of innovations. "So the question is not one of whether or not to innovate but rather of how to do so successfully."[3]

2 Fundamentals on Innovation

Innovations are vital to the success of any company. Without innovation to at least some extent, a corporate strategy is unlikely to result in significant additional growth or the gain of a competitive edge. There are ample initiating triggers for innovations:

Megatrends	Markets	Technology/ materials management	Efficiency pressure	Political framework
• Demography/life style	• Competition	• Technological changes	• Capacity utilization	• Environmental legislation
• Life sciences	• Life cycle	• Material availability	• Experience curve	• Social provisions
• Public private partnership	• New knowledge	• Bioscience	• Rising quality requirements	• Terms of trade
• ...	• Change of consumer habits	• Inventions	• ...	• ...
	• ...	• ...		

Fig. 1 Initiating triggers for innovations

To support these depicted "driving forces" has always been a key to success for a company. The increasing dynamic nature of the environment has raised the pressure on companies to act, the periods for new competitive products have been significantly shortened and the rate of innovation has noticeably increased.

---

[1] Krüger (2000), p. 45.
[2] Thomaschewski (2002), p. 108.
[3] Cf. Hauschildt and Salomo (2007), p. 32.

2.1 Strategic Planning and Innovation Strategy

Innovations and the innovation portfolio of a company are directed by strategies. In connection with the corporate objectives as the foundation of the corporate strategy, the innovation strategy must be developed and derived in a second step.
Vital strategic questions should be answered in the course of this process:[4]

- What does the company want? How is the portfolio to be shaped in the future with regard to product, region and customers? In which fields of business does the company intend to operate?
- How does the company want to tackle competition? What are the important core competencies as the longer-term basis for success?
- How does the company intend to achieve the desired positioning? What has to be done for the needed expansion/restructuring/cutback? What are the implications for the growth strategy (Growth by the company's own strength? Growth by acquisition?)

The answers to these questions then form the necessary basis for the company's innovation goals. The innovation strategy and an innovation portfolio defined on[5]

- products, services and markets
- technologies, processes and materials
- social system, environment and information

are then used for the optimum allocation of resources. The innovation strategy embraces all the strategic statements for the development and marketing of new products and processes, for opening up new markets, for the introduction of new organizational structures and social relations in the company.[6] No matter what innovation the company focuses on, one conclusion remains: Meeting customer demands and delivering more value than the competition is crucial for success. This is the only way to ensure that an invention turns out as innovation and at the same time initiates a sustainable and economically feasible market cycle.

2.2 Features of the Innovation Strategy

Strategic planning differs from operative planning through the maturity of the developed plans. Whereas the strategic planning process is applied long term (doing the right things), operative planning (doing things right) is substantially shorter term. Furthermore, the strategic measures for safeguarding the long-term success

---

[4] Schreyögg and Koch (2007), S-72.
[5] Gaubinger, Werani, and Rabl (2009), p. 34.
[6] Pleschak and Sabisch (1996), p. 58.

of a company have a series of specific framework conditions that also apply to the innovation strategy without restrains:[7]

- The numerous interactions in the company are extremely complex
- The long-term character and continuity require consistency in entrepreneurial behavior
- Planning and designing innovations require cross-company coordination to avoid conflicts of targets

These conditions lead to the evident conclusion that formulation and definition of innovation is definitely a top-management task: setting the direction of the company, defining the corporate goals and describing the corporate vision cannot be delegated.

Innovation projects usually tie up long-term resources. They need large investments in research and development and in production and marketing. If a project fails, not only are there massive financial losses that have to be dealt with, even harder to handle are possible negative effects on image for the company. Also, another fact calls for top-management decision making: Innovation is always risky, either due to appearing me-too products, new competitors, new technologies or weak market acceptance.

2.3 Success and Failure Rates

High development costs and long development periods characterize (at least radical) innovations. For example, the Motorola Company needed almost 15 years and US$ 150 million in order to achieve wide acceptance of the mobile phone. Corning Glas invested 10 years and US$ 100 million in the development of the fibre-optic-cable.[8] Gillette spent US$ 75 million to develop and launch a new safety razor. This list can easily be extended: an "innovative" cigarette needed about € 20 million, a new pharmaceutical about € 600 million, a new car model ranges at about € 1.5 billion and the Airbus A 380 approximately € 15 billion.[9]

These investments in strategic renewal that are initially made to sustainably improve market positions and deliver the basis for success,[10] are frequently not reflected in an adequate return: Amortization is often not achieved. Numerous studies provide evidence on this. *Meffert* states a failure rate of 65% for the food industry in Germany and 70–90% for the packaging industry in the United States. *Homburg*, based on Nielsen evaluations, shows flop rates of about 30% in the consumer industry, *Kerka* shows a success ratio of 6 per 100 ideas, and *Stevens/Busley* one of 1 per 3000.

---

[7] Vahs and Burmester (2005), p. 101.
[8] Vahs and Burmester (2005), p. 134.
[9] Meffert, Burmann, and Kirchgeorg (2008), p. 411.
[10] Krüger (2000), p. 45.

Here are some more examples: Volkswagen needed to write off € 700 million due to insufficient sales of the luxury limousine Phaeton, Texas Instruments lost US$ 660 million before withdrawing from the home computer sector, Dupont made a loss of US$ 100 million with an artificial leather called "Corfam", the investments in the "Concord" were never regained, and finally Motorola and partner companies wasted US$ 6 billion on the satellite telephone system Iridium.[11]

Numerous studies and their findings can be recited, the quintessence remains the same.

The main goal is to safeguard innovations. In this respect, one of the main challenges and tasks of innovation management lies in reducing the probability of failure.[12]

In annual reports companies state their R&D spendings. Knowing that simply spending the money does not automatically lead to a market success, the validity of this figure as a key performance indicator on innovation should be questioned. Efficient usage of R&D-spendings can only be achieved by reducing failure rates.[13] For this, the approach that a company takes towards innovation management is key.

3 Innovation Management

The underlying principles of innovation management are basically the same as in general management. Management in its general form consists of goal-oriented steering tasks carried out by all instances of a company that have managerial authority.[14] Planning, directing and controlling in innovation management are just as much classic functions of management as are organizing, staffing and information management. However, in innovation management, these functions are focused on three task segments:

- Generating, developing and applying knowledge to satisfy customer needs and create competitive advantages.
- Implementing this knowledge in a marketable, novel range of offerings (products, services, systems and processes).
- Successful introduction of these innovations in the market and ensuring their sustainable economic success.

---

[11] Kotler, Keller, and Bliemel (2007), p. 437.
[12] Meffert, Burmann, and Kirchgeorg (2008), p. 408.
[13] Cooper (1999), p. 116. Today's new product teams and leaders seem to fall into the same traps that their predecessors did back in the 1970s; moreover there is little evidence that the success rates or R+D productivity have increased very much.
[14] Schreyögg and Koch (2007), p. 8.

3.1 Elements of Innovation Management

Market success and added customer value is of critical importance for innovations. All the company's activities need to be aligned to achieve this goal. Successful innovation follows a structured approach that has three vital elements: (a) a systematic and consistent preparation, (b) boundaryless and seamless execution and (c) efficient coordination of all activities and operations connected with them.[15] The planning, organization and control of innovation processes and the creation of the necessary framework conditions is embodied in the main components of innovation management:[16]

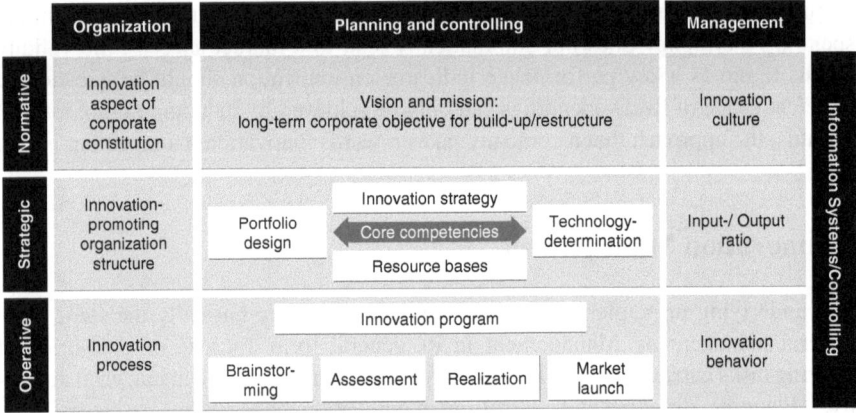

Fig. 2 Components of innovation management

Following the above, the main tasks of innovation management are:

- Definition of innovation goals derived from corporate strategy
- Determination of the investment strategy including reconciliation of needed resources and potentials
- Definition of the innovation program including profitability performance indicators
- Development of an innovation-prone organizational culture and -structure
- Planning and steering of the innovation process and execution of defined measures
- Establishment of a process-spanning information system and controlling.

Thus, the function of innovation management consists ultimately of perceiving the opportunities in the innovation process/transformation process and reducing risks.[17] It is one of the primary and original functions of top management. The

---

[15] Vahs and Burmester (2005, p. 47).
[16] Based on König and Völker (2002), p. 10.
[17] Macharzina and Wolf (2008) p. 751.

innovation agent in a company is not solely one unit or one function: Innovation is cross-departmental; the identification with innovations can pervade the whole company.

3.2 The Innovation Process

On the way to innovation, an invention passes through a multistage, multiphase process. The phase structure results from the allocation of specific tasks and problems to units of the organization. Each of these phases requires different instruments and methods. The permanent orientation along the company's strategic goals and the market demand is of crucial importance. In theory and practice numerous different partitions of the innovation process can be found.[18] A more detailed format of the process is depicted here and should help to better understand the factors for success and failure.

Innovations need time and resources. If it is possible to coordinate the innovation-driven activities more precisely with each other and reduce the needed time to run through each phase, then the return on innovation investments starts earlier and the probability of success rises. Clearly assigned tasks and responsibilities promote this process. The phases of the processes are dependent on each other. This is often not reflected in the real world, where processes rarely flow seamlessly and the missing interface efficiency of organizational units leads to a "throw-it-over-the-wall mentality": employees from different units neglect that they need to coordinate and crosslink information adequately and fit their work-results into the overall project activities.

As a result of the considerably higher degree of uncertainty (multistage complex decisions, resistance, market risk, etc.) in comparison to routine processes, it is of crucial importance to monitor innovation from different perspectives in order to safeguard innovation success. The main perspectives in this regard are

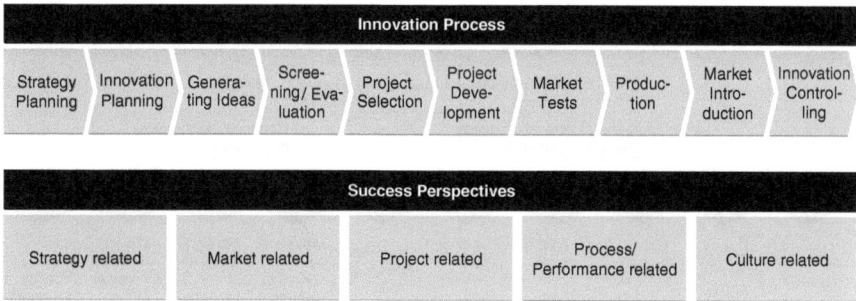

Fig. 3 Phase model of the innovation process

---

[18]Cf. Thom (1980), Brockhoff (1994), Will (1996), Pleschak and Sabisch (1996) and Cooper (1999).

- strategy related,
- market related,
- product/project related,
- process/performance related and
- culture related.

These perspectives reflect the most important areas that need to be taken into account while planning and implementing innovation.

3.3 Roadblock to Success

Numerous barriers have to be torn down to finally reach innovation success. Some of these are determined by external factors. The willingness of investors to provide funding is just as much one of these factors as is fiscal legislation, delayed and complex approval procedures or the risk-averse attitude of politicians. However, apart from these external factors, there are also numerous internal causes for a lack of innovativeness.

Many companies have deficits regarding effective and efficient innovation management. In various studies,[19] the following problem clusters have been identified as significant in the course of the innovation process:

Roadblocks on the way to success

Overview roadblocks per phase

Strategy Planning	Innovation Planning	Generating Ideas	Screening/ Evaluation	Project Selection	Project Development	Market Tests	Production	Market Introduction	Innovation Controlling
						▪Too many changes ▪No objective market feedback			
					▪Lack of comm. between R&D and marketing ▪Kill early/ kill cheap neglected				
				▪Fear to build early business case ▪Lack of resources					
			▪Underestimation cost of development ▪No clear evaluation criteria						
		▪Limited out-of-the-box thinking ▪Too many ideas				▪Missing techn. strategy ▪Old technology			
	▪Undefined general innovation needs ▪No proper market intelligence						▪Incomplete selling story ▪Long delays		
▪Lack of company strategy ▪Sparkling eyes = no systematic approach								▪Not reflected ▪Input not output driven	
Unfocused approach	We know already the answer	Always the same way	Lack of project prioritization	Too many projects for reduced ressources	No cross functional teams	Customers not sufficiently involved	Old technology for new products	Missing resources for market launch	Unclear expected performance

Fig. 4 Roadblocks on the way to success – overview of roadblocks per phase

---

[19]Brockhoff (1993), Cooper (2001), Vahs and Burmester (2005), Albrecht, Bauer, and Kühnl (2006), IBM Global CEO Study (2006) etc.

These roadblocks are examples of factors that ultimately lead to the failure of innovation projects. Numerous other ones can be enumerated[20]

- practiced forms of management are obstacles to innovation
- centralistic, power-retaining, rigid organization structures
- high proportion of standards and rules hostile to innovation
- projection of past successes

However, these roadblocks have one thing in common: They must – if present in the company – be removed in order not to block the way to success. They are "most significant management challenges for innovation performance". Let's blast them away.

3.4 Key Questions for Innovation Management

New business systems, new products and services, new processes and habits need to be developed in innovation management. In a nutshell: "Economics of Innovation" are the needs of the moment. Change is not understood as adaptation to external developments, but as idea-driven evolution.[21]

However, this change and renewal always involves special risks. Failure rates are high. To prevent this, management must permanently deal with certain vital questions of innovation strategy:

Some key questions for the innovation management

Questions according to perspective

Perspective	1 Strategy related	2 Market related	3 Product/Project related	4 Process/Performance related	5 Culture related
Key questions	• Innovation pipeline balanced through all phases?	• Qualitative marketing in place?	• Resources allocated according to project necessities?	• Acceptable time to market?	• Instruments to determine innovation success in place?
	• Required competencies present to follow the strategy?	• Market size sufficiently and verifiable determined?	• Expected return meets requested return?	• Methods like Target-costing, investment calculations applied?	• Innovation for all executives major factor in target agreement?
	• Sufficient financing provided for projects?	• Customer benefit as strategic orientation secured?	• Clear go/ no go criteria defined?	• Process directed by dedicated "leader" in cross-functional team?	• Share of regulations opposing innovation relatively small?

Fig. 5 Key questions for innovation management according to perspectives

---

[20]Wahren (2004), p. 32.
[21]Krüger (2000), p. 42.

It is unrealistic, even in the case of successful innovations, that all of these questions can be answered positively. However, there is no doubt that at least the majority needs to be addressed accordingly in order to safeguard success.

The most important topics in this regard are:

- strategic considerations (goals, resources and competencies)
- market principles (market valuation/assessment, benefit promises and competitive advantages)
- product/project development (business cases and kill criteria)
- process/performance orientation (pacing and leadership)
- level of innovation culture (agreed goals and degrees of freedom).

4 Determinants for Success

Investigating the success factors of innovations has a long history. In determining these it is essential to define what goals are being aimed at with an innovation and furthermore with which strategies and measures the goals are ultimately to be achieved. The driving forces behind success and failure can then be defined using the components of innovation management and the various phases of the innovation process. Ultimately, humans are the guaranteeing success factor with their necessary competencies and habits.

4.1 Performance Measurement

Innovation means 'improvement' and 'progress' compared to an existing status. These improvements and advances refer to the following dimensions: content, intensity, time and space. This dimensioning necessarily leads to individual focal-points of interest. Producer and consumer interests, economic and ecological interests and different philosophies of life define an interface where a consensus on success can hardly be depicted.[22] However, measuring success in an economic sense needs a tight definition. An innovation is successful if it positively contributes to profits and returns of a company.

In this regard it is difficult to state what the optimal length of the planning period is. Some innovations require a significantly longer market introduction and acceptance period, others are generating payback quickly.[23] An ideal amortization period needs to compensate for the invested capital: e.g. the costs for research and

---

[22] Hauschildt and Salomo (2007), p. 28.

[23] A breakdown of profits into the various engines such as volume, prices, variable costs, fixed costs, tied-up capital etc. is immanently included.

development, for production, for the market launch and marketing as well as the cost of capital. The innovation only then creates 'added value' for the company, shareholders and stakeholders if the internal rate of return of the project lies above the market interest rates for investments with a comparable risk profile. An interest yield must at least be achieved with the innovation that is (hopefully significantly) above the company's weighted capital cost rate.

Managerial steering in this setting means to constantly monitor the development of the innovation's business case to safeguard the innovation payback and limit eventual losses. This can be accomplished via an ex ante and ex post comparison of the innovation business case in each critical decision phase. The constant business case monitoring also ensures that the "modification" of the success factors is measureable in the economic parameters of company's success.

In summary, innovation projects need to be treated as investments. For this reason analog controlling mechanisms need to be in place to safeguard the invested capital for innovation.

4.2 Empirical Studies on Success Factors

The question of the causes of the innovation success for companies is linked to the question which instruments ultimately support innovation success. *Hauschildt/ Salomo*[24] cite more than 60 empirical studies that are dealing with the question if there is one single recipe for success for companies operating successfully in the market. Several meta-analyses – with different methodological approaches – condense and concentrate these success factors. *Bauer/Albrecht/Kühnl*,[25] too, have processed various important dependency relationships of company operations and corporate success from a marketing-oriented point-of-view based on the results of metastudies.[26]

Balachandra/Frias place special emphasis on market-related, technology-related and organization-related factors. *Montoya-Weiss/Calantone* see the strategic factors, the development process, the market environment and the organization as the essential elements. *Henard/Szymanski* refer to product properties, strategic factors, process properties and market properties. Finally, *Van der Penne/van Beers/Kleinschmidt* prioritize company-related, project-related, product-related and market-related factors.

---

[24] Hauschildt and Salomo (2007), p. 35 ff.
[25] Bauer, Albrecht, and Kühnl (2006), p. 10 ff.
[26] For the megastudies cf. Balachandra and Frias (1997), Montoya-Weiss and Calantone (1994), Henard and Szymanski (2001), van der Penne, van Beers, and Kleinschmidt (2003), Hauschildt and Salomo (2007), p. 18.

All these findings are plausible and comprehensible:

- Provision of benefits for the customer
- Technologically novel
- Professional market research/assessment
- Innovation-friendly corporate culture
- etc.

Irrespective of the different methodological approaches and weaknesses of scientific explanations,[27] one important insight remains: without doubt there are certain success perspectives. Within these perspectives, specific success factors can be identified that positively influence the innovation process. The consideration of these success factors should significantly improve innovation success. It was not the intention of the authors to publish the next version of success factor research. The task was, in fact, to create a simple taxonomy for successful innovation management that delivers implementable advice on what to do to successfully innovate, differentiating between the different involved units in a company. The foundation for the identification and definition of success factors is taken from existing and practice-oriented empiricism.

4.3 Perspectives of Success Factors

Looking at the different reasons for success there are five different perspectives concerned that either lead to satisfying innovation results or, if neglected, to inefficiently used budgets. The approach here was to cluster similar types of success factors in a common perspective. They indicate certain requirements that need fulfillment in order to lead to success. Thus the perspectives are not independent of each other but are to be considered interacting with each other.

In the case of the *strategy-related perspective*, it is clear that innovations which are relevant to corporate growth are understood to be investments in the future. The innovation strategy should be crafted in a way that it harmonizes with the goals and strategies of the whole company and is also the basis for a targeted, concentrated allocation of resources.[28] As a result of the innovation strategy specification, the company has a portfolio of innovation initiatives that it needs to manage.

The *market-related perspective* addresses first and foremost the customer. Awareness in the company needs to be strengthened so that the market (investigated transparently, consistently and completely), the customer and its visible and

---

[27] Hauschildt and Salomo (2007), p. 18.
[28] Pleschak and Sabisch (1996), p. 58 ff.

latent needs should be permanently in the center of innovation planning. Genuine, sustainable, and long-term competitive edges can only be achieved by thoroughly dealing with the customer's needs and embracing in a productive dialogue with the customer target groups.

The *product/project-related perspective* touches aspects of project preparation, project planning and project realization. For the target-oriented, complex project (interaction of various functions, various competencies, and various people acting under time pressure), which is limited in time and for which only limited resources are available, a clear planning (performance requirements), specific direction (allocation of responsibility) and permanent controlling (business case) is necessary.

In the *process/performance-related perspective*, the focus lies on the effectiveness and efficiency of the innovation cycle. Innovations are processes of preparation, decision, execution and control of novel combinations of means and ends. As processes, these are characterized by activity sequences, time limits and outputs/results. The efficient throughput of the innovation process from idea generation to market introduction is a mandatory area of attention.

The *culture-related perspective* combines all aspects that enable a company from within to initiate, foster, support and complete an innovation process successfully. Generating permanent innovations and growth from innovations needs to be accepted as an imperative in the company. The structures and processes of the organization, the behavior and the motivation of the employees must be aligned with this guideline. Innovation is driven by a collective and collaborative attitude of all functions involved.

To create, maintain and develop an innovation culture is one of the most important but at the same time hardest to achieve success factors in innovation management.

4.4 Determinants for Success

Success and failures factors are analyzed, processed and published in numerous studies. In many companies, the essential critical success factors are recorded in handbooks. The communication of these factors does not seem to be successfully absorbed and implemented despite an existing awareness. However, these factors are manageable and can be controlled. The challenge for management consists "merely" in converting words into action.

4.4.1 Challenges and Goals by Perspectives

Each perspective has its own determinants for success. Furthermore, each perspective has challenges and goals to be achieved. The deduction of the specific challenges that have to be met will ultimately help to differentiate the success factors, giving support for the implementation:

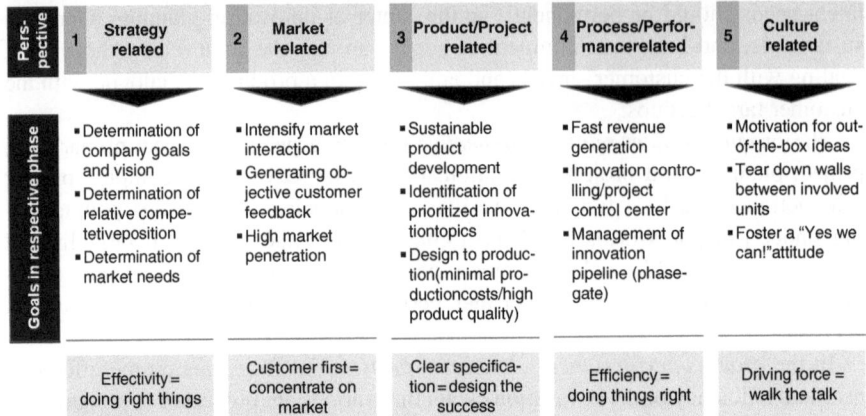

Fig. 6 Challenges and goals by perspective

The development and management of the success factors for each perspective must be supported by goals that the company ultimately wants to achieve. These can be condensed into goal clusters.

- In the strategic context, effectiveness is clearly in the focus: Doing the right things. Innovation success following corporate objectives is guided by the questions: What does the company want and how does the company intend to achieve this?
- In the case of market environment, the challenge is to concentrate on markets, customer needs and wants. Innovations serve to satisfy the needs of the market users with an offer that is significantly superior to what competitors are offering (so called "delighters"). "Built-in voice of the customer" requires constant interaction with the same.
- In the product-project perspective, the focus is on consistent planning. Before development begins, the concept, benefits, requirements, features und specs need to be defined. A lack of planning is a major cause of both new product failure and serious delays in time-to-market.[29] A clear and distinct product/project definition is key to success. In the advanced stages of the process, this saves expensive modifications and adjustments of the innovation work.
- Challenges in the process/performance related category are concerned with process efficiency and fast revenue generation. To achieve this, a seamless process flow is needed. Furthermore, a structured monitoring and controlling of the process is a precondition. Successful innovation projects aim at constant monitoring of project costs, timelines and project results. The goal is very clear: timely escalate eventual breaches of defined threshold values.

---

[29]Cooper (1999), p. 119.

- The culture perspective finally is the driving force in the innovation process. Even though it is a highly intangible topic, the importance cannot be overstated. Looking at the culture-related challenges, striving for an organization that fosters innovation and collaboration as well as co-evolvement of ideas is the main goal. To reach that goal, the motivational system should be closely examined.

4.4.2 General View of the Factors

Despite various individual studies on successes and failures of innovations, the authors still believe that there is a need to systematically assess factors decisive for success and failure of innovation. Despite different priorities, the heterogeneity of innovations, numerous industries and services and finally despite various suggestions for measuring success, an important conclusion remains when comparing the success factors.

There is a consensus that innovation success is not gut- but system-driven, is the results of hard and consistent work and is based on a manageable number of success factors:

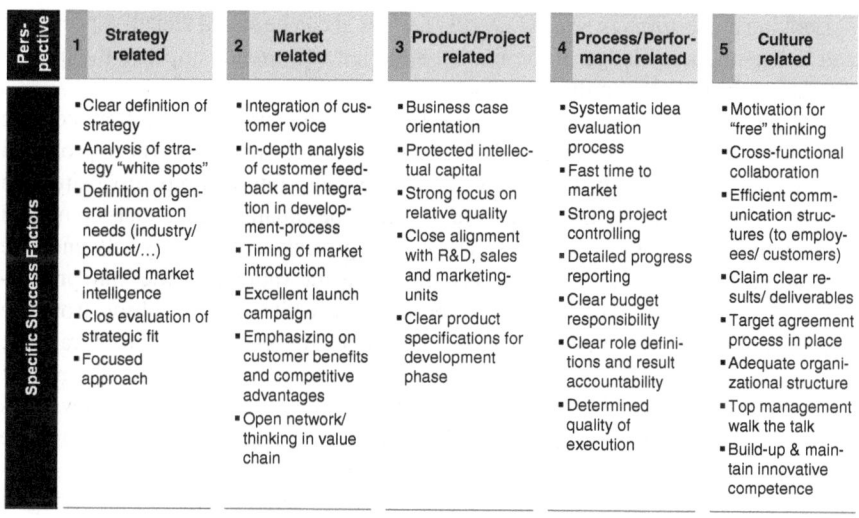

Fig. 7 Overview of success factor in respective perspectives

Knowledge of the contents and effects of these success factors helps to operate innovation management in an even more efficient way, since complexity in decision making is significantly reduced. The complexity of the cause-and-effect relations does not permit a one-dimensional allocation to a single success factor. The interdependencies of the relationships of cause and effect require multidimensional consideration and simultaneous development of these factors in order to make companies more effective and more efficient in innovation management.[30] The authors have developed a qualitative review of

- innovation-specific parameters
- internal structuring forces
- external claims
- person-immanent motivation elements

in order to arrive at a more systematic assessment.[31] Ultimately, this shall lead to necessary improvements in the company through creating a higher awareness of innovation barriers. This article cannot cover and discuss all elements in detail, if, however, it is possible to direct attention to the fact that "the critical factors are noticeably absent from the typical new product project",[32] this article has achieved an important objective.

4.4.3 Some Considerations by Perspectives

The scope of this article does not permit in-depth treatment, nevertheless some supplementary remarks are designed to stimulate further discussion in regard to the defined perspectives.

Looking at the *strategy related success factors*, it can be stated that all aspects that lead to a sound knowledge of the market-situation and -trends improve the quality of strategy formulation in the first place.

The information gathered is the input for a focused strategy definition, coming from clear corporate vision and goals. Hence, the markets that need to be approached with the company's products as well as the recent and future competitive situation are to be reviewed and defined. The result is the to-be situation that then is compared to the actual set-up of the company. The resulting weak spots greatly influence the innovation strategy so they can focus and direct the innovation activities into promising strategic fields of interest. More precisely, the white-spots resulting from the strategy are to be identified. With this knowledge, successful innovators translate the white-spots directly into areas for innovation focus and thus give the innovation process more direction. In this regard it is of great importance that the innovations that are expected will fit to the company's resources. Innovations that overstretch the company's competencies have a higher probability of failure. In this case two strategies are observed: Either the idea is abandoned after determination of a missing fit or a plan to further develop the company's set-up is defined and implemented.

---

[30] This statement is independent of the intensity of the innovation or the size of the company. The parameters are present in each case although their attributes and weighting may differ.
[31] Van der Penne, van Beers, and Kleinschmidt (2003), p. 3.
[32] Cooper (1999), p. 116.

- Strategic direction of the innovation program
- Long-term, balanced innovation portfolio
- Competencies and resources as essential basis.

Evaluating the *market related factors* it is no question that 'the pull' of the market is decisive for any innovation success. Successful innovation starts and ends with the customer's needs and wants. The perceived value of any idea in the eye of the customer is halfway to success. The customer therefore always is first, everything needs to be focused on this precondition. Innovation is successful if customers buy the product; therefore all innovation management activities need to take customer satisfaction into account. This does not necessarily mean that customers recognize the improvements of a new product or service right away. Successful innovators will anticipate an acclimatization trajectory that has an impact on the development of sales and revenues. The goal for innovators is to shorten the time span that customers need to learn the improvements of a new product or service. Therefore the quality and timing of the launch campaign is of great importance for the success of innovation projects. This does include a specific innovation marketing (pre-marketing, internal selling story, pilot marketing) to first open up the markets.

But success is even higher if the voice of the customer is integrated into the development process early. This can be achieved by customer focus groups that can test and appraise ideas at a prototype stage. By integrating the customers in the idea generation stage, e.g. through moderated forums on the company's internet site, valuable impulses for the innovation process can be gained.

Successful innovation needs a strong interaction with the user of the new and innovated products or services to get a feeling for the value perception. This includes cooperative ventures with market partners (also in the sense of an open network), in order to ensure a comprehensive knowledge base also beyond vertical value-adding chains. Synergy potentials can thus be exhausted over the whole value-adding chain and individual shortages of resources can be overcome.[33]

- Sound and comprehensive market analysis and assessment
- Dedication to the benefits/values for the customer, customer inputs throughout that project
- Creation of important, perceivable and lasting competitive advantages.

Product/project related factors have some technical, economical, organizational challenges. Successful innovation projects are those that are not exposed to moving targets and shifting management ideas. The key to efficient innovation therefore is clear specification of what is needed. This is important throughout the whole process, beginning with the briefing of the innovation team and ending with clear production specifications (that are reliable for the production set-up).

---

[33] Vahs and Burmester (2005), p. 387.

Project-related innovation success also plays an important economic role. This means that everything needs to be planned in a business case which needs constant updating every time new aspects that are involved with the profitability of an innovation idea appear. Coming from the business case, clear go and no go criteria need to be in place to let only those ideas pass into the development stage that prove to be economically interesting. Successful innovators decide for or against an idea based on numbers; if the numbers change in the course of the development process they still bury ideas regardless of the investments that went into it in the past.

If an idea passes the test of economic feasibility another aspect grows in importance: the protection of the intellectual capital that went into idea generation. If this only has second priority, the door is open for early adopters and me-too-producers that skim the company's profits.

Also, a clear focus on product quality leads to success. At the time the promising prototype hits the production phase and the first customers buy the product, the quality needs to meet at least expectations so that the customers positively talk about the product and praise it. At that time, resources are spent to correct the mistakes in the production process and the faulty products, above that the product image has deteriorated and sales are dropping. To avoid this, a close alignment between R&D, production, sales and marketing units needs to be guaranteed in order to only release the product if production delivers the right quality, sales staff is trained, distribution is set-up and the launch campaign is well prepared.

Sustainable product development, proper production and a successful market launch can only be achieved if all potentially involved parties are integrated into the innovation process and are also motivated to positively interact and drive the product to ultimate success.

- Clear, plausible and consistent definition for the innovation project
- Sustainable and permanent cross-functional cooperation with clear project responsibility
- As-early-as-possible orientation to profitability criteria with a clear business case and definite project-kill criteria.

Challenges in the *process/performance* related category are focused on efficiency. A set of targets and strong controlling are tools to enable it. Innovation process controlling is concerned with two aspects: the project costs and the project timelines. Successful innovation projects aim to constantly monitor these aspects and escalate eventual breaches of defined threshold values in time. After the project plan and timeline are agreed and the financial resources are budgeted, a controlling mechanism that tracks the due dates and budgets should be in place for management to see the early warning signs that could evolve in the course of the project. The reaction to projects that move out of the defined bounds can either be to search for countermeasures, or, in case this is not possible, terminate the project. The difficulty is to not stick to the decision to develop the idea but rather to stick to the process to eliminate the risk of overinvesting into one idea. The risk is that opportunities will be missed, but the advantage is that on a larger scale the innovation portfolio is balanced and risk adjusted. In sum, the goal is to effectively manage the innovation portfolio and the innovation pipeline to stay in the track defined by the company and innovation strategy.

In order to have a functioning mechanism, successful innovators define clear responsibilities for all involved parties. These responsibilities include accountability for budgets and results, and include clear and unquestionable milestones to be mandatorily observed.

- Consistent process orientation and determination of definite milestones
- Structured process monitoring and reporting for results-, progress- and premise-control
- Determination of responsibilities and definition of accountability.

The *culture success factors* are the driving force in the innovation process. The cultural aspect is elusive by nature. Still the importance for successful innovation is undeniable. The main success factor is a culture that fosters cross-functional collaboration and motivates for "out-of-the-box" thinking. It needs to encourage organization to take controlled risks.

This can be achieved by good leadership that "walks what it talks" and encourages the units to act respectively. To support this, an aligned motivational system needs to be set-up that rewards this behavior. Even though innovation is to a certain extent not always plannable, successful innovation projects do use target agreements to give direction and rewards in case of target conformance.

Next to the motivational aspects, culture is also positively influenced by efficient communication. If there is a forum, there is also encouragement for employees to talk about their innovation success and also, which is even more important, exchange ideas with other colleagues in different units and from different backgrounds.

All this needs to be accompanied by the right organizational structure. In companies in stable environments and with more formal cultures it is sometimes difficult to reach a culture with inspiration for innovation. So an organization has to open up for new ideas.

Above that, culture has to encourage employees to target the formal mental walls between different units and demand a collaborative solution to upcoming problems in the course of the innovation process. Summing this up, the challenge is to have a leadership that walks the talk and constantly positively encourages innovation.

Finally: Innovation is not a one-time event. Innovation is a permanent experience. Therefore a company should steadily work on building up and maintaining competence. That is the ground where success can be achieved.

In conclusion: Experience shows that the right tools and methods will not deliver value if the organization, employees and culture are not aligned.

- Design of incentive-system for all involved employees to generate motivation and commitment.
- Efficient and ongoing communication of project-progress and -results
- Top management dedication to innovation and "walking the talk".

4.5 The "Human Side" of Innovation

Organizations can be understood as integrating, target-oriented, communicative social systems. The correct organizational structure/ management structure (e.g. for dealing with product ideas, their development and implementation) promotes the efficiency and effectiveness to fulfill tasks in the company and influences the behavior of employees and executives. This shows on the one hand that the level for analysis is the organization but on the other hand the promotion of innovation in an organization resides on the individual level.

Ultimately, the success of innovations depends on individuals. Success is closely related with structuring, activating and utilizing the full power and the full potential of human resources. Integral part in achieving this is long-term support of the employee's ability and willingness to innovate. The willingness to innovate is largely determined by the motivation of the individual and the personal risk perception. The ability to innovate is fundamentally linked to individual qualification, attitude and behavior.

Looking into the innovation process, it is often possible to distinguish between certain types of employees that influence the implementation of innovation. In this context it is important to point out that usually the implementers of innovation projects are often exposed to an "acceptor bias".[34] Because they are surrounded by employees which either participate directly in the development of new ideas or promote innovation processes they have the impression that resistance towards innovation is low. This does not have to be reality; therefore an analysis of the stakeholders in the innovation implementation is useful. The types of employees that can be observed differ on the one hand by the energy they bring into the innovation process and on the other hand by their acceptance of the new ideas (relevant types are the "creators", the "rebels", the "sufferers" and the "followers").

Main goal is to positively affect the mentioned influencing factors supporting innovation success. If this is not done successfully, any of the parameters can be a potential barrier in the cause of an innovation project.

The possibility to influence innovation success begins at the point where employees are selected. When selecting, the relevant competencies, such as entrepreneurial spirit, know-how, open-mindedness and creativity are to be considered. At the end the innovation manager needs to make sure that the innovation team has the necessary skills and the required knowledge.

The innovation manager himself has to be equipped with thorough professional/technical competencies (e.g. innovation knowledge and organizational potential), conceptional competencies (e.g. problem-structuring and skills) and social competencies (communication skills and networking abilities).

Eventually, with decision making, decision execution and decision revision innovation management is a process focusing on power. Realization of innovative solutions to problems requires influencing interaction partners inside and outside

---

[34]Maurer and Austin (1996), S. 172 f.

the company. With an information supply, high motivation and conflict management, individuals and groups can be incited to interact. Even in an efficiently functioning organization this can also require the use of power. Power can help safeguarding innovation success and overcoming resistance by applying escalation procedures and making use of sanctions or, more positively, supporting charismatic leadership abilities.[35] Innovations are realized when people act, and this is the top management's most ambitious task.

5 The "Ten Commandments" of Innovation Management

Successful innovation management is not a 'miracle' nor is it an obscure phenomenon. Successful innovation management is sound, hard, sustainable and consistent work. This work is directed by 10 essential guidelines that should always strictly govern innovation management and the innovation process in every company.

1. *Place the focus of innovation on the market and the customer.* The market success of an innovation depends on how successful the customer can be convinced of the benefits of the innovation. The market, the customer and the success of the innovation are the sole sources of company's income. Sustainable customer benefit is the alpha of the innovation's success: Listen to the customer.
2. *Trust on a permanent, superior performance/competitive advantage that is clearly perceivable by the market.* The unique benefit of the provided solution with a clear advantage in value is vital for market success. Innovation means differentiating oneself from competition and not imitating it. Honest appraisal is the basis for this.
3. *Oblige the organization to an economically rational behavior at an early stage and constantly review the business cases.* Invention does not mean innovation: Only the economic success increases the company's value. The amortization of an innovation project is to be ensured at any time. This means also "loosing" innovation ideas: Kill early, kill cheap.
4. *Closely monitor the process with in all phases of innovation development.* Milestones secure a clear and comprehensible progression of the innovation process. Milestones make room for autonomous actions but also enforce critical assessments after the end of each phase.
5. *Focus your resources on what the company can handle.* Companies and their employees are highly creative in an appropriate innovation culture. Diversity of ideas is a desirable state but can also overstrain companies. Concentration on valuable ideas in sync with the available human and financial resources is key to successful innovation.

---

[35] Hauschildt and Salomo (2007), p. 44.

6. *Create innovation competence in the company and maintain it.* Innovating needs to be learnt: A company necessitates permanent innovation. This requires competence. Building up and maintaining this competence using outstanding knowledge management is an extraordinary asset.
7. *Promote cross-functional co-operation in the organization.* Innovation is neither an exclusive domain of the engineer or natural scientist, nor is it a monopoly of the market expert or the market maker, or an exclusive right of the manager: Innovation success is only obtained when these forces are streamlined.[36]
8. *Safeguard consistent "product" development using performance standards and functional specifications.* Before the development starts, market and customer expectations need to be defined and documented– in comparison with what the company is able to do. A stable, market-driven definition sets the innovation on the right track.
9. *Do not bank on the free play of forces in an organization, but build on an innovation-friendly management structure.* Successful innovation development calls for an effective management structure to deal with innovations. The proactive shaping of such a structure facilitates to overcome conflicts, to eliminate resistance and to ensure implementation.
10. *Develop and secure an innovation culture that pervades the whole company.* Enthusiasm, pleasure, tolerance in failures and room for development enable success. It is the natural dealing with people in an organizational structure that determines success.

Adherence to these 10 commandments never completely eliminates the possibility of failures since there still is imminent risk and the complexity of the new, but it can help to reduce the failure rate significantly.

References

Ansoff, I. H. (1965). *Corporate strategy*. New York.
Bauer, H. H., Albrecht, C.-M., & Kühnl, Ch. (2006). Aspekte der Einführungsstrategien als Erfolgsfaktorenvon Produktinnovationen. Reihe Wissenschaftl. Arbeitspapiere W109 des Institutes für Marktorientierte Unternehmensführung, Universität Mannheim.
Coenenberg, A. G., & Salfeld, R. (2003). *Wertorientierte Unternehmensführung*. Stuttgart.
Cooper, R. G. (1999, April). From experience: The invisible success factors. *Journal of Product Innovation Management, 16*(2), pp. 115–133.
Gaubinger, K., Werani, T., & Rabl, M. (2009). *Praxisorientiertes Innovations- und Produktmanagement*. Wiesbaden.
Hauschildt, J., & Salomo, S. (2007). *Innovationsmanagement*. München.
Homburg, C., & Krohmer, H. (2003). *Marketingmanagement*. Wiesbaden.
Jaworski, J., & Zurlino, F. (2007). Innovationskultur: Vom Leidensdruck zur Leidenschaft, Frankfurt a. Main.

---

[36]Hauschildt and Salomo (2007), p. 48.

König, M., & Völker, R. (2002). *Innovationsmanagement in der Industrie.* München/Wien.
Kotler, P., Keller, K. L., & Bliemel, F. (2007). *Marketing management.* München.
Krüger, W. (2000). Strategische Erneuerung: Probleme, Programme und Prozesse. In: Krüger, W. (Ed.), *Excellence in change.* Wiesbaden.
Macharzina, K., & Wolf, J. (2008). *Unternehmensführung.* Wiesbaden.
Maurer, R. (1996). *Beyond the wall of resistance.* Austin.
Meffert, H., Burmann, C., & Kirchgeorg, M. (2008). *Marketing.* Wiesbaden.
Micic, P. (2006). *Das Zukunftsradar.* Offenbach.
Pleschak, F., & Sabisch, H. (1996). *Innovationsmanagement.* Stuttgart.
Rohe, C. (1999). Erfolgreiches Management von Innovation und Wachstum. In: Rohe, C. (Ed.), *Werkzeuge für das Innovationsmanagement.* Frankfurt a. Main.
Schreyögg, G., & Koch, J. (2007). *Grundlagen des Managements.* Wiesbaden.
Thomaschewski, D. (2002). Strategische Erfolgsfaktoren des internen Wachstums. In: Glaum, M., Hommel, U., & Thomaschewski, D. (Eds.), *Wachstumsstrategien internationaler Unternehmungen.* Stuttgart.
Vahs, D., & Burmester, R. (2005). *Innovationsmanagement.* Stuttgart.
Van der Penne, G., van Beers, C., & Kleinknecht, A. (2003). Success and failure of innovation: A literature review. *International Journal of Innovation Management, 7*(3), 1–30.
Wahren, H. K. (2004). *Erfolgsfaktor Innovation.* Berlin, Heidelberg, New York.

Innovation Generating and Evaluation: The Impact of Change Management

Dietmar Vahs, Verena Koch, and Michael Kielkopf

> Innovation has nothing to do with how many R&D dollars you have ... it's not about money. It's about the people you have, how you're led, and how much you get it.[1]

Contents

1 Change Management and Innovation Management 151
   1.1 Change Management: How to Define? 151
   1.2 Innovation Management: How to Define? 153
   1.3 Change Management and Innovation Management: Differences and Similarities 154
2 The Ability to Change and Innovate 156
   2.1 Ability to Change 156
   2.2 Ability to Change as a Prerequisite for Innovation, Innovation Generating and Innovation Processes 160
3 Evaluation for Innovation Projects 163
   3.1 Stage Gate Process 164
   3.2 Criteria for Evaluation 168
4 The Impact of Change Management on the Innovation Process 170
References 172

1 Change Management and Innovation Management

1.1 Change Management: How to Define?

The advancement of technology, economic pressures, and globalization are only three of many other factors forcing companies to review the way to operate.[2] As a

---

D. Vahs (✉)
Esslingen University of Applied Sciences, Esslingen, Germany
e-mail: Dietmar.Vahs@hs-esslingen.de

[1] Steve Jobs, interview with Fortune Magazine (1998); quoted after Tidd (2005), 467.
[2] Oakland and Tanner (2007), 1.

result, "change is an ever-present element that affects all organizations. There is a clear consensus that the pace of change has never been greater than in the current continuously evolving business environment".[3] In the past, change referred to the adoption of a new idea or behaviour by an organization, leading to its repositioning in terms of performance – now it becomes increasingly important to *manage* change rather than let it happen. Or, to use the words of former General Electrics CEO Jack Welch: "Drive change or it will drive you".[4]

Organizational change can happen planned or unplanned, but it always happens through people. There are three different ways to classify different kinds of change.

The first distinguishes between planned and unplanned. *Unplanned change*, on the one hand, happens coincidentally without people actually realizing it. This phenomenon of change is also referred to as *"panta rhei"*, a term that goes back to *Heraklit*, the great Greek philosopher, who recognized that everything – even the whole world – is in constant movement. *Planned change*, on the other hand happens intentionally. It is directed, controlled and organized by management to improve the effectiveness and the efficiency of an organization.[5] A second dimension in assessing change is its degree of novelty: *Incremental change* is evolutionary, comprising a series of continuous events that maintain the organization's equilibrium and often affecting only part of the organization.[6] It is therefore low in intensity and complexity, and seems logical and rational. As incremental change occurs in established structures and processes, people tend not to fear it. By contrast, *radical change* is rule-breaking. It is a drastic, high-risk, paradigmatic change in the organization's mode of operation and it often transforms the entire organization itself.[7] Radical change is revolutionary, abrupt, it seems irrational and incomprehensible. Due to that fact, people tend to fear it.

As a third, DiBella distinguishes *desirable* from *undesirable* change, which is a key factor for the motivation for change of the members of an organization.[8] It influences whether people engage themselves productively in a change initiative or not: "This perception provides a foundation upon which resistance or participation ultimately rests".[9]

For most authors *change management* means a goal-oriented analysis, planning, realisation, evaluation and continuous improvement of the integrated methods of variations in companies and prepares the people concerned for the future situation. It contains different management tools and methods, which support the change process and lead to a successful change.[10]

---

[3] By (2005), 378.
[4] Quoted after Picot et al. (1999), 1.
[5] Krüger (2006a), 25; Vahs (2007), 267.
[6] Daft (2004), 400.
[7] Daft (2004), 400; Krüger (2006a), 25 pp.; Staehle (1999), 900.
[8] DiBella (2007), 233.
[9] DiBella (2007), 233.
[10] Cacaci (2006), 41p.; Heberle and Stolzenberg (2006), 5; Vahs (2007), 281.

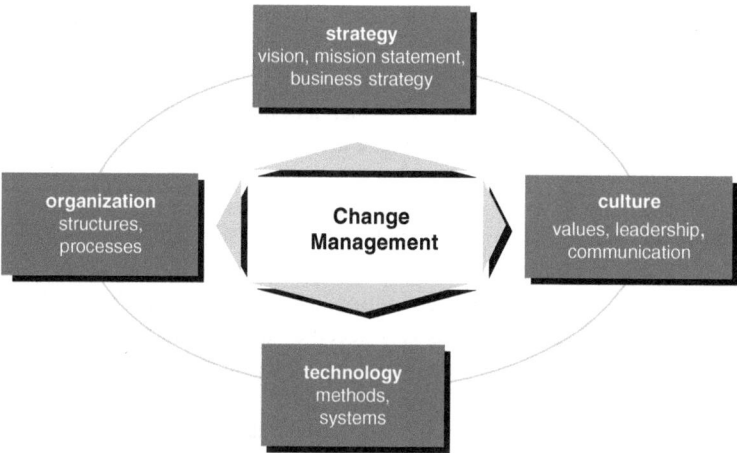

Fig. 1 The four dimensions of change management (Vahs, 2007, 320)

Change management is the consideration of *four dimensions* in an organization, *strategy*, *organization*, *technology* and *culture*, which should be seen in a dynamic general view and have to be coordinated with each other in the best possible way.[11] While strategy, organization and technology are the so-called "hard dimensions" (see Fig. 1), which can be easily measured, corporate culture is a "soft dimension" which cannot be measured easily but has a tremendous effect on the organization in the whole. For this reason successful change requires an integrated approach considering the hard factors as well as the soft factors permanently.

Change is always associated with a loss in stability and security. The old and well-known is replaced by something new and ambiguous. Employees experience uncertainty about organizational change.[12] They associate fear and threats within change, instead of challenges and opportunities. Therefore change often leads to resistance of the people concerned. But change becomes increasingly important for decision-making processes in high levels of turbulence, especially in the case of disruptive technological change or discontinuous innovation. A close relation to the management of innovations can be seen here, where "novelty" and hitherto unknown elements, too, are always involved, which has to assert itself against traditional products, methods and processes and thus also against human resistance.[13]

1.2 Innovation Management: How to Define?

As this book treats innovation and innovation management, this subchapter will be very short. Only the main statements will be taken up.

---

[11] Vahs (2007), 319 and 328.

[12] Allen, Jimmieson, Bordia, and Irmer (2007).

[13] Christensen (2006), XV and 3pp.; Antoniou and Ansoff (2004), 275; Ansoff (1977), 77; Tidd et al. 18pp.; see subchapter 4.2.

Innovation management is the systematic generation, development and implementation of an innovation.[14] It is the successful way to lead new ways of doing things from the generation and evaluation of novel ideas over the whole research and development process to the market launch and last but not least to the long-term adoption of a new idea by the customers. Further innovation management includes leadership, organization, financing and controlling.

According to Siguaw et al. a broader view of innovation is important to create a long-term survival or a competitive advantage.[15] It is insufficient to concentrate only on products and processes. Also the importance of the organization system and the corporate culture have to be mentioned: "..., there must be a collective set of understanding and beliefs, pervasively accepted throughout the firm and likely to occur at all levels and functions, that facilitates continual process to insure long-term competitive advantage".[16]

What do change management and innovation management have in common – and what differentiates them?

1.3 Change Management and Innovation Management: Differences and Similarities

Today, every company must change and innovate not only to prosper but merely to survive in a world of increased competition. To manage the threats and take advantage of the opportunities, the top management has to handle dramatic changes in all areas of operations. As change and innovation, rather than stability, are the norm today, the difference between change management and innovation management is primarily that change management, on the one hand, focuses on the corporation itself, and innovation management, on the other hand, concentrates on the customers and the markets. Thus, change management is a method to reorientate and restructure a company, whereas innovation management represents a method to establish new products, new processes or new services on existing markets, or even a method to establish a new market for an innovative idea.

Change is mainly driven by today's turbulent, unpredictable environment. "Even when internal reasons were given for the need of change; these could be related to some form of external pressure on the organization."[17] In contrast to that, innovation can be driven by the market (market-pull-innovation) or by an innovator inside

---

[14] Thom and Müller (2006), 251.
[15] Siguaw, Simpson, and Enz (2006), 570.
[16] Siguaw et al. (2006), 570.
[17] Oakland and Tanner (2007), 12.

the company (technology-push-innovation).[18] Especially pushed innovations have to create a market demand to be successful.

As "every innovation causes change",[19] there are a lot of similarities between change management and innovation management: Both types of management differentiate their respective subjects regarding their effects. On the one hand, there are incremental changes and innovations and, on the other hand, there are radical changes and innovations. Radical innovations which are "new to the world", in the majority of the cases will lead to drastic or radical changes, whereas incremental innovations only will lead to smaller but often continuous improvements.[20] These changes affect not only the customers but all persons who are involved in the innovation process. As often employees are the main barrier for innovation and change,[21] it is necessary to achieve the acceptance of the persons concerned. "Innovation management is constantly fighting against resistance."[22] The groups and types of resistance in an innovation process and in a change process are quite similar.[23] An open-minded internal communication and information policy, shared values and clear and goal-oriented leadership support the elimination of uncertainty and resistance.[24]

Management commitment in an innovation process as well as in a change process is of great importance for its respective success. The company's top decision makers have to confess subjectively to change and innovation, and should anchor innovation in their value system or their corporate agenda. They have to be the convincing promoters and enablers of change.[25]

Summarizing the facts, one can say that change management needs sufficient participation, comprehensive communication and, as a result, highly motivated employees, whereas innovation management needs a supremely good idea, a systematic approach, fearless consequence and clear customer focus. The intention of both is to create new possibilities to enable the company to meet the challenges of the global environment. This is why there is a constant need for the ability to change and innovate.

---

[18] Vahs and Burmester (2005), 80p.
[19] Seidenschwarz (2003), 54; translation by the authors.
[20] Tidd et al. (2005), 17p.; see next subchapter and Fig. 4.2.
[21] Stern and Jaberg (2007), 20 and 27.
[22] Hauschildt and Salomo (2007), 206; translation by the authors.
[23] Hauschildt and Salomo (2007), 183pp.; Vahs (2007), 329pp.
[24] Stern and Jaberg (2007), 84.
[25] Hauschildt and Salomo (2007), 29; Krüger (2006a), 41p.; Manimala, Jose, and Thomas (2006), 56; Stern and Jaberg (2007), 21.

2 The Ability to Change and Innovate

2.1 Ability to Change

2.1.1 What does Ability to Change Mean?

While flexibility represents the ability to adopt the existing game rules in an established system with less effort in reaction, the ability to change is the skill to transform an established system structurally and with fast reaction.[26]

The ability to change relies on the knowledge and the skills of individuals in an organizational unit or whole organization. Coping with drastic change requires specific experience in such situations and specialized technical and methodical skills. The human factor is often a barrier in this field. This barrier should be eliminated by suitable methods, like coaching, mentoring etc.[27] The necessity for change, the willingness to change and the ability to change are the three main factors, which build the framework of change in the *3-W-Modell* by Krüger. "Only enterprises with the ability to change are sustainable."[28] This is the reason, why top management has to orchestrate these three coordinates of change to fit in tune.[29]

The participation of the employees of a company plays a tremendous role regarding the ability to change. It is important to reduce the uncertainty of the employees. This is while successful change processes must activate and empower the people concerned. This in turn requires transparent and well-timed information and communication processes as well as the toleration of faults.[30] "Such interventions (actions) allow employees to 'see' and influence the process, assert a measure of control over their futures, and understand their role in the new change environments."[31]

2.1.2 Why do Companies Need the Ability to Change?

Success in the long run always comprehends innovation, thus an adaption to the changing environment. Without innovation, there will be no long-term success. The impact of the ability to change on innovation management is shown by Tidd et al. (see Fig. 2):[32]

- In *Zone 1* no change happens, which means that "the rules of the game are clear". Only incremental innovation, like steady-state improvements, in products or processes takes place. Knowledge about process and product core components is accumulated.

---

[26] Spath, Hirsch-Kreinsen, and Kinkel (2008), 11.
[27] Krüger (2006a), 34pp.; Vahs (2007), 375pp.
[28] Krüger (2006a), 34; translation by the authors.
[29] Krüger (2006b), 135.
[30] Vahs (2007), 340p.
[31] Fugate, Kinicki, and Prussia (2008), 31.
[32] Tidd et al. (2005), 17p.

- In *Zone 2* the overall architecture remains the same, but there is a significant change in one element: "There is a need to learn new knowledge but within an established and clear framework of sources and users".
- In *Zone 3* discontinuous innovation happens, which is abrupt, the whole set of "rules of the game" changes and there is the opportunity for new entrants. In this zone, "neither the end state nor the way in which it can be achieved are known about".
- In *Zone 4* architectural innovation happens and new combinations emerge. Existing knowledge sources and configurations have to be reconfigured. Either the existing knowledge is used and recombined in different ways, or old and new knowledge are combined.

Only in Zone 1, the incremental innovations, is the necessity of the ability to change not shown. In all the other zones, the company needs the ability to change, especially in Zone 4, where the "steady-state" innovation conditions are punctuated by occasional discontinuities which are shifting the basic conditions dramatically.[33]

Change management helps the company to steer through the different situations in the four zones of innovation, especially "when radical change takes place along the technological frontier or when completely new markets emerge".[34] In this case, Oakland and Tanner identified several key points for successful change management: First, companies should scan their external environment to improve the organization's ability to implement improvements. Second, the company's leaders should set a clear direction and communicate success criteria. Third, companies

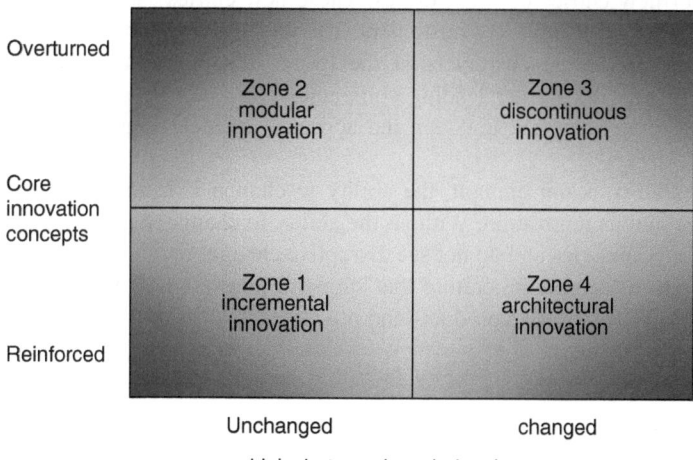

Fig. 2 Component and architectural innovation (Tidd, Bessant, & Pavitt, 2005, 17)

---

[33] Tidd et al. (2005), 18.
[34] Tidd et al. (2005), 18.

should align the need for change to their operational issues, because people have to understand how they will be affected and what must to be done to cope with the challenge. Fourth, they should translate the high level strategic change into a process approach, because process thinking is central to successful change. Fifth, companies should align their culture to support changes in peoples' behaviour, because this will help to minimise resistance. Sixth, companies should support change by performance measurement, as this will allow target-areas for improvement to be identified and will play a key role in communication. Seventh, companies should continuously review priorities, structure and programme metrics of the chosen change approach to be sure that the required benefits are really delivered.[35]

As already stated, the ability to change is essential. If companies or managers do not monitor business discontinuities and changes that are associated with them, they become enemies of their own success.[36] Christensen mentions that especially great innovative companies tend to fail, when they only follow their customer's next-generation needs.[37] This represents one of the innovator's dilemmas: "Blindly following the maxim that good managers should keep close to their customers can sometimes be a fatal mistake".[38] Customer orientation and/or lead-user orientation is definitely a good method for push-innovations, but sometimes the established customers or lead-users cannot look behind the boundaries and do not see that times are changing, and that new markets emerge. Christensen illustrates this by the case of the hard disk drives, where the leading companies, like IBM, did not see the necessity of smaller drives, because their established customers did not see this necessity either. Nor did the leading companies see that there was a new market emerging, with different customers, and this was the chance for new entries of small drives. These new entries satisfied the needs and wishes of the non-established customers. The new market increased, and suddenly the former established customers of the former leading companies also wanted smaller drives. The former leading companies could not satisfy their customers' wishes, because they had not seen the approach of weak signals of a disruptive change.[39]

As flexibility is not enough, the ability to change is a fundamental necessity for discontinuous innovation. Without the ability to change, companies do not look behind the boundaries and do not see disruptive changes and innovations approaching. Therefore a corporate culture that supports innovation as well as change is a prerequisite for innovative products and processes.

---

[35] Oakland and Tanner (2007), 12pp.
[36] Atoniou and Ansoff (2004), 275; Ansoff (1977), 77.
[37] Christensen (2006), 3p.
[38] Christensen (2006), 4.
[39] Christensen (2006), 3pp.

2.1.3 The Effect of Organizational and Innovation Culture on the Ability to Change

What describes an innovative organization? Which culture does it comprehend? Why is culture of central importance in the field of innovation? An innovative organization "implies more than a structure; it is an integrated set of components which work together to create and reinforce the kind of environment which enables innovation to flourish".[40]

As an organizational culture affects how organizations do things, it is of great importance for the success of an innovation and the change process involved. The organizational culture of a company has to take innovation into account. Vahs and Burmester therefore characterize an innovation friendly organizational culture, which encompasses the following: (1) Place great importance on innovation in the value system of the company. (2) Integrate the employees and give them security. (3) Support employee qualification. (4) Make innovation-champions visible; give them the information they need, communicate directly and well-timed, create individual freedom, be tolerant about and learn from failures and disappointments. (5) Avoid territorial and hierarchical thinking, information hiding, monitoring and supervising.[41]

Normally one of the main problems in change and innovation processes is the "not invented here" symptom, which means that people avoid ideas they did not have themselves. Therefore it is important to integrate employees and especially the middle managers in the process of generating ideas and putting them into action. Risk acceptance and failure tolerance are also important for a culture that leads to successful innovations.[42] This is the reason while some organizations encourage and reward risk-taking behaviour even if it is not successful. According to Mintzberg this leads to treating the enterprise as a community of engaged members: "Corporations are social institutions which function best when committed human beings ... collaborate in relationships based on trust and respect. Destroy this and the whole institution of business collapses".[43] As a result a "good" corporate culture has a coordinating, stabilising and integrating character and leads to a motivating team spirit as well as to the will to innovate, both of which are main prerequisites for the ability to change.

---

[40] Tidd et al. (2005), 468.
[41] Vahs and Burmester (2005), 360pp.
[42] Tidd et. al. (2005), 468pp.
[43] Mintzberg (2007), 25.

2.2 Ability to Change as a Prerequisite for Innovation, Innovation Generating and Innovation Processes

2.2.1 Why is Change Important for Innovations and Innovation Generating?

Seidenschwarz notices that innovation and change are closely related.[44] From the intra-corporate sight, fundamental or radical change always has an innovative character. As innovation in general "is inherently uncertain and will inevitably involve failures as well as successes", a successful innovation thus requires "that the organization be prepared to take risks and to accept failure as an opportunity for learning and development".[45]

With reference to product and technology management, Cooper and Kleinschmidt determined *four key success factors* in new product development which have the strongest effect on the business's new product performance: (1) A high-quality new product process with sharp and early product definition and tough Go/Kill decision points, (2) a defined new product strategy for the business unit with areas of focus delineated, (3) adequate resources of people and money and (4) R&D spending for new product development as a percentage of sales.[46]

They also uncovered *five other success factors*, with a modest effect on performance: (1) High-quality new product project teams, (2) senior management committed to and involved in the development of new products, (3) an innovative climate and culture, (4) the use of cross-functional project teams and (5) senior management accountability for new product results.[47]

The four determined key driving factors may implicate that only "hard" factors and structures lead to successful innovations. But as an essential prerequisite for the ability of the employees to create new ideas and to generate innovation, in our opinion the five other factors, which are all "soft" factors, are very important too (even though in the benchmarking study of Cooper and Kleinschmidt they had far less effect on performance). The hard factors may lead an innovation successfully to the market and help to make go/kill decisions, but without the soft factors there will be no breeding-ground for idea and innovation generating.

In consequence, an innovative corporate culture influences the company's strategy and thus the way in which companies do things and how they deal with change, because without an innovation culture people do not develop the ability to change or the ability to look behind boundaries.

---

[44] Seidenschwarz (2003), 60.
[45] Tidd et al. (2005), 471.
[46] Cooper and Kleinschmidt (2007), 57.
[47] Cooper and Kleinschmidt (2007), 57.

2.2.2 What does an Innovation Culture Encompasses?

The correlation of innovation success and organization culture was the subject of an empirical study by Ernst.[48] The study is based on the typology of companies in four dimensions according to Cameron and Freeman. They distinguish four types of culture: clan, adhocracy, hierarchy and market. The *hierarchy culture* is characterized by administrative leadership, order, uniformity, rules and regulations. It is based on mechanistic structures and an internal orientation. The *clan culture* is also oriented internally, but has more organic structures and processes. It is characterized by participation, teamwork and sense of family. Employees confess to their company and to their parent-figure leader. The main elements of the *market culture* are competitiveness, goal achievement, environment exchange and a decisive and achievement-oriented leadership. It is externally oriented with mechanistic structures and processes. The *adhocracy culture* has organic structures and processes and is externally oriented. It is characterized by entrepreneurship, creativity, adaptability and an innovative and risk-taking leadership (see Fig. 3).[49]

The study dealt with the effects of organizational culture on new product development. Which culture fosters innovation success?

The posted negative correlation of technology dynamic and hierarchy culture and the posted positive correlation of technology dynamic and adhocracy culture were approved. It was also shown that companies with the highest technology dynamic in their environment, instead of companies with the lowest technology dynamic in their environment, show more often significant adhocracy culture and rarely hierarchy culture. But it was even shown that too much "adhocracy" can contribute to a loss in innovation success. This non-linearity of the relationship means in its turn that an optimum degree of "adhocracy culture" exists.[50]

This result leads to the conclusion that only externally oriented and organic cultures lead to successful innovations. However, it cannot be concluded that only organic organizations lead to successful innovations.[51] Most times the flexibility of mechanistic structures is underestimated. Perhaps mechanistic structures lead to more resistance, but mechanistic structures also include specific techniques to negotiate resistance.[52] So it cannot be said that only companies with an adhocracy culture will lead to successful innovation, but it can be said that companies with an external, customer oriented culture will be more successful in innovation than companies with an internal oriented culture. What does this mean to innovation processes?

---

[48] Ernst (2003).
[49] Cameron and Freeman (1991), 27.
[50] Ernst (2003), 37pp.
[51] Hauschildt and Salomo (2007), 111pp.
[52] Hauschildt and Salomo (2007), 113.

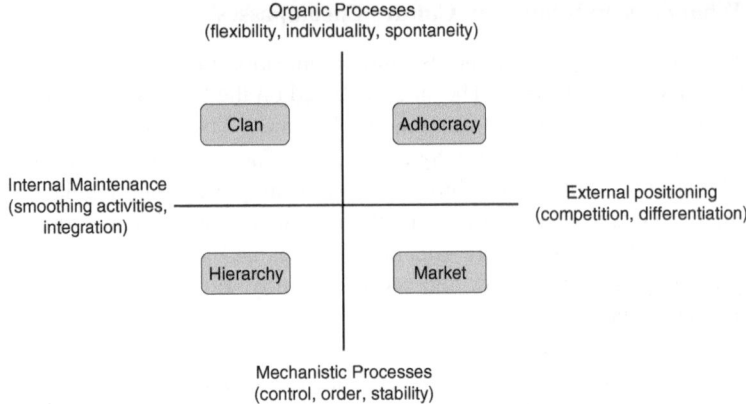

Fig. 3 Typologies of organization culture (Cameron & Freeman, 1991, 27)

2.2.3 What Requirements or Standards do Innovation Processes Need?

Innovations cannot be decreed. Innovation processes are unique. There are scores of different effects and influences and their results are insecure. Therefore, on the one hand, innovation processes are complex and difficult to control; on the other hand, some degree of freedom and flexibility is an essential ingredient to innovation processes operated by cross-functional teams.[53]

According to Tidd et al. innovation processes comprehend the enabling of a continuous searching and scanning of the corporate environment, strategy-making, effective knowledge acquisition, implementation, innovation launch and last but not least learning and re-innovation.[54] Innovation processes should not forget to look beyond the steady-state and beyond the boundaries to make innovation happen under discontinuous conditions.[55] Hauschildt and Salomo present an overview of empiric results for process-steering and its impact on the innovation process and conclude, that (1) a mixture of taut process steering and coexisting autonomy is promising for success. (2) Well defined targets and milestones, intermediate deadlines and intermediate results permit actions of the people concerned to be controlled. (3) Formalisation and documentation are not prejudicial to innovation. They make the processes' structure transparent, define responsibilities, document the initiators and complement reviews. (4) A formal monitoring of the projects progress by supervisors is helpful.[56]

A high-quality innovation process is judged to be a success factor for winning businesses in product development.[57] Because of changing surrounding conditions,

---

[53] Bonner et al. (2001), 233, Stern and Jaberg (2007), 8.
[54] Tidd et al. (2005), 349pp.
[55] Tidd et al. (2005), 405pp.
[56] Hauschildt and Salomo (2007), 495pp.
[57] Cooper and Kleinschmidt (2007), 57.

a continuous process improvement and evaluation of the innovation process is important and necessary. Processes have to be optimized by time, costs and quality. According to Schmelzer and Sesselmann there are different alternatives to improve and optimize processes. "If you don't go forwards you go backwards, steadiness is regression, if achievements are not improved continuously."[58] Quality gates with precise target-settings are an important tool to prove achieved results and make go/kill-decisions.[59]

As "steering is essential to the success of the change process" and as "every innovation causes change", change and innovation processes need a systematic performance measurement.[60] Permanent evaluation and reviewing will help to keep the process updated and to avoid making mistakes twice. However, could be taken that not too much emphasis is placed on the hierarchical-mechanistic elements, particularly where there is high technology dynamism. As has already been described above with reference to the study by Ernst,[61] an optimum degree of adhocracy contributes to the success of the innovation. For the design of the innovation process, this in its turn means a balancing act between stability and standardization on the one hand and flexibility and entrepreneurial scope on the other.

3 Evaluation for Innovation Projects

Innovation processes are important for successful innovation management. But without evaluation and reviewing of the whole innovation project, improvements and lessons learned cannot happen.

As the majority of innovations do not achieve marketability, the creation of new products is a business that includes tremendous risks.[62] Reichwald et al. point out that 20–40 percent of innovations in the industrial goods industry and 30–90 percent of innovations in the consumer goods industry fail.[63] Innovation management therefore needs substantial commitment by the company. Apart from soft factors, such as culture and human resources management, the company has to take the hard factors into account as well, i.e. structure and evaluation.

The development of new products is characterised by four major points: (1) The degree of innovation varies by improvement of products to global basic innovations. (2) The risk to fail is high as there are technical, social and economic risks. (3) The innovation process depends on different departments and people, which

---

[58] Schmelzer and Sesselmann (2008), 371; translation by the authors.
[59] Schmelzer and Sesselmann (2008), 311p.
[60] Pendlebury, Grouard, and Meston (1998), 119; Seidenschwarz (2003), 54; Vahs (2007), 393.
[61] Ernst (2003).
[62] Ozer (1999), 77.
[63] Reichwald, Meyer, Engelmann, and Welcher (2007), 15.

increases complexity. (4) As many parties are involved in the new product development, conflicts may crop up.[64] In order to manage this complexity, researchers suggest understanding innovation, including all activities around the innovation, as one process.[65]

To manage the new product development process effectively, evaluation is essential. Hauschildt and Salomo understand evaluation of innovation projects as the critical assessment of the overall success of an innovation project.[66] In contrast to that, the authors of this article focus on evaluation as a constant monitoring of the new product development process. The aims of this evaluation are constant output controls and target-performance comparisons which include setting deadlines, budgets and objectives, in order to react when a deviation of the project occurs and to directly influence the desired end. Adequate measures can then be taken to adapt the process or to stop the project, if necessary. Further, continuous reviews and evaluation at the end of the process help to decrease failures in future processes.

3.1 Stage Gate Process

Many researchers have adopted this idea and "most best-practice companies have implemented a robust idea-to-launch system, such as Stage-Gate".[67] In order to manage the innovation process efficiently, the necessity of a conceptual and operational game plan is mentioned. Cooper introduces the *stage gate process*,[68] which is "a widely accepted process for new product development".[69] This means an effective as well as an efficient approach, so that the new product can be moved from idea to launch in a systematic way. The stage gate process divides the new product development (NPD) process into different stages and gates. Usually five, six or seven stages are defined. Nevertheless it is not a linear process, nor is it a rigid system to handle different types and sizes of new product projects.[70]

Every stage is preceded by one gate. At each stage, information is gathered to reduce project uncertainties and risks, which is then evaluated at the following gate. Gates represent decision points with deliverables (what the project team brings to the decision point) and must-meet/should-meet criteria, where the company can decide, if it will proceed with the NPD project or if it is to be killed, held or recycled. Thus, gates are also referred to as "Go/Kill check points", where a decision to invest more

---

[64] Thom and Müller (2006), 252.
[65] Tidd et al. (2005), 87; Cooper (1998).
[66] Hauschildt and Salomo (2007), 527.
[67] Cooper (2008), 213.
[68] Cooper (1998), 95; also see Hart et al. (2003); Birkenmeier and Brodbeck (2005), 306; Lühring (2006).
[69] Canez (2007), 49.
[70] Hart et al. (2003), 27; Ozer (1999), 78; Cooper (1998), 109; Cooper (2008), 213.

Innovation Generating and Evaluation

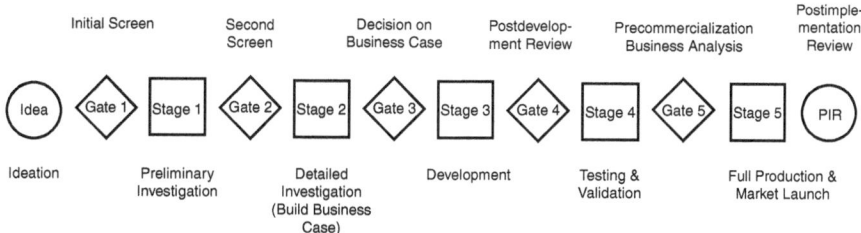

Fig. 4 A generic stage-gate new product process (Cooper, 1998, 108)

or not is made.[71] Every gate encompasses mandatory and desirable characteristics which evaluate the information about the project.[72] The stages are multifunctional and encompass parallel activities which are carried out by employees from different departments (e.g. R&D, marketing, production, or engineering).

The majority of researchers typically divide the new product development process into five or six stages.[73] A complete stage gate process is illustrated in the following Fig. 4.

3.1.1 How does the Stage Gate Process Work and What does It Encompass?

The first gate is preceded by an *idea*. Idea generation is an important activity which triggers the process. External sources, e.g. customers and suppliers, may be valuable for information gathering. In the end, it is the customer who has to buy the new product. Group discussions with customers, customer panels or surveys may help to find out the preferences of customers, their wishes, needs, possible problems and solutions.[74] It can also be helpful to identify lead users. Herstatt and von Hippel found that the joint development of products with lead users is successful because it enables the company to get a very intense understanding of customers needs.[75] Cooper puts emphasis on the fact that some progressive firms redesigned their stage-gate processes in order to include open innovation and to handle externally derived ideas, intellectual property, technologies, and even fully developed products.[76]

At *gate 1* a first screening explores whether the idea is according to the strategy of the company, and the project is feasible and attractive to customers. The company has to define beforehand which criteria represent "must meet" or "should meet" criteria. When the decision is in favour of the idea, it passes to *stage 1*. This stage

---

[71] Cooper (1998), 107; Cooper (2008), 215.
[72] Cooper (1998), 109p.
[73] Cooper (1998), 108; Hart et al. (2003), 27.
[74] Cooper (1998), 124p.
[75] Herstatt and von Hippel (1992).
[76] Cooper (2008), 231.

represents the preliminary investigation of the innovation project. Here the scope of the project is determined and a market and technical assessment is carried out.

At *gate 2* the project is re-evaluated based on the criteria of gate 1 and additional variables like market potential. A first financial calculation measures the potential return on investment of the new product. After that, a more detailed investigation follows in *stage 2*. The project is defined in detail, opportunities and threats are assessed and a project plan, i.e. a business case, is elaborated. After this stage the actual spending commences. Stage 2 is therefore seen as critical; however it is often neglected. The business case includes a clear definition of the new product: What are the target markets? What is the product positioning strategy? What are the product benefits? Then, the market and competition are analysed. Market research studies are undertaken. Further, the technical aspects of the product are assessed. Is the product actually feasible? And a detailed financial analysis is carried out. In order to get the necessary information for the business case, employees from different departments have to work closely together. Due to the multifunctional team members and the importance of the business case for the continuation, it is suggested that the business case is done by the core group of the project.

The following *gate 3* decides on the business case. If the decision is positive, substantial financial investments commence. In *stage 3* the product is finally developed physically. The result of this stage is a tested prototype. Apart from technical and qualitative aspects, it is important to involve the customers or users for feedback in order to better understand their unmet and unspoken needs, problems and benefits sought in the new product. Cooper nowadays describes this "spiral development" as a "series of 'build-test-feedback-revise' iterations or loops" which are deliberately built in from the front-end stages through the development stage to the testing stage.[77]

Gate 4 assesses the progress of the product development and reviews if it is incessantly attractive to the customer. Economic data and detailed plans, e.g. production and marketing plans, are reviewed. Based on this, in *stage 4*, the product is tested again for overall operability. This includes testing the product in the market.[78] Cooper suggests field trials, pre-tests or test markets in order to assess customers' reactions and calculate approximate market share or revenues.[79]

Gate 5 assesses the product a last time before its launch. The gate reviews the quality of the measures taken in stage 4. The last *stage 5* implements the production and marketing plans of the new product. The new product is launched on the market. After 6–18 months the new product project is finished. It becomes a regular product. At this point the product is assessed once more. Actual performance is compared to forecasts. Further, strengths and weaknesses of the project as well as customer acceptance and satisfaction and unit sales are primary considerations

---

[77] Cooper (2008), 225.
[78] Cooper (1998), 108pp.
[79] Cooper (1998), 116.

for the evaluation in a *postimplementation review*, in order to improve following projects.[80]

3.1.2 What are the Key Success Factors of the Stage Gate Process?

An important point of the process is *strategy formulation*.[81] This means the company has to define the framework in which it will operate and according to which it will select its innovation projects. Strategy depends on the resources of the company and it should take the external forces of the company like markets, competitors, customers, etc. into account.[82] The strategy formulation has to precede any process.

The *integration of customers* in the new product development process is stressed by several researchers. Cooper mentions the integration of customers especially during stages 1, 3 and 4.[83] Yet, the integration should not be organized selectively but as a continuous interaction as pointed out in the description of stage 3 above.[84]

As already mentioned, gates represent the *information gathering points* where quality is controlled and Go/Kill decisions are taken. This is important for several reasons: Quality is controlled constantly. Further, the company decides about the potential of the new product. This enables its resources on projects which are according to the strategy to be bundled. Several researchers[85] refer to innovation projects that were rejected too late or never and consequently failed. Companies had already invested a large amount of money and thought it would be false to stop this investment. When the new product failed, more money and resources were wasted. According to the motto "kill your darlings", the company has to evaluate constantly whether the project is feasible and realistic. A common example is Apple. After Steve Jobs joined Apple again as CEO in 1997, he stopped several new product development projects radically, in order to focus on those having the highest potential.[86] This strategy paid off regarding the success of iPod, iTunes etc.[87]

The cycle time of the stage gate process is not prolonged by the different stages, as activities can be carried out *simultaneously*. The overlapping of key activities and even entire stages leads to more flexibility and accelerates the process itself. As the simultaneous execution adds risk to a project, this risk has to be calculated by weighing the cost of delay of market entry against the cost and profitability of being wrong.[88]

---

[80] Cooper (1998), 108pp.
[81] Tidd et al. (2005), 98, 362pp.
[82] Tidd et al. (2005), 362p.
[83] Cooper (1998), 108pp., 124p.
[84] Tidd et al. (2005), 357.
[85] For example Grün (2005); Apenburg (2006), 19.
[86] Chmielewski and Guynn (2009).
[87] Goldbrunner (2006), 31.
[88] Cooper (2008), 224.

This way of doing the innovation process is enabled by a project team, formed by employees form different departments creating a *multifunctional team*. The team approach is necessary for the implementation of the stage gate process as it is mandatory that the different departments (marketing, R&D, engineering etc.) work together.[89] Each team member can contribute his expertise and share his experience, in order to develop the new product holistically. If the team is formed only by one department, e.g. R&D, the team might loose its market focus. The team structure is variable. People join the team or quit the team when their task is fulfilled. However, a core team should exist, whose members stay during the whole process, so that a certain responsibility for the process exists. Furthermore, a *strong leader* is required, who has the formal authority, in order to take the decisions within the team. Before development starts in stage 3, the first steps of the process, which are the most risky ones, are evaluated constantly.[90]

After having presented the stage gate process, it has to be mentioned that this process should be adapted to the special needs of the company to make it a efficient, lean and rapid system. "Processes (...) are tailored to the immediate requirements of their environments."[91] If the requirements change, the company has to *adapt the process* as well, that is to say the company needs a continuous change management. This applies particularly when what is required of the process logic (e.g. on account of the integration of completely new stages or gates), the process speed (e.g. on account of the appearance of competitors in the target market) or process flexibility (e.g. on account of unforeseen events in the course of the process) changes. It is possible that the overlapping of individual phases and gates is then no longer sufficient. For this reason Cooper suggests a "next generation" stage gate process which is scalable to suit very different types and risk levels of projects, e.g. high risk major new product projects as well as moderate risk modification projects and marketing requests.[92]

3.2 Criteria for Evaluation

After having designed an individual stage gate process for the new product development of a company, the criteria of evaluation for each gate have to be determined, in order to maintain a continuing assessment of the process. This enables the company to react immediately, so that the company may adjust its process and take adequate measures. The necessity of continuing assessment is shown by Goldbrunner.[93] When analysing R&D expenditures, he found out that no definite relation between

---

[89] Cooper (1998), 98p.
[90] Cooper (1998), 96pp.
[91] Harmancioglu et al. (2007), 422.
[92] Cooper (2008), 223.
[93] Goldbrunner (2006).

the total of R&D expenditures and the performance of a company existed. In order to manage its resources efficiently, it is necessary to monitor the NPD process.

What criteria should be selected for a certain gate? It is not reasonable to measure e.g. the technological success of a product at the development stage. And if the economic success of a new product is indisputable, it is less meaningful to start evaluating the technical elegance of the new product. This leads to the conclusion that different measures and different criteria are important at the variable gates of the new product development process[94] and it is not one criterion that may be sufficient, but multiple criteria should be considered.

A study by Hart et al. examines the evaluative criteria which are used most frequently by different manufacturing companies.[95] The study analyses 134 companies in the UK and the Netherlands. Based on a literature review, Hart et al. selected *20 evaluative criteria*. These criteria were allocated to four dimensions: *market acceptance* (e.g. customer acceptance, customer satisfaction, revenue growth and market share), *financial performance* (e.g. break-even tine, margin goals and profitability goals), *product performance* (e.g. development cost, launched on time and product performance),[96] *additional indicators* (e.g. technical feasibility and intuition).[97] The findings demonstrate that, although companies vary in their choice of evaluation criteria at different gates, companies apply a customer and market orientation at all stages, i.e. they fulfil the requirement of customer integration.[98] The companies participating in the study estimated financial performance to be especially important during business analysis and after product launch. This enables management to judge the outcome of the product and to assess if resources are allocated properly. The next dimension, product performance, is basically applied during product development and market testing. And additional factors are notably applied during idea screening. The authors conclude that idea screening is assessed holistically. At Procter & Gamble, management introduced a new strategy to increase the number of ideas generated. They decided to use external resources, e.g. external laboratories, other companies or customers (open innovation), in order to foster new product development. Approximately 35 percent of Procter Gamble's products contain external elements.[99] However, this strategy implies management and evaluation of the generated ideas, so that the resources of the company can be concentrated on promising new products. The ideas are evaluated, e.g. the manager of a certain business unit assesses if the product idea fits in with the goals of the business unit,

---

[94] Hauschildt and Salomo (2007), 568p.

[95] Hart et al. (2003).

[96] The first three dimensions are based on the research of Griffin and Page ((1993), 294). They did a secondary analysis of 77 published articles concerning new product success measures (Griffin and Page (1993), 292).

[97] Hart et al. (2003), 28.

[98] Desouza et al. (2008), Reichwald et al. (2007).

[99] Sakkab and Huston (2006), 24.

analyses the technical infrastructure and economic potential, etc. This tight assessment leads to the fact that only one of 100 external ideas is moved to market launch at Procter & Gamble.[100]

Further, the allocation of the different dimensions was relatively stable, disregarding company size, market share, innovation strategy, type of product or driver for the new product development process[101],[102].

Researchers assume that the process evaluation has to be adapted to each stage. "If a firm applies the same metrics throughout the R,D&E process it does not get the most out of its technological efforts."[103] This is supported by the study of Hart et al. as companies attribute importance to the fact that specific dimensions in evaluation are more important in certain stages than others.[104] Thus, evaluation criteria have to be selected according to the needs of the special stage.

4 The Impact of Change Management on the Innovation Process

The implementation of an innovation process in the company requires certain characteristics, in order to manage the NPD projects successfully. These characteristics seem to be concordant with the basic ideas of change management as innovation involves change. These are presented below.

Organizational culture influences strategy. *Strategy formulation* represents the framework for new development projects and is a prerequisite for the stage gate process and its operations. It constitutes the first selection of projects as the company defines resources and parameters of competition.[105] Tidd et al. point out that innovation needs a strategy tolerating failure and fast learning.[106] Failures are seen as a possibility or a chance for learning, and tolerance of failure encourages idea generation of employees.[107] After having finished the new product development project the company should evaluate its process and learn from prior experience.[108] This involves questioning the new product development process all the time and adapting it continuously.

---

[100] Sakkab and Huston (2006), 30.

[101] Innovation strategy: technological innovator, fast imitator, cost reducer; product type: line addition, improvement, completely new; new product development driver: market or technology driven (Hart et al. (2003), 26).

[102] Hart et al. (2003), 33pp.

[103] Hauser and Zettelmeyer (1997), 33.

[104] Hart et al. (2003).

[105] Tidd et al. (2005), 362p.

[106] Tidd et al. (2005), 73.

[107] Vahs and Burmester (2005), 365; Schoemaker and Gunther (2006).

[108] Tidd et al. (2005), 400pp.

Apart from strategy, *organizational culture* influences the code of conduct within the company. Employees are a source of idea generation. The company should therefore actively involve employees and their knowledge in order to support the generation of ideas.[109] As innovation means change, innovation can create uncertainty and resistance. *Effective communication* can help avoiding dysfunctions as already benign information reduces employee's insecurity and uncertainty feeling.[110] It is therefore advisable to communicate information regarding innovation at an early stage.[111] Clear, honest and consistent communication fosters the transparency of the process for the participating members from the beginning.

Flexibility and *value stream orientation* seem to be another prerequisite for innovation processes. Waste of time and money as well as inefficiency have to be removed from the NPD process at every opportunity. Cooper emphasizes the fact that the stage gate process is not a rigid, bureaucratic system, but an adaptable process with the possibility of a simultaneous execution of key activities or even entire stages in order to accelerate the process itself and to improve its results. The organization of the process as well as the organization of project teams should be able to be changed according to the needs of the new product development. This also concerns employees, who have to adapt when new product development projects are stopped because a kill decision is taken.[112] One should also consider the fact that too much formal control may constrain the creativity of the NPD team, impede their progress, and even adversely affect the ultimate performance of the team and the process.[113]

The formulation of a clear strategy, integration of employees, comprehensible and consistent communication and flexibility seem to be the main prerequisites for an effective stage gate process and at least for a rapid commercialization of new ideas. These factors are not only important for the new product development, i.e. innovation (see Sect. 3), but also are prerequisites for change management (see Sect. 2). The management of an innovation process and the management of a change process are intertwined as the innovation process involves constant adaptation as well as the way people cope with each other in this process. Companies therefore need people who are able to drive change and innovation. However, they cannot take the right measures, if management does not monitor deviances. Therefore an evaluation of the change and innovation processes is essential. Only a continuous, holistic evaluation enables the planned and actual situation to be compared, in order to adapt, if necessary. And this adaptation, again, is the task of a successful farsighted management.

---

[109] Cooper (1998), 128p.
[110] Ashford, Lee, and Bobko (1989), 821.
[111] Balmer, Invesini, Planta, and von Semmer (2000), 130.
[112] Cooper (2008), 224.
[113] Bonner, Ruekert, and Walker (2001), 234.

References

Allen, J., Jimmieson, N. L., Bordia, P., & Irmer, B. E. (2007). Uncertainty during organizational change: Managing perceptions through communication. *Journal of Change Management, 7*(2), 187–210.

Ansoff, H. I. (1977). Strategy formulation as a learning process: An applied managerial theory of strategic behavior. *International Studies of Management & Organization, 7*(2), 58–77.

Antoniou, P. H., & Ansoff, H. I. (2004). Strategic management of technology. *Technology Analysis & Strategic Management, 16*(2), 275–291.

Apenburg, E. (2006). Innovation – wo ist das Problem? *Wissenschaftsmanagement, 6*, 14–20.

Ashford, S. J., Lee, C., & Bobko, P. (1989). Content, causes, and consequences of job insecurity: A theory-based measure and substantive test. *Academy of Management Journal, 32*(4), 803–829.

Balmer, R., Invesini, S., Planta, A., & von Semmer, N. (2000). Innovation im Unternehmen. Leitfaden zur Selbstbewertung von KMU, vdf Verlag, Zürich.

Birkenmeier, B., & Brodbeck, H. (2005). Marktleistungsentwicklung. In: Hugentobler, W., Schaufelbühl, K., & Blattner, M. (Eds.), *Integrale Betriebswirtschaftslehre*. Zürich: Orell Füssli.

Bonner, J. M., Ruekert, R. W., & Walker, O. C. (2001). Upper management control of new product development projects and project performance. *The Journal of Product Innovation Management, 19*, 233–245.

By, R. T. (2005). Organizational change management: A critical revue. *Journal of Change Management, 5*(4), 369–380.

Cacaci, A. (2006). *Change Management – Widerstände gegen Wandel, Plädoyer für ein System der Prävention*. Wiesbaden: Deutscher Universitäts-Verlag.

Cameron, K. S., & Freeman, S. J. (1991). Cultural congruence, strength and type: relationships to effectiveness. In: Woodman, R. W., & Passmore, W. A. (Eds.), *Research in Organizational Change and Development, 5*, 23–58.

Chmielewski, D. C., & Guynn, J. (2009). Apple CEO Steve Jobs Takes Medical Leave, in: Los Angeles Times, 15 January 2009. http://www.latimes.com/business/la-fi-stevejobs15-2009jan15,0,7042254.story , Accessed 28 January 2009.

Christensen, C. M. (2006). *The innovator's dilemma: The revolutionary book that will change the way you do business*. New York: Harper Collins.

Cooper, R. G. (1998). *Winning at new products*. Reading, MA: Perseus Books.

Cooper, R. G. (2008). Perspective: The Stage-Gate Idea-to-Launch Process – Update, What's New, and NexGenSystems. *The Journal of Product Innovation Management, 25*, 213–232.

Cooper, R. G., & Kleinschmidt, E. J. (2007). Winning businesses in product development: The critical success factors. *Research Technology Management, 50*, 52–66.

Daft, R. L. (2004). *Organization theory and design*. Mason, OH: Thomson South-Western.

Desouza, K. C., Awazu, Y., Jha, S., Dombrowski, C., Papagari, S., Baloh, P., et al. (2008). Customer-driven innovation. *Research Technology Management, 51*, 35–44.

DiBella, A. J., (2007). Critical perceptions of organisational change. *Journal of Change Management, 7*(3/4), 231–242.

Ernst, H. (2003). Unternehmenskultur und Innovationserfolg – Eine empirische Analyse. *Zfbf, 55*, 23–44.

Fugate, M., Kinicki, A. J., & Prussia, G. E. (2008). Employee coping with organizational change: An examination of alternative theoretical perspectives and models. *Personnel Psychology, 61*, 1–36.

Goldbrunner, T. (2006). Mehr hilft nicht mehr! Hohe F&E Ausgaben sind kein Garant für Efolg. *Wissenschaftsmanagement, 1*, 30–33.

Griffin, A., & Page, A. L. (1993). An interim report on measuring product development success and failure. *Journal of Product Innovation Management, 10*, 291–308.

Grün, O. (2005). Risiken und Erfolgsfaktoren von Großprojekten: Lehren für das Innovationsmanagement. *zfo, 74*, 207–210.

Hart, S., Hultink, E. J., Tzokas, N., & Commandeur, H. G. (2003). Industrial companies' evaluation criteria in new product development gates. *Journal of Product Innovation Management, 20*, 22–36.

Hauschildt, J., & Salomo, S. (2007). *Innovationsmanagement*. München: Vahlen.

Hauser, J. R., & Zettelmeyer, F. (1997). Metrics to evaluate R, D&E. *Research and Technology Management, 40*, 32–38.

Heberle, K., & Stolzenberg, K. (2006). Change Management. Veränderungsprozesse erfolgreich gestalten – Mitarbeiter mobilisieren, Springer, Heidelberg.

Herstatt, C., & von Hippel, E. (1992). From experience: Developing new product concepts via the lead user method: A case study in a "low-tech" field. *Journal of Product Innovation Management, 9*(3), 213–221.

Krüger, W. (2006a). Topmanager als Promotoren und Enabler des Wandels. In: Krüger, W. (Ed.), *Excellence in Change, Wege zur strategischen Erneuerung* (pp. 21–46). Wiesbaden: Gabler.

Krüger, W. (2006b). Das 3 W-Modell: Bezugsrahmen für das Wandlungsmanagement. In: Krüger, W. (Ed.), *Excellence in Change, Wege zur strategischen Erneuerung* (pp. 125–170). Wiesbaden: Gabler.

Lühring, N. (2006). *Koordination von Innovationsprojekten*. Wiesbaden: Deutscher Universitäts-Verlag.

Manimala, M. J., Jose, P. D., & Thomas, K. R. (2006). Organizational constraints on innovation and intrapreneurship: insights from public sector. *Vikalpa: Journal for Decision Makers, 31*(1), 49–60.

Mintzberg, H. (2007, July–August). Productivity is killing American enterprise. *Harvard Business Review, 85*, 25.

Oakland, J. S., & Tanner, S. (2007). Successful change management. *Total Quality Management, 18*(1/2), 1–19.

Ozer, M. (1999). A survey of new product evaluation models. *Journal of Product Innovation Management, 16*, 77–94.

Pendlebury, J., Grouard, B., & Meston, F. (1998). *The ten keys to successful change-management*. Chichester: Wiley.

Reichwald, R., Meyer, A., Engelmann, M., & Welcher, D. (2007). *Der Kunde als Innovationspartner*. Wiesbaden: Gabler.

Sakkab, N., & Huston, L. (2006). Wie Procter & Gamble zu neuer Kreativität fand. *Harvard Business Manager, 28*, 21–31.

Schmelzer, H. J., & Sesselmann, W. (2008). Geschäftsprozessmanagement in der Praxis, Kunden zufrieden stellen – Produktivität steigern – Wert erhöhen, Hanser, München.

Schoemaker, J. H., & Gunther, R. E. (2006). Machen Sie mehr Fehler: Es lohnt sich! *Harvard Business Manager, 28*, 72–81.

Seidenschwarz, W. (2003). *Steuerung unternehmerischen Wandels*. München: Vahlen.

Siguaw, J. A., Simpson, P. M., & Enz, C. A. (2006). Conceptualizing innovation orientation: A framework for study and integration of innovation research. *Journal of Product Innovation Management, 23*, 556–574.

Spath, D., Hirsch-Kreinsen, H., & Kinkel, S. (2008). Organisatorische Wandlungsfähigkeit produzierender Unternehmen, Unternehmenserfahrungen, Forschungs- und Transferbedarfe, Fraunhofer IAO, Stuttgart 2008.

Stern, T., & Jaberg, H. (2007). *Erfolgreiches Innovationsmanagement, Erfolgsfaktoren – Grundmuster – Fallbeispiele*. Wiesbaden: Gabler.

Thom, N., & Müller, R. (2006). Innovationsmanagement in KMU. In: Bruch, H., Vogel, B., & Krummaker, S. (Ed.), *Leadership – Best Practices und Trends*. Wiesbaden: Gabler.

Tidd, J., Bessant, J., & Pavitt, K. (2005). *Managing innovation: Integrating technological, market and organizational change*. West Sussex: John Wiley & Sons.

Vahs, D. (2007). *Organisation. Einführung in die Organisationstheorie und –praxis*. Stuttgart: Schäffer-Poeschel.

Vahs, D., & Burmester, R. (2005). *Innovationsmanagement. Von der Produktidee zur erfolgreichen Vermarktung*. Stuttgart: Schäffer-Poeschel.

Implementing Change Management Successfully – Reinventing an Innovative Corporation: The Bayer Case

Alexander Moscho, Lydia Bals, Matthias Kämper, and Stefan Neuwirth

Contents

1 Introduction 175
2 Times of Change for Bayer: Yesterday, Today and Tomorrow 177
3 Conceptual Background: Change Management 179
4 The Case: Change Management at Bayer 181
5 Summary and Conclusion 188
References 189

1 Introduction

Globalization has intensified competition throughout industries.[1] Product lifecycles have shortened dramatically.[2] As pharmaceutical firms are characterized by particularly high R&D upfront costs, long before any revenues can be expected,[3] the consequences of shorter periods to reap the rewards is particularly high.

Bayer is an innovators' company, striving for technology leadership. As a company committed to innovation, with annual R&D expenditures of approx. € 2.65 billion in 2008, Bayer has been continuously confronted with the challenge to sustain innovation and success over its corporate history.

From a strategic management perspective, Miles and Snow (1978) differentiated between different types of firms with characteristic behavior: Prospector, Analyzer, Defender and Reactor. Prospector companies emphasize product and market effectiveness. These companies tend to have access to markets with large potential and seek to innovate and initiate changes within their industry.

A. Moscho (✉)
Bayer Business Consulting, Leverkusen, Germany
e-mail: alexander.moscho@bayerbbs.com

[1] Czinkota and Ronkainen (2005).
[2] Fenwick (1999).
[3] DiMasi, Hansen, and Grabowski (2003) and Moscho, Hodits, Friedemann, and Leiter (2000).

As a prospector, Bayer has strived to actively shape its own future during its history. The quote "The precondition for success is not adaptation, but the will to design changes yourself"[4] by Werner Wenning, highlights this philosophy.

After Mr. Wenning became Chairman of the Board in 2002, Bayer was fundamentally transformed, encompassing changes in organization, portfolio, cost structures and the remuneration system. Business was consistently aligned towards innovation and growth.[5] Already during the years prior to that, Bayer had been characterized by an increasingly dynamic growth with varying degrees throughout the business divisions.[6] Today, life sciences within its portfolio and Asia from a regional perspective represent Bayer's major growth areas. Regarding innovation, taking Bayer Healthcare as an example, Fig. 1 gives an overview of the innovative products' pipeline.

In order to put innovation and change management into the right conceptual context, the article is structured as follows. In the next section, we are going to provide more background information regarding Bayer. The third section provides a conceptual background on organizational change and change management. In the fourth section we analyze how change management is pursued at Bayer, drawing on the

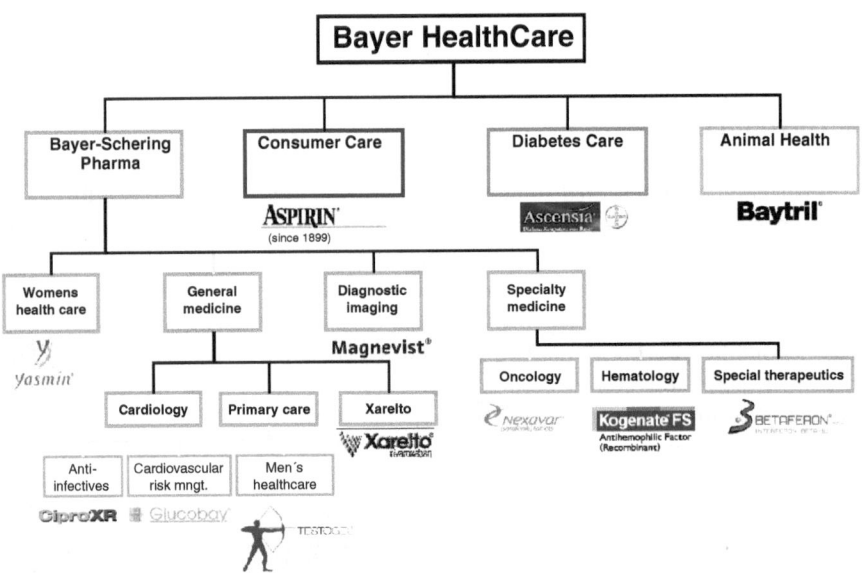

Fig. 1 Innovative Bayer HealthCare products

---

[4]Focus (2008), p. 12.
[5]Focus (2008).
[6]Metelmann and Neuwirth (2002).

example of the Schering Integration in 2006. Last but not least, the last section summarizes the main findings, provides conclusions and gives an outlook on the future of change management at Bayer.

2 Times of Change for Bayer: Yesterday, Today and Tomorrow

Founded in 1863, Bayer looks back on a long corporate history as a big player in the chemical and pharmaceutical industries. Nowadays, Bayer is structured into three operative divisions Bayer MaterialScience, Bayer HealthCare and Bayer CropScience organized under a strategic holding. In 2008 it had net sales of over € 32.9 billion and capital expenditures of approx. € 1.98 billion. Figure 2 provides an overview of the different phases as well as the changes in structure.

When this structure was established, the goal was to concentrate resources on those business areas, in which Bayer possessed its major strengths/core competencies and saw greatest growth opportunities, leading to the portfolio of the three divisions.[7] Since 2002, Bayer has moved a transaction volume of € 42 billion. It acquired businesses for € 28 billion (e.g. Schering and Aventis CropScience) and sold parts of its portfolio for € 14 billion (e.g. divesting its classical chemical business).

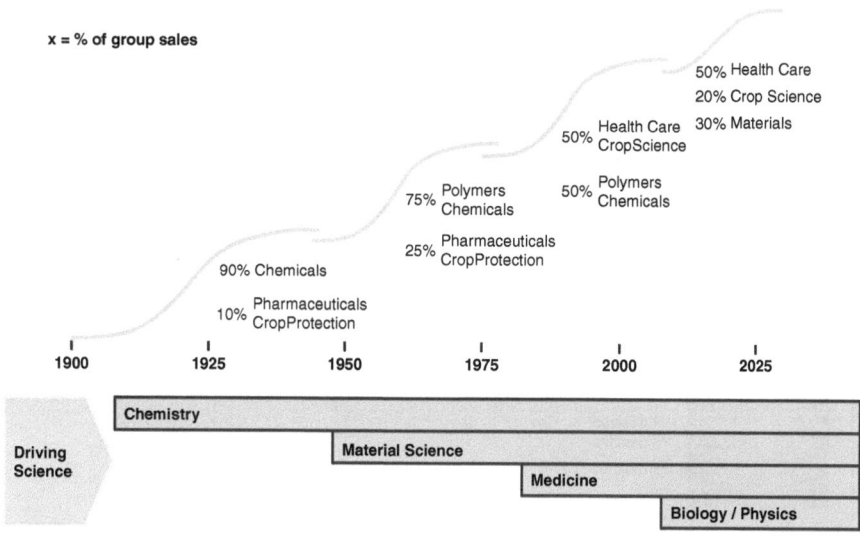

Fig. 2 History of Bayer business

---

[7] Focus (2008).

Today, Bayer HealthCare accounts for approximately 50 percent of the business, Bayer MaterialScience for approx. 30 percent and Bayer CropScience for approx. 20 percent (as also shown in Fig. 2).

Looking at the decade from 1996 to 2008, overall sales were increased from € 24.9 to 32.9 billion (please see Fig. 3). The significant reorganization enabled the operatively acting subgroups to operate close to their market and to orient themselves better towards the critical success factors of their respective industries.[8]

This process helped Bayer Schering Pharma, for example, to intensify its research focus and concentrate on four central research areas: Cardiology, oncology, women's healthcare and diagnostic imaging. Taking the oncology business of Bayer Schering Pharma as an example, the product Nexavar has set the foundation for the global oncology portfolio.

Furthermore, Bayer is extending its position as a truly global player, e.g. with investments in China. Taking the latter as an example, China is expected to become Bayer's top market within the next 10 years. Bayer has been in China since 1913, and today China is Bayer's third largest market globally. Bayer is the largest healthcare company in China, and CropScience holds a leading market position. Bayer's largest fully integrated polymer site has been built up in China. The growth has been tremendous – headcount from 2004 to 2007 developed from 2,278 to 7,100 FTE.

Considering which future challenges await Bayer on a global level, the trends of growing world population and climate change[9] are only two of the issues currently being tackled.

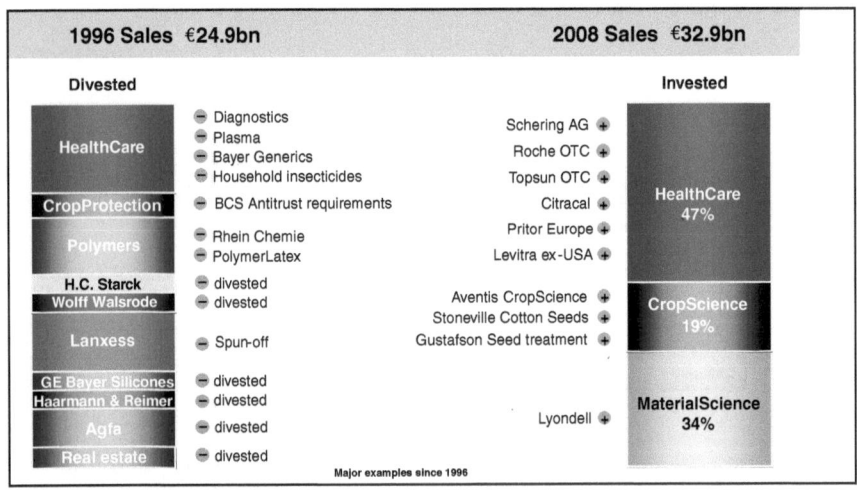

Fig. 3 Investments and divestments

---

[8] Ibid.
[9] Ibid.

Given the shift from a centrally managed cross-industry business as well as the size of its transactions, the change impact of Bayer's reorganization was dramatic. During the whole process, approximately 85,000 employees joined or left the company due to acquisitions and selling parts of the business.[10] In 2008, about 50 percent of Bayer's 106,200 employees have been less than five years with the company,[11] putting particular strain on maintaining one global corporate culture.

3 Conceptual Background: Change Management

In a literature review Pardo del Val and Fuentes (2003, p. 148) define organizational change as follows: "Organizational change is an empirical observation in an organizational entity of variations in shape, quality or state over time, after the deliberate introduction of new ways of thinking, acting and operating. The general aim of organizational change is an adaptation to the environment or an improvement in performance."

This definition highlights that deliberate actions are undertaken in order to achieve an improvement, and that corresponding changes in organizational structure and processes, as well as performance are a result. These changes can then be empirically observed.

In this context "change management subsumes all the measures that are necessary to initiate and implement new strategies, structures as well as systems and behaviours",[12] therefore resembling what is done in order to achieve that deliberate organizational change.

Top management usually is the dominant coalition within the organization which determines organizational goals and objectives, influences decision-making processes, shapes the organizational structure, formal procedures and influences the reward systems.[13] It designs the organizational structures which in turn facilitate or hinder certain processes. With its preferences for certain approaches it moreover gives strong implicit and explicit signals to all employees as to which practices are favored. Explicitly or implicitly, it communicates its preferences as vision, intent or during social interactions with employees. Therefore it shapes a great part of the content of the shared mental models existent in the organization, determining how people communicate and what is communicated.

Therefore, the communication of a clear vision of the future state is important to initiate change successfully.[14] To overcome possible resistance to cultural change, employees have to be well and timely informed about the changes as well

---

[10] Ibid.
[11] Ibid.
[12] Al-Ani and Gattermeyer (2001), p. 14.
[13] Hofstede (1981).
[14] Also confer Schein (1995) and Schreyögg (2000).

as the goals to be achieved, and should ideally have a strong individual performance orientation.[15]

In order to explicitly foster change processes, management can take measures on the organizational and the individual level.

On the organizational level, structures, processes and personnel have to be prepared, and aligned with appropriate and management systems in place that set the right signals.[16] Organizationally, the desired behavior and attitudes can be visibly promoted with the help of change agents within the organization.

Another important topic is the establishment of the necessary formal and informal communication channels.[17] Here, technological communication channels are not only referred to but also create opportunities for direct face-to-face interaction. This personal interaction, especially in groups,[18] can serve as a powerful way to maintain the necessary trust and to facilitate the evolvement of the desired shared mental models.

On the individual and group level, individuals and groups already within the company have to internalize the change content, while newly incoming members of the organization have to demonstrate a basic fit with its requirements. The latter can be accomplished by appropriately defined HR hiring profiles. Newcomers can receive training to ensure they have the necessary methodological and analytical know-how, are informed about the organization and processes within the company.[19]

In considering which instruments have to be applied for successful change, gaining an overview over the situation at hand in which to apply these instruments is vital. For this purpose the "consensus matrix" is shown in Fig. 4.[20]

The first dimension refers to how much consensus exists among employees about what they want to achieve (e.g. which goals do they have, which priorities do they set, how much effort do they invest in achieving the intended results?). The second dimension refers to how they want to achieve it (e.g. which cause and reaction chains exist and which measures can be taken).

Christensen et al. (2006) propose that it depends on the situation of the company which instruments will work best, and they differentiate between the four general approaches power (e.g. rules and role definitions), leadership (e.g. charisma and role modeling), management (e.g. measurement systems and standardized processes), corporate culture (e.g. democracy and behaviors) and mixed/overlapping instruments (such as visions, negotiations and financial incentives).

The issue of establishing consensus is amplified, when it is not only one company with potentially different subcultures, but two companies, as in the case of

---

[15] Miller, Johnson, and Grau (1994).
[16] Sackmann (1999).
[17] Grudowski (1998).
[18] Schreyögg (2000).
[19] Koch (2008).
[20] Christensen et al. (2006).

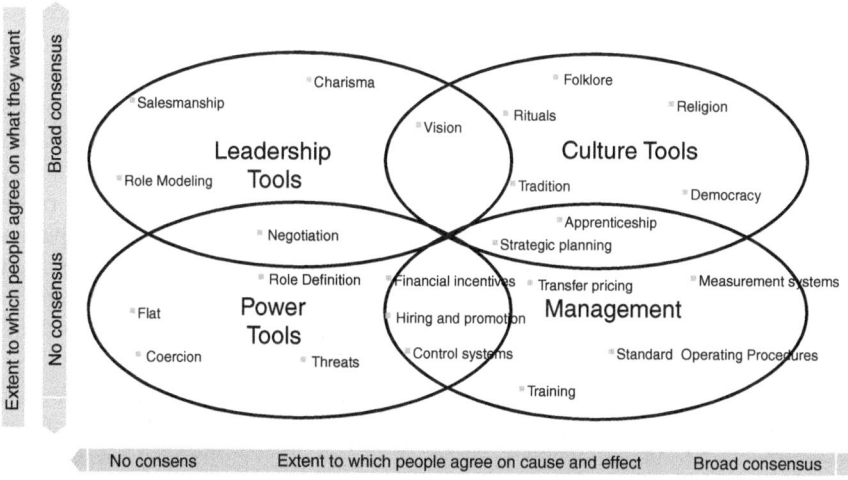

Fig. 4 Cooperation instruments within the consensus matrix, adapted from: Christensen, Stevensen, and Marx (2006) p. 30

post-merger integration. In that sense, post-merger integration resembles the ultimate change management challenge.[21] Such a situation puts a particular strain on a careful analysis and well-designed transition process towards establishing consensus.

4 The Case: Change Management at Bayer

As stated earlier, over the years Bayer has accumulated vast experience with change management throughout various situations, such as the general reorganization into the three subgroups, or for example the Aventis CropScience and Schering acquisitions.

Regarding the reorganization of the company, the importance of creating understanding for the key messages at managerial level was considered crucial, and that managers would become living examples and bring the change to their organizations. Therefore, the issues were always discussed very openly with the management levels of the "Global Leadership Circle" (GLC), comprising top managers worldwide. Communication was essential during the restructuring phase and proves to be a competitive advantage in the ever increasingly changing global economy, still. In this context, within Regional Executive Conferences it was continuously communicated to managers and interactively discussed with them, what the aims are, what the self-image is and how business management is envisaged for the future.[22] It was

---

[21] Galpin and Robinson (1997).
[22] Focus (2008).

essential to maintain the cultural core of Bayer with the values: A will to succeed; a passion for its stakeholders; integrity, openness and honesty; respect for people and nature; as well as sustainability of its actions.[23]

In order to master the insecurity that comes with such change for the employees, this process was accompanied by intensive discussions with the employee representatives. As a consequence, plans could be turned into reality relatively smoothly, also because established processes could be maintained during integration of the new companies, establishing more security. At the same time and in order to permanently learn, it was thoroughly analyzed which processes could be taken over from the newly acquired companies. By doing so, during the time of transition, insecurity for the new employees could be replaced by the recognition of being part of a better solution.[24]

What is essential to fostering this change sustainably is proof that the new situation is beneficial for the company – or expressed as "The best driver is always success"[25] by Mr. Wenning with regards to how improved performance was a result of the changes. Indeed, for five and a half years Bayer was able to improve its quarter to quarter performance and employees were given the chance to participate in this by changing the remuneration system to become more performance-based (only altered recently by the world economic crisis).

Also, surveys are regularly performed to establish an overview regarding how employees perceive the situation and to get their feedback on topics such as "The values and leadership principles of the Bayer Group are quite clear to me/have my support/are very important for my daily work" or "The compensation system is transparent" and results of these surveys have been continuously improving.[26]

In the following paragraphs, the Schering integration will now be taken to illustrate Bayer's approach to change management in more detail. As already mentioned earlier, post-merger integration represents the ultimate change management challenge,[27] making it a very interesting scenario for illustrating the approach in more detail. The Schering acquisition is quoted as Bayer's largest acquisition in the company's history until today.

Like Bayer, Schering was also a traditional German company operating internationally. The merger of Bayer and Schering started in March 2006, when Bayer had the chance to act as a "white knight" for Schering, after Merck KGaA had launched a hostile takeover. Financial markets were surprised how fast Bayer proceeded not only with the takeover itself but also with the preparations for the Post Merger Integration (PMI), which started in parallel.[28]

---

[23] Bayer (2008b).
[24] Focus (2008).
[25] Focus (2008), p. 12.
[26] Focus (2008).
[27] Galpin and Robinson (1997).
[28] Courth et al. (2008).

To illustrate the complexity of the consolidation and optimization effort, here are a few numbers concerning this undertaking:[29]

- Approximately 450 associated companies in about 150 countries had to be analyzed;
- The supply chain had to be optimized considering approx. 250 warehouses and approx. 30 production plants worldwide;
- The harmonization of heterogeneous IT infrastructures had to be tackled up to a detailed level into the operative processes of both Bayer and Schering.

Less regarding the complexity, but more the sheer dimensions of this undertaking, here are a few key figures: The consolidated sales of the new Bayer Schering Pharma AG were approx. € 7.5 billion in 2006. The number of employees involved was the sum of about 16,900 employees from Bayer's pharmaceutical division and about 25,000 employees from Schering.[30] The phases of the Post Merger Integration project are shown below (see Fig. 5).

The project organization reflected a number of deliberate considerations for successful change management (see Fig. 6 for an overview of the project organization). For example, an Integration Champion was appointed.

The program management decided to treat "Day One" (the legal entity implementation in each country) as a key milestone, and in order to prepare this, weekly Global PMO (Project Management Office) lead calls with 60 participants were

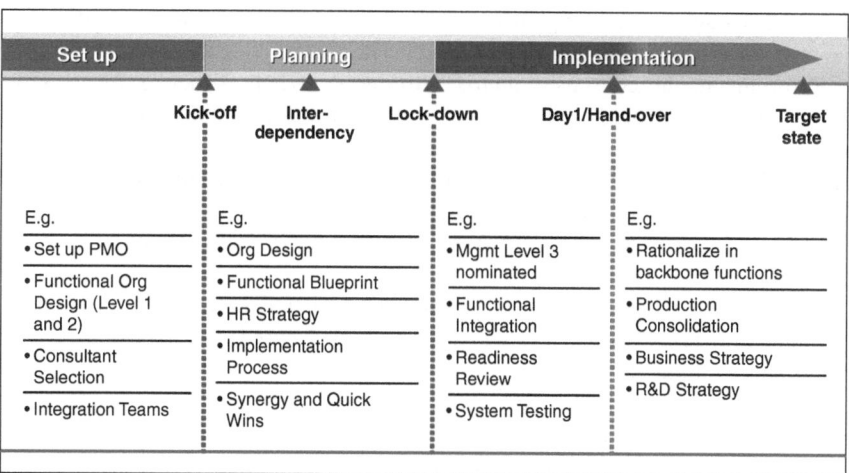

Fig. 5 Phases of the post merger integration. Source: Courth, Marschmann, Kämper, and Moscho (2008) p. 10

---

[29]Courth et al. (2008).
[30]Courth et al. (2008).

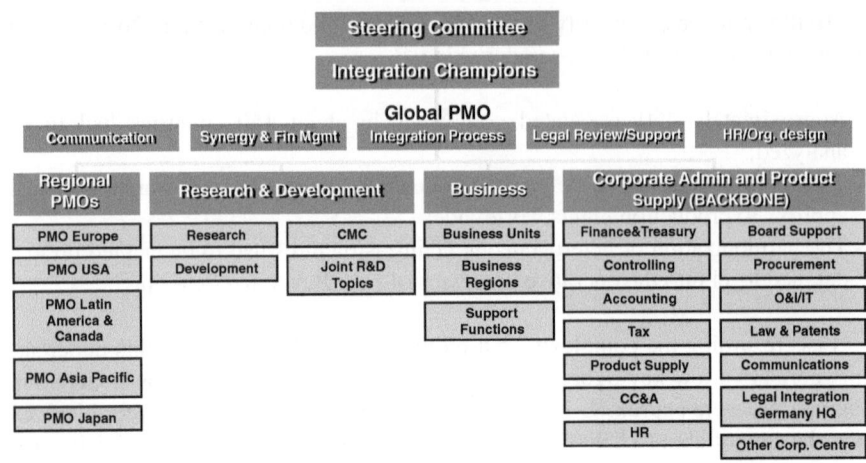

Fig. 6 PMI project organization at Bayer Schering Pharma. Source: Courth et al. (2008) p. 11

performed, followed by regional project management office (PMO) calls with 60 countries and by country calls with several hundred teams, as well as functional calls with 17 teams and 100 sub teams.[31] Also, all countries were given a toolbox, for which a download platform was established, e.g. featuring presentations, and Questions & Answers.[32] After "Day One" responsibility for continuous tracking of the integration process was with the line management in each country, for which a handover document, outlining customized milestones and specific action items for the respective synergy targets and the target state was provided to them.[33]

Returning to the consensus matrix introduced earlier, particular focus is put below on how a shared corporate culture was fostered within the post-merger integration project.

In order to foster a shared corporate culture of the newly emerging entity, the "STAR" (Success Together Achieves Results) project was conducted as part of the overall post-merger integration (PMI) project. This project was structured as shown in Fig. 7.

It was divided into two phases, the "Successful Integration Through Cultural Alignment" (SITCA) phase and the second phase focused on establishing a high performance culture "Creating the Driving Force".

During the analysis phase, the cultures of the two companies Bayer and Schering were analyzed. Then followed a definition phase, in which the future behaviors within the integrated new entity Bayer Schering Pharma were developed. The established Bayer principles and values as shown in Fig. 8 guided this phase.

---

[31] Tuschke, Müller, Kämper, Marschmann, and Moscho (2009).
[32] Tuschke et al. (2009).
[33] Tuschke et al. (2009).

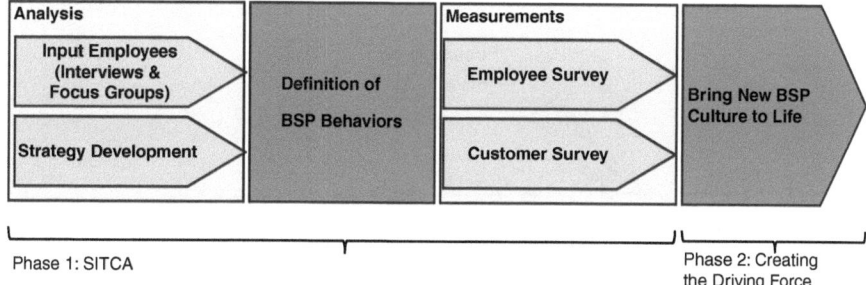

Fig. 7 Phases of the STAR project. Adapted from: Bayer AG (2009b)

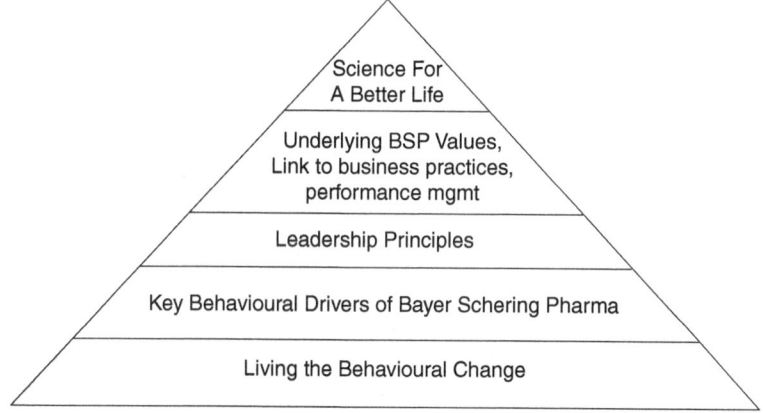

Fig. 8 Integrated approach to behavior development. Source: Bayer AG (2009b)

In general, Bayer fosters an environment favorable to prospector behavior also by promoting the Bayer Values and specific Leadership and Management Values. In this project, as well as in others, an important factor for the development of a shared corporate culture was that Bayer would consistently stick to and promote its leadership principles (these are shown in more detail below in Fig. 9). The leadership principles provide a framework for talent management and are the basis for performance management.

These leadership principles are translated into a holistic, stringent set of leadership instruments, such as performance assessment, 360° feedback, talent assessment, and leadership development. In this context, it should also be mentioned that Bayer managers are prepared for change by systematically getting opportunities to obtain international experience and to work in various functions.[34]

As part of the STAR project, the two companies' current cultures were analyzed according to the developed profiles. The following profiles emerged, which showed

---

[34] Focus (2008).

Focusing on
Customer Needs
Delivering Results
Overcoming
Complexity
Thinking
Strategically
Partnering
Leading People
Developing Yourself
& Others
Living our
Values

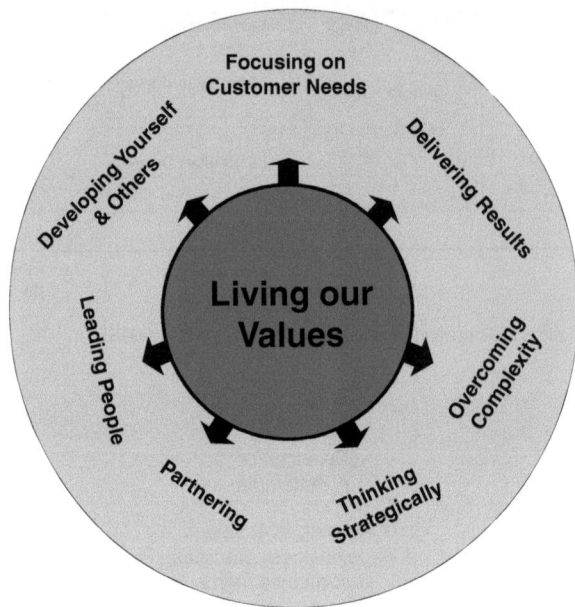

Fig. 9 Bayer leadership principles. Source: Bayer (2008b)

the cultural spread between the companies (Fig. 10). "B" stands for Bayer, whereas "S" shows the respective scoring of Schering.

As can be seen, there was a considerable need for changing to a single shared corporate culture. The one trait most consistent between them at that time was that both showed a similar customer focus.

Then, the implementation phase followed, which tackled turning the shared BSP culture into reality. The behaviors that were defined and guide BSP's behavior today

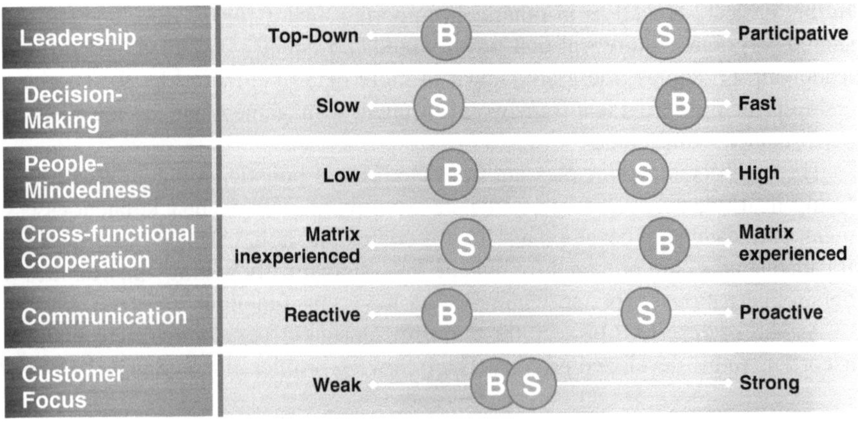

Leadership
Top-Down
B
S
Participative
Decision-
Making
Slow
S
B
Fast
People-
Mindedness
Low
B
S
High
Cross-functional
Cooperation
Matrix
inexperienced
S
B
Matrix
experienced
Communication
Reactive
B
S
Proactive
Customer
Focus
Weak
B S
Strong

Fig. 10 Results of measurements. Source: Bayer AG (2009b)

are: (1) Be Decisive, (2) Customers First, (3) Communicate with Integrity, (4) Live our Commitments, (5) Challenge the Status Quo, (6) Partner across Boundaries, (7) Strive for Success.[35]

Figure 11 summarizes the change process for the cultural transition from two distinct cultural profiles towards one shared BSP culture. This comprised the implementation of vision, values, principles and expected behaviors. On the one hand, processes, structures and the allocation of resources were aligned, and, on the other, cultural change was managed. The new community was built by enhancing competencies and adjusting behavioral patterns.[36]

Considering what was said earlier regarding the consensus matrix it has been crucial to establish a shared understanding regarding the "what to do" and "how to do" of the newly established entity Bayer Schering Pharma. The thorough approach chosen helped to prepare a holistic and sound basis for future management.

Concluding on the Bayer-Schering integration case, it is important to mention that not only Bayer has had the opportunity to experience and learn from several significant acquisitions and integrations during the past years. Moreover, it deliberately chose to retain this knowledge by promoting its conservation within specific organizational units, like the Business Unit Business Consulting.

In these terms Bayer's inhouse consultancy, Bayer Business Consulting, acts as an internal service provider and knowledge pool for capabilities crucial to the Group. One business unit of the service company Bayer Business Services GmbH (Fig. 12 provides an overview of its organization) Business Consulting has been growing significantly during the last 3 years and now comprises about 100 employees, in three offices – Leverkusen, Shanghai, Wayne (NJ) and Pittsburgh.

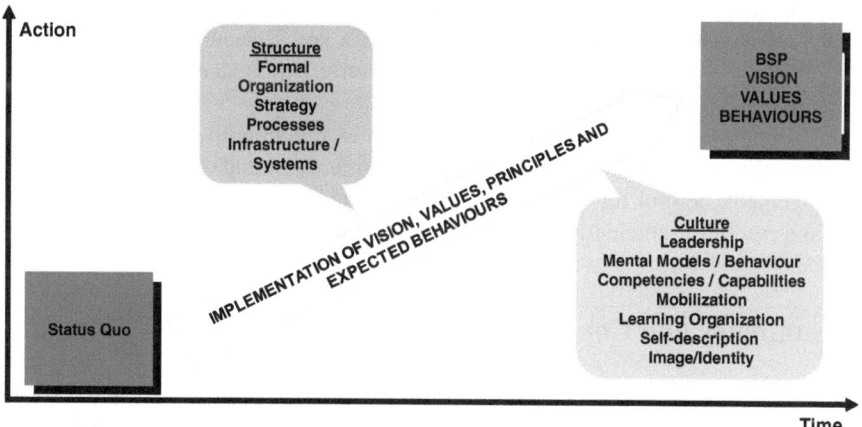

Fig. 11 The cultural change. Source: Bayer HealthCare (2007)

---

[35]Bayer (2009a).
[36]Bayer HealthCare (2007).

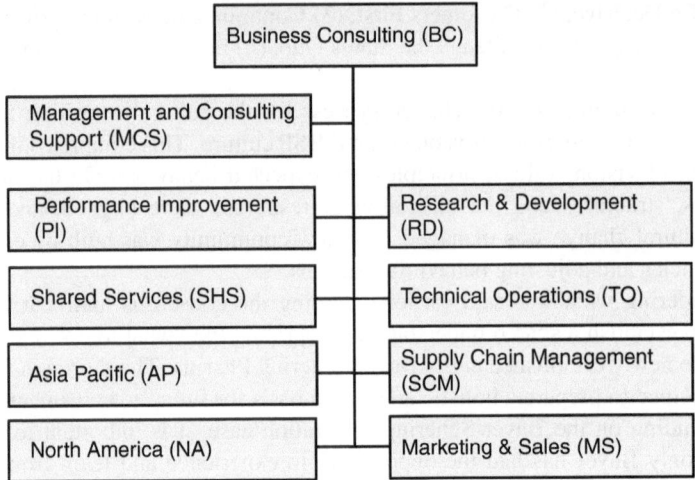

Fig. 12 Organization of business consulting. Source: Bayer AG

Within Business Consulting, different departments work to bundle knowledge relevant to Bayer along the value chain (e.g. R&D, Technical Operations, Marketing and Sales) and "Shared Services". In additional areas of special importance, practices are formed as in the case of "Post-merger Integration" (PMI). The latter one is completely dedicated to this type of radical change, ensuring that the experience and knowledge gathered during previous integrations is conserved and further developed for similar future undertakings. Then, Bayer can reach back to the services of its internal consultants on demand and, compared to external providers, with lower costs. As acquisitions have become a key element of Bayer's growth, Bayer Business Consulting today offers a collective experience in this area unique even in the external consulting industry.

All in all, the described frameworks as well as retaining the knowledge how to perform these tasks enable Bayer to maintain and further build capabilities in change management, helpful for maintaining the company in a condition, in which it can act and react fast, efficiently and effectively to market opportunities.

5 Summary and Conclusion

During the past years, Bayer has lived through a major transformation. With the consolidation of its business to three subgroups, the organization was subjected to a tremendous change structurally. At the same time, the employee base was greatly transformed, putting even more strain on the successful management of maintaining a shared corporate culture. Today, the company is successfully positioned to face the relevant challenges of the 21st century.

Considering how the experience and knowledge regarding successful change management is retained within the organization, it affords an interesting opportunity for further research to study this deliberate creation of a change management capability in more detail. In particular, an analysis showed how such internal change management capabilities compare with externally integrated change management capabilities performance-wise and, as could be reasonably assumed, if this is contingent upon the type of project (e.g. acquisition versus internal reorganization), as well as industry and the size of operations.

Concluding, Bayer strives to be a dynamic, learning organization and has created certain "repositories" for knowledge and abilities to react fast in handling challenges as efficiently and effectively as possible. In creating this collective knowledge and developing it further, Bayer is able to strengthen its ability to behave strategically as a prospector.

As a company committed to value creation and rigorous value management, portfolio management will continue to play a crucial role within the Bayer Group, aiming to further strengthen its market positions through strategic acquisitions and alliances.[37] With upcoming challenges such as the growing world population and climate change, decreasing arable area per person worldwide,[38] and demographic change, change management capabilities broadened over the past years have much potential to be put to good use.

References

Al-Ani, A., & Gattermeyer, W. (2001). Entwicklung und Umsetzung von Change Management-Programmen. In: Gattermeyer, W. & Al-Ani, A. (Eds.), *Change Management und Unternehmenserfolg – Grundlagen, Methoden, Praxisbeispiele*. Gabler Verlag, pp. 13–40.

Bayer. (2008a). *Policy: Future, Goals, Strategy, Values – The Mission Statement of the Bayer Group*.

Bayer. (2008b). *Policy: Values and Leadership Principles – Living our Values*.

Bayer. (2009a). *BSP Today: Our Behaviors*. Bayer Intranet.

Bayer. (2009b). Making Acquisitions Work: The Bayer/Schering Case, Presentation at University of Wuppertal, January 2009.

Bayer HealthCare. (2007). Towards One BSP Corporate Culture: Creating the Driving Force, Internal Presentation.

Christensen, C., Stevensen, H., & Marx, M. (2006). Die richtigen Instrumente für den Wandel. In: *Harvard Business Manager, 12*, 26–38.

Courth, L., Marschmann, B., Kämper, M., & Moscho, A. (2008). Spannungsfeld zwischen Geschwindigkeit und Best-in-Class-Ansätzen – PMI am Beispiel der Bayer-Schering-Übernahme. *M&A Mergers and Aquisitions Review, 1*, 8–14.

Czinkota, M. R., & Ronkainen, I. A. (2005). A forecast of globalization. International business and trade: Report from a Delphi study. In: *Journal of World Business, 40*(2), 111–123.

DiMasi, J. A., Hansen, R. W., & Grabowski, H. G. (2003). The price of innovation: New estimates of drug development costs. In: *Journal of Health Economics, 22*(2), 151–185.

Galpin, R. J., & Robinson, D. E. (1997), Merger integration: The ultimate change management challenge. In: *Mergers & Acquisitions, 32*(1/2), 24–28.

Fenwick, S. (1999). Will the Mirror Crack? In: *Drug Discovery Today, 4*(1), 3.

---

[37] Bayer (2008a).
[38] Focus (2008).

Focus. (2008). Interview with Werner Wenning, Focus, 02/2008, pp. 5-15.
Grudowski, S. (1998). *Informationsmanagement und Unternehmenskultur – Untersuchung der wechselseitigen Beziehung des betrieblichen Informationsmanagements und der Unternehmenskultur.* Stuttgart.
Hofstede, G. (1981). Culture and organizations. In: *International Studies of Management and Organization,* X(4), 15–41.
Koch, C. (2008). *Die Kunst des Erfolgs: Wie Market-Based Management das weltweit größte Familienunternehmen aufgebaut hat.* Wiley-VCH.
Metelmann, K., & Neuwirth, S. (2002). Wachstum und Organisation im Bayer-Konzern. In: Glaum, M., Hommel, U., & Thomaschewski, D. (Eds.), *Wachstumsstrategien internationaler Unternehmungen. Internes vs. externes Wachstum* (pp. 123–157). Stuttgart.
Miles, R. E., & Snow, C. C. (1978). *Organizational strategy, structure, and process.* New York: McGraw-Hill.
Miller, V. D., Johnson, J. R., & Grau, J. (1994), Antecedents to willingness to participate in a planned organizational change. *Journal of Applied Communication Research, 22,* 59–80.
Moscho, A., Hodits, R., Friedemann, J., & Leiter J. (2000). Deals that make sense. *Nature Biotechnology, 18,* 719–722.
Pardo del Val, M. P., & Fuentes, C. M. (2003). Resistance to change: A literature review and empirical study. *Management Decision, 41*(2), 148–155.
Sackmann, S. (1999). Cultural Change – eigentlich wäre es ja ganz einfach...wenn da nicht die Menschen wären. In: Götz, K. (Ed.), *Cultural change* (vol. 4, pp. 15–37). München/Mering: Managementkonzepte.
Schein, E. H. (1995). *Unternehmenskultur: Ein Handbuch für Führungskräfte.* New York: Frankfurt am Main.
Schreyögg, G. (2000), Neuere Entwicklungen im Bereich des Organisatorischen Wandels. In: Busch, R. (Ed.), *Change Management und Unternehmenskultur: Konzepte in der Praxis* (vol. 20, pp. 26–44). Forschung und Weiterbildung für die betriebliche Praxis, München, Mering.
Tuschke, A., Müller, S., Kämper, M., Marschmann, B., & Moscho, A. (2009). Bayer and Schering – The Integration of Two Global Players in the Pharmaceutical Market, Case Study.

Ambidextrous Leadership in the Innovation Process

Kathrin Rosing, Nina Rosenbusch, and Michael Frese

Contents

1 Introduction 191
2 Ambidexterity 192
3 An Example of an Ambidextrous Organization 194
4 The Innovation Process 195
5 Leadership in the Innovation Process 196
6 Ambidextrous Leadership: The Behavioral Flexibility of Leaders 198
7 How to Become an Ambidextrous Leader? 200
8 Conclusion 202
References 202

1 Introduction

Innovation research is full of paradoxes. Bledow, Frese, Anderson, Erez, and Farr (2009) summarize several kinds of conflicting demands inherent to the innovation process and demonstrate the commonness of tensions within this process. The main paradoxes of innovation are probably achieving a balance of new and old activities, of structured and chaotic activities, and of uncertain and reliable activities. All these activities map onto ambidexterity – the ability to achieve a balance of exploration and exploitation. In this chapter, we will argue that ambidexterity is required within the innovation process, not only on the organizational level but also for each individual person involved in an innovation process. Leaders in the context of innovation need to be able to support subordinates in their attempts to act ambidextrously – by ambidextrous leadership.

We will begin this chapter by explaining the concept ambidexterity and illustrating its application in one highly innovative company. Afterwards we will demonstrate its importance within the innovation process. Subsequently we will

---

K. Rosing (✉)
University of Giessen, Giessen, Germany
e-mail: kathrin.rosing@psychol.uni-giessen.de

give a short overview of the existing literature on leadership and innovation. We will conclude by introducing the term ambidextrous leadership and explaining the importance of this concept for the innovation literature and practice.

2 Ambidexterity

Ambidexterity literally means the ability to use both hands equally well. In management science the label ambidexterity has been linked to the balance of explorative and exploitative organizational strategies, i.e. the ability to engage in exploration and exploitation equally well.[1] Exploration and exploitation were originally defined by March (1991) as two different forms of organizational learning. In this respect, exploration is connected to increasing variance, experimentation, search for alternatives and risk taking. In terms of innovation, exploration is linked to radical innovation, entering new product markets and new technology. On the other hand, exploitation is linked to reducing variance, adherence to rules, alignment and risk avoidance. In the innovation context, exploitation means rather implementation, incremental innovation and refinement of existing products. Both exploration and exploitation have their benefits and their costs. For example, exploration may lead to radically new products, but the success of these products may be very uncertain. The outcome of exploitation in turn is rather predictable, but will be unlikely to lead to competitive advantage in the long run. Thus, for firms to be successful in the short and also the long run it is necessary to be both explorative and exploitative – i.e. to be ambidextrous. Several studies already have shown that organizations that are able to achieve a balance of exploration and exploitation are more successful than organizations that do not achieve such a balance.[2]

Raisch and Birkinshaw (2008) reviewed the literature on ambidexterity and its different connotations. They conclude that ambidexterity is not limited to the balance of exploration and exploitation, but also refers to several other pairs of contradictory concepts, e.g. incremental vs. radical innovation, continuity vs. change, induced vs. autonomous organizational strategies, and organic vs. mechanical organizational structures. Thus, the central feature of organizational ambidexterity is the integration of conflicting or even contradictory activities, strategies or features.

Gupta, Smith, and Shalley (2006) discuss several possibilities of handling the contradictory explorative and exploitative organizational strategies. They argue that it is important to be clear about the question whether exploration and exploitation are actually mutually exclusive (i.e., two ends of a continuum), or whether exploration and exploitation are two orthogonal dimensions (i.e., theoretically independent). In the first case, an organization may be more explorative *or* more exploitative and a balance of exploration and exploitation would be the middle of the continuum. In the latter case, an organization may be more or less explorative *and* more or

---

[1] Bledow et al. (2009).
[2] For example He and Wong (2004).

less exploitative and, for greatest success, organizations should be high on both exploration and exploitation. Gupta et al. further derive two different mechanisms to achieve an organizational level balance of exploration and exploitation from the conceptualization of exploration and exploitation as either a continuum or as orthogonal dimensions. They argue that a temporal separation of exploration and exploitation makes sense in the case of a continuum, i.e., a temporal shift alongside the dimension from exploration to exploitation and vice versa. In contrast, they suggest that a structural separation of exploration and exploitation is superior in the case of orthogonal dimensions. Structural separation equates the concept of structural ambidexterity as defined by Benner and Tushman (2003). Structural ambidexterity is achieved by dual structures, i.e., separate units that are specialized in either exploitation or exploration.

The two mechanisms proposed by Gupta et al. (2006) have in common the separation of exploration and exploitation on an organizational level (either temporal or sturctural). However, we suggest that exploration and exploitation can be integrated into a complex strategy, following the reasoning of Bledow et al. (2009) that exploration and exploitation might be perceived as thesis and antithesis that can be integrated into a higher order synthesis that combines aspects of both.

To understand the interplay of exploration and exploitation, it is important to have a deeper look at the organization. The concept of ambidexterity has been developed for the organizational level. But what ambidexterity actually means on lower levels of an organization – especially in behavioral terms for teams and individuals – has not yet been elaborated in the existing literature. We argue that exploration and exploitation are interwoven activities that cannot be separated. For example, in highly innovative R&D teams, explorative activities are doubtlessly very important. But even those R&D teams would not be able to produce any results without exploiting their existing competencies and relying on well-learned routines. Therefore, individuals and teams need to able to both explore and exploit and additionally integrate both activities for meaningful behavior to take place. That is, an individual switching from exploration to exploitation and vice versa needs to integrate both activities within an overall plan to be able to reach his or her goals. We argue that the either-or-logic of separating exploration and exploitation (whether leading to temporal or structural separation) may be a reasonable strategy on the organizational level. But on lower levels we do not think that exploration and exploitation can be completely separated. Working in a unit that is specialized in exploitation does not mean that absolutely no exploration activities are needed. In turn, working on mainly exploratory tasks does not imply that no routine or exploitative activities are appropriate at all. Therefore, we propose that a both-and-logic,[3] i.e., an integration of exploration and exploitation, is a more reasonable strategy on lower levels of the organization.

Similar to this idea, Gibson and Birkinshaw (2004) introduced the concept of contextual ambidexterity. Contextual ambidexterity implies that the individuals of

---

[3] Lewis (2000).

an organization have to engage in both exploration and exploitation and management systems are designed in such a way that individuals are able to decide by themselves when to use which kind of activity. Ambidexterity needs to be already incorporated in the organizational context in such a way that it is conducive to both exploration and exploitation.

3 An Example of an Ambidextrous Organization

At this point, we would like to illustrate our arguments by analyzing one of the most successful ambidextrous organizations: the Toyota Motor Corporation. In *Business Week's* 2009 ranking of the 25 top innovative companies, Toyota ranks third after Apple and Google and thus appears to be the most innovative non-IT company. Over the past few decades, Toyota has managed to become the leading automotive company in the world using a variety of different measures. Until the recent global financial crisis hit companies worldwide, Toyota had never reported an operating loss in its history. The company's products were often rated as the best with respect to quality, and Toyota's market share has risen dramatically. Researchers and practitioners have frequently asked what it is that makes Toyota so successful. Whereas the Toyota Production System has often been viewed as the reason for success, Takeuchi, Osono and Shimizu (2008) argue that Toyota's unique strength is the ability to deal with contradictions embedded in a corporate culture that facilitates continuous improvement (*kaizen*) and radical change (*kaikushin*) at the same time.

In an interview with Thomas Stewart and Anand Raman, Katsuaki Watanabe, the president of Toyota stated: "Fifteen years ago I would have said that as long as we had enough people, Toyota could achieve its goals through kaizen. In today's world, however, change can be produced by *kaizen*, but it may also need to be brought about by *kaikushin*. When the rate of change is too slow, we have no choice but to resort to drastic changes or reform: *kaikaku*".[4] Dealing with the contradictory tasks of exploitation and exploration requires a unique organizational setting. Thus, in order to learn from Toyota, one needs to identify factors that enable the company to be ambidextrous.

Toyota's corporate culture is characterized by a high tolerance for failure that fosters radical change. The company's leaders set impossible goals to create breakthrough innovations.[5] At the same time, the deeply embedded respect for people, one of the pillars of the Toyota Way (in particular the "Customer First" principle), encourages continuous improvements. The regional strategy of Toyota also aims to serve the contradictory demands of radical and incremental change. Whereas Toyota utilizes the home market for experimentation, its local customization strategy for

---

[4] Katsuki Watanabe in Stewart and Raman (2007, p. 81).
[5] Takeuchi, Osono, and Shimizu (2008).

international operations facilitates incremental improvements of products and processes and especially aims at increasing efficiency.[6] A third factor that enables Toyota to manage the contradictory tasks of exploration and exploitation relates to human resources. The unique combination of a strict hierarchical structure with freedom in employee's decision-making facilitates both exploration and exploitation. Information flows freely across hierarchical levels and functions leading to a diffusion of knowledge in all directions[5]. Employees are encouraged to participate in networks to generate even more knowledge. Toyota also encourages employees to challenge current state of the art and find solutions for problems[5]. Even first-line employees are part of the innovation process.[7] Toyota heavily invests in improving the capabilities of their employees at all levels and functions. In addition, the company trains managers to become T-type people. T-type managers intensify their knowledge in their own area while at the same time learning other jobs. It takes Toyota up to 20 years to train T-type managers.[8] Perhaps most importantly, Toyota sticks to its long-term focus. There are no deadlines for innovation projects. The idea of a never-give-up culture leads to continuous improvements and revolutionary change at the same time.

After introducing the general concept of ambidexterity and explaining its application in a highly successful organization, in the next section we will now describe the meaning of ambidexterity within the innovation process. We will argue that for being innovative, it is important to be ambidextrous and act ambidextrously.

4 The Innovation Process

West and Farr (1990) define innovation as

> the intentional introduction and application within a role, group or organization of ideas, processes, products or procedures, new to the relevant unit of adoption, designed to significantly benefit the individual, the group, organization or wider society (p. 9).

Thus, to be innovative it is not enough to be just creative. Creative ideas need to be implemented to create value for the organization. Following this idea, the innovation process can be split up in two rough stages: the creativity stage and the implementation stage.[9] The creativity stage includes the identification of a problem or an opportunity and the generation of ideas to solve the problem or use the opportunity. To be creative, explorative behavior as defined by March (1991) is necessary. The implementation stage compromises the evaluation of the generated ideas, the selection of one or more ideas, and finally the actual implementation of the idea/s.

---

[6] Stewart and Raman (2007).
[7] Hamel (2006).
[8] Watanabe in Stewart and Raman (2007).
[9] Farr, Sin, and Tesluk (2003).

To implement ideas, it is necessary to exploit. The distinction between the creativity and implementation stage of innovation is widely accepted in the literature.

High performance in the creativity and implementation stages of innovation should be fostered by very different conditions, i.e., conditions that stimulate explorative and exploitative activities respectively. Research has shown that for creativity, intrinsic motivation, a divergent thinking style and autonomy are important antecedents. In contrast, implementation is predicted not as much by individual variables, but more by variables that reside on higher organizational levels, such as management support and organizational support for innovation.

What makes innovation a really challenging endeavor is that the creativity stage and the implementation stage cannot be clearly separated. Rather, the innovation process is chaotic and nonlinear.[10] That means creativity and implementation are not bound to specific phases but are needed throughout the whole innovation process. Therefore, exploration and exploitation are needed not only in specific phases but also within the whole process. For this reason, individuals and teams need to be able to switch between these two kinds of behaviors. Thus, as we argued before, ambidexterity is necessary in the innovation process. We suggest that it is an important question to ask how this balance of exploring and exploiting may be best achieved within the innovation process since, for the individual employee or for the individual work group, ambidexterity means balancing behaviors that are quite different. Below, we will discuss possibilities of how leaders can support their subordinates in their attempts to achieve ambidexterity in the innovation process. To this end we will now give a short overview on the existing literature on leadership and innovation to see what we already know about the influence of leaders within the innovation process.

5 Leadership in the Innovation Process

The link between leadership and innovation has attracted increasing attention in the literature. In fact, some researches argued that leadership is one of the most important predictors of innovation.[11]

Transformational leadership. Transformational leadership has been defined as "moving the follower beyond immediate self-interests through idealized influence (charisma), inspiration, intellectual stimulation, or individualized consideration". [12] Thus, transformational leadership motivates people to reach higher goals beyond the primary work task. Transformational leadership is by far the most frequently studied leadership behavior in the innovation context. However, the literature on the relationship of transformational leadership with innovation and creativity does not yield a consistent picture. Some studies did find a positive relationship, but others

---

[10] For example Anderson, De Dreu, and Nijstad (2004).

[11] Mumford, Scott, Gaddis, and Strange (2002).

[12] *Bass* (1999, *p.11*).

did not find such a relationship. A few studies even reported a negative relationship. More likely than not, the relationships between transformational leadership and creativity/innovation are contingent on other variables, such as the type of dependent variable (e.g., creativity vs. innovativeness), the level of analysis, the work tasks (e.g., research vs. development projects) and the several features of the individuals, teams and organizations studied (e.g., climate for excellence, centralization, etc.). Nevertheless, it appears to be difficult to draw clear conclusions from the existing results on the transformational leadership – creativity/innovation link. We suggest that the particular nature of innovation processes calls for more situational leadership behaviors rather than for the broad cluster of stable leadership behaviors such as transformational leadership. We will outline this approach in more detail later in this chapter.

Transactional leadership. In contrast to transformational leadership, transactional leadership establishes an exchange-based relationship by clarifying goals and rewarding goal achievement and by intervening only when necessary.[13] Very few studies have looked at the relationship between transactional leadership and creativity/innovation. Experimental studies suggest a positive relationship. However, a field study by Jansen and colleagues[14] showed a positive relationship of transactional leadership with exploitative innovation, but a negative relationship with explorative innovation. Thus, no simple conclusion may be drawn from the existing results concerning the transactional leadership – innovation/creativity link.

Initiating Structure. Initiating structure is defined as leader behavior that structures tasks, defines goals and controls goal attainment.[15] Unfortunately, only a few studies have been done on the relationship between initiating structure and innovation. The scarce evidence suggests that initiating structure has a positive relationship with performance in R&D teams, i.e., innovation, especially when the innovation is incremental rather than radical.[16]

Consideration. Consideration[15] is the concern and respect of the leader for subordinates' feelings and the leader's appreciation and support of subordinates. To our knowledge, only one study has analyzed the relationship between consideration and innovation. Keller (1992) found a positive relationship between consideration and performance in R&D teams.

Leader-Member Exchange. In contrast to the above mentioned leadership styles that describe overall leader behaviors, leader-member exchange (LMX) focuses on the leader-follower dyad and the quality of this relationship.[17] High-quality exchange relationships are characterized by mutual trust and respect. Leader-member exchange has been studied several times on the individual level. Apart from a few exceptions, the literature on LMX and innovation is quite consistent

---

[13] *Bass* (1999).
[14] *Jansen, Vera, and Crossan* (2009).
[15] *Fleishman* (1953).
[16] *Keller* (2006).
[17] *Graen and Uhl-Bien* (1995).

in the result that a high quality leader-member exchange has a positive relationship with innovation.

Supervisor support. Supervisor support is not a clearly defined leadership style, but rather a cluster of leader behaviors that are supportive of subordinates' innovative behaviors. Supervisor support has been studied quite frequently in the innovation context. About half of these studies show a small, but significant positive relationship of supervisor support and creativity/innovation. However, nearly as many studies did not find such a relationship. The results suggest that supervisor support is more important for actually acting on ideas (i.e., implementation) rather than just having ideas (i.e., idea generation). This may be due to the fact that the implementation of ideas is more of a social process that involves other persons than idea generation that can be more easily kept to the individual. Therefore supervisor support may be more helpful when it comes to selling and actually acting on ideas.

Summary and conclusions. It seems to be difficult to draw clear conclusions from the existing literature on the leadership – creativity/innovation link. Very different and even opposing leadership behaviors are important for the innovation process. At the same time, the same leadership behaviors are in some studies related to innovation, but in others they are not. We suggest that the traditionally studied leadership behaviors such as transformational leadership and initiating structure may be too general to accurately predict innovativeness. In addition, the specific conditions under which leadership behaviors are successful have to be outlined, i.e., research has to look for moderating conditions. Therefore, a contingency theory that is able to explain what specific leadership behavior is effective in which situation and that additionally considers the particularities of the innovation process is called for.

6 Ambidextrous Leadership: The Behavioral Flexibility of Leaders

It seems to be very unlikely that relatively stable and broad clusters of leadership behaviors such as transformational leadership or initiating structure are suitable for predicting very variable and specific behaviors such as exploration and exploitation. We argue that for a reliable prediction we need equally variable and specific behaviors on the leadership side. That is, the best way to predict specific follower behavior is to predict it by specific leader behavior. Therefore, we will introduce two sets of leader behaviors that we suggest for predicting exploration and exploitation respectively.

A good way to characterize the leadership behaviors needed for exploration might be the term "opening". To foster idea generation and exploration, the leader needs to create an open atmosphere. One prerequisite of being creative is the increase of variance. Diversity and different approaches lead to creativity. For being explorative, employees need to know that they are not only allowed but are actually required to experiment and play with ideas. Opening leader behaviors encourage employees to break up rules and search for solutions outside the safe ground. They also allow for different approaches to work. And opening leader behaviors mean being critical of the ways things have been done in the past. Thus, we define **opening**[18] as *a set of leader behaviors that includes encouraging doing things differently and experiment, giving room for independent thinking and acting, and supporting attempts to challenge established approaches.*

In contrast, when it comes to leadership and exploitation, "closing" might describe the necessary leader behaviors quite well. When implementing ideas, the reduction of variance is crucial. Employees need to be in line and adhere to rules as errors and failures must not be risked in this stage of the process. Exploiting means reliance on well-trained competencies and being efficient in acting. Closing leader behaviors mean establishing routines that subordinates have to follow and giving exact instructions how to carry out tasks. In addition, they mean defining specific work goals and pre-structuring tasks. Thus, we define **closing**[19] as *a set of leader behaviors that includes taking corrective action, setting specific guidelines and monitoring goal achievement.*

We argued earlier that to be innovative, it is not enough to be able to explore and exploit, but it is also necessary to switch flexibly between the two kinds of behaviors. For leaders, this means that they need to be not only able to show opening and closing leader behaviors, but they need to be able to flexibly switch as well. In conclusion, what we describe as an **ambidextrous leader**[19] is *a leader that is able to foster exploration by opening behaviors and exploitation by closing behaviors and flexibly switch between these behaviors according to situational and task demands.* Figure 1 summarizes our theoretical model of ambidextrous leadership.

In the example of Toyota earlier in this chapter we already learned about ambidextrous leaders. Toyota's leaders are able to foster both radical change and incremental improvements, thus, exploration and exploitation. The description of Toyota's characteristics includes hints for opening and closing leader behaviors. The high tolerance for failure that is embedded in the company's culture encourages leaders to show opening leader behaviors. Strict hierarchies lead to closing leader

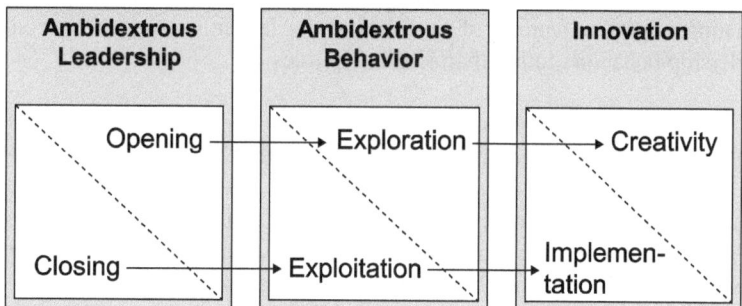

Fig. 1 Theoretical model of ambidextrous leadership

---

[18]Rosing and Frese (2009).
[19]Ibid.

behaviors. The T-type managers are another example for ambidextrous leaders as they are trained to both exploit their existing knowledge within their own job and explore new knowledge in different areas. A more detailed analysis of the organization would be necessary to make specific leader behaviors visible. Nevertheless, the description of Toyota gives an indication of organizational level characteristics that are supportive of ambidextrous leadership.

An approach to leadership in the innovation process similar to ambidextrous leadership has also been taken by Van de Ven, Polley, Garud, and Venkataraman (1999). On the organizational level, they distinguish four top management leadership roles that they find are important in the "innovation journey": the sponsors who champion an innovation, the mentors who coach and supervise the innovation team leader, the critics who test the innovation against hard business criteria, and the institutional leaders who balance the power of the other three leadership roles. In the innovation projects Van de Ven et al. studied, these roles were – contrary to assumptions – not accomplished by different persons, but individual members of the top management carried out several of these roles. Thus, managers need to be able to carry out more than one leadership role and must be able to switch between them. For example, a manager supports an innovation project by gathering resources and at the same time takes on the role of a critic questioning the progress and the success of the project. Paradoxical use of different leadership styles or roles (e.g., as sponsor and critic) therefore seems to be necessary for innovation success. Van de Ven et al. claim that "the odds of organizational learning and adaptability increase when a balance is maintained among dialectical leadership roles throughout the innovation development".

In contrast to Van de Ven et al. (1999) we do not propose distinct leadership roles that need to be balanced within the innovation process, but rather specific leadership behaviors that foster exploration and exploitation – namely opening and closing leader behaviors as defined above. The advantage of our approach is that these leader behaviors are more specific on the one hand and more flexible to be used on the other hand. Roles generally include a broader cluster of behaviors and are rather stable and inflexible. As we argue that flexibility and situational adaptability are the most important features of an ambidextrous leader, we suggest concentrating on leadership behaviors rather than leadership roles.

7 How to Become an Ambidextrous Leader?

The next important question is: How might a leader achieve being an ambidextrous leader? We argue that there are two main routes to ambidextrous leadership. First, leaders may directly influence their subordinates' behavior by opening and closing leader behaviors. For this strategy, we need to know what the prerequisites are to do so. Second, leaders may instead indirectly influence their subordinates in establishing a culture or climate that is beneficial for both exploration and exploitation. Thus, we need to know what kind of culture or climate might be suitable. In the following paragraphs we will elaborate both routes to ambidextrous leadership in more detail.

Zhou and George (2003) argue that emotional intelligence enables leaders to understand and channel the emotions of subordinates connected to the innovation process. Emotional intelligence therefore might be helpful to be sensitive to what kind of leader behaviors are called for in a given situation. A leader with high emotional intelligence might be able to have an intuition for what kind of behavior his or her subordinates need to show and be able to adapt his or her leader behavior appropriately. For example, when an individual generates a lot of ideas of how to solve a problem it might be necessary for a leader to find the right point in time to intervene before this individual gets overwhelmed by too many ideas, but not before enough useful ideas have been generated. Emotional intelligence should enable a leader to find this "right point in time".

Another more cognitive prerequisite for being an ambidextrous leader might be the ability of integrative thinking. In his book "The Opposable Mind", Martin (2007) gives several examples of leaders capable of integrative thinking. These leaders have in common that they are able to simultaneously hold in mind two contradictory ideas. Leaders who are convinced that exploration and exploitation or opening and closing leader behaviors are not mutually exclusive, but are able to combine and integrate both into a higher order plan should be able to foster both exploration and exploitation and to frequently switch between opening and closing leader behaviors.

The second route to ambidextrous leadership is a more indirect way of fostering both exploration and exploitation in subordinates by influencing the organizational and team culture. That means, a leader may establish an organizational or team culture that is beneficial for both explorative and exploitative activities. We propose a climate for initiative[20] to be helpful in advancing ambidexterity. Personal initiative is a self-starting, proactive, and persistent approach to work. Climate for initiative fosters such an approach to work. Personal initiative is positively linked to exploration[21] as the self-starting and proactive approach motivates individuals to search for new possibilities of accomplishing tasks. Being persistent helps to overcome barriers that frequently arise within the innovation process. On the other hand, personal initiative also fosters exploitation, as it encourages individuals to use their existing knowledge to continuously improve their work on a self-starting, proactive and persistent base. Thus, establishing a climate for initiative is supportive of innovation in fostering both exploration and exploitation, i.e., ambidexterity.

In addition, an error management culture[22] should be favorable for both exploration and exploitation. An error management culture combines the acknowledgement of the positive consequences of errors and the prevention of the negative consequences of errors. An organization or team with a high error management culture is able to utilize the diversity errors can create for exploration. It is also able to avoid the negative consequences of errors in quickly responding to errors in exploiting existing knowledge about handling errors.

---

[20] Baer and Frese (2003).
[21] Frese, Teng, and Wijnen (1999).
[22] Van Dyck, Frese, Baer, and Sonnentag (2005).

Thus, to conclude, we suggest there are two main routes that ambidextrous leaders may use to positively influence the innovation process. The first route is to directly influence the exploration and exploitation activities of subordinates by combining opening and closing leadership behaviors. For being able to do so emotional intelligence and integrative thinking may be conducive. The second route is to indirectly influence subordinates by affecting the organizational or team climate by establishing an error management culture and a climate for initiative. The two routes complement each other in their influence on ambidexterity.

8 Conclusion

Leadership in the innovation context is complex. This is due to the fact that innovation itself is complex as it demands very different activities, i.e., exploration and exploitation. Unfortunately, the innovation process cannot be easily split into distinct stages, but is rather nonlinear and difficult to predict. Leaders of an innovative workforce need to be able to deal with this complexity by ambidextrous leadership. We argued that established concepts of leadership cannot explain subordinates' performance in the innovation process, as these concepts are too broad to predict specific explorative and exploitative behaviors. Therefore, we introduced two sets of leader behaviors that directly foster exploration and exploitation: opening and closing leader behaviors respectively. We called the flexible application of these leader behaviors according to situational and task demands ambidextrous leadership. Leaders willing to achieve ambidextrous leadership may use two different routes: first, by directly influencing their subordinates in interaction and, second, by indirectly influencing the organizational culture and climate to support ambidexterity. Doubtlessly, ambidextrous leadership is not easy to achieve. But given the importance of innovation for organizational success and viability, it is an objective worth pursuing.

Acknowledgments This research was supported by a research grant by the Volkswagen Foundation (II/82 408). We would like to thank our colleagues Andreas Bausch, Nataliya Baytalskaya, Ronald Bledow, James Farr, Verena Mueller, Alexander Schwall, and Shaker Zahra for discussions on initial ideas from which this chapter emerged.

References

Anderson, N. R., De Dreu, C. K. W., & Nijstad, B. A. (2004). The routinization of innovation research: A constructively critical review of the state-of-the-science. *Journal of Organizational Behavior, 25*(2), 147–173.

Baer, M., & Frese, M. (2003). Innovation is not enough: Climates for initiative and psychological safety, process innovations and firm performance. *Journal of Organizational Behavior, 24*(1), 45–68.

Bass, B. M. (1999). Two decades of research, and development in transformational leadership. *European Journal of Work and Organizational Psychology, 8*(1), 9–32.

Benner, M. J., & Tushman, M. L. (2003). Exploitation, exploration, and process management: The productivity dilemma revisited. *Academy of Management Review, 28*(2), 238–256.

Bledow, R., Frese, M., Anderson, N. R., Erez, M., & Farr, J. L. (2009). A dialectic perspective on innovation: Conflicting demands, multiple pathways, and ambidexterity. *Industrial and Organizational Psychology: Perspectives on Science and Practice, 2*(3), 305–337.

Farr, J. L., Sin, H.-P., & Tesluk, P. E. (2003). Knowledge management processes and work group innovation. In L. V. Shavinina (Ed.), *The international handbook on innovation* (pp. 574–586). NY, USA: Elsevier Science.

Fleishman, E. A. (1953). The description of supervisory behavior. *Journal of Applied Psychology, 37*(1), 1–6.

Frese, M., Teng, E., & Wijnen, C. J. (1999). Helping to improve suggestion systems: Predictors of making suggestions in companies. *Journal of Organizational Behavior, 20*(7), 1139–1155.

Gibson, C. B., & Birkinshaw, J. (2004). The antecedents, consequences, and mediating role of organizational ambidexterity. *Academy of Management Journal, 47*(2), pp. 209–226.

Graen, G. B., & Uhl-Bien, M. (1995). Relationship-based approach to leadership: development of Leader-member Exchange (LMX) theory of leadership over 25 years: Applying a multi-level multi-domain perspective. *The Leadership Quarterly, 6*(2), 219–247.

Gupta, A. K., Smith, K. G., & Shalley, C. E. (2006). The interplay between exploration and exploitation. *Academy of Management Journal, 49*(4), 693–706.

Hamel, G. (2006). The why, what, and how of management innovation. *Harvard Business Review, 84*(6), 140–140.

He, Z.-L., & Wong, P.-K. (2004). Exploration vs. exploitation: An empirical test of the ambidexterity hypothesis. *Organization Science, 15*(4), 481–494.

Jansen, J. J. P., Vera, D., & Crossan, M. (2009). Strategic leadership for exploration and exploitation: The moderating role of environmental dynamism. *Leadership Quarterly, 20*(1), 5–18.

Keller, R.T. (1992). Transformational leadership and the performance of research and development project groups. *Journal of Management, 18*(3), 489–501.

Keller, R. T. (2006). Transformational leadership, initiating structure, and substitutes for leadership: A longitudinal study of research and development project team performance. *Journal of Applied Psychology, 91*(1), 202–210.

Lewis, M. W. (2000). Exploring paradox: Toward a more comprehensive guide. *Academy of Management Review, 25*(4), 760–776.

March, J. G. (1991). Exploration and exploitation in organizational learning. *Organization Science, 2*(1), 71–87.

Martin, R. L. (2007). *The opposable mind: How successful leaders win through integrative thinking*. Boston, MA: Harvard Business School Press.

Mumford, M. D., Scott, G. M., Gaddis, B., & Strange, J. M. (2002). Leading creative people: Orchestrating expertise and relationships. *Leadership Quarterly, 13*(6), 705–750.

Raisch, S., & Birkinshaw, J. (2008). Organizational ambidexterity: Antecedents, outcomes, and moderators. *Journal of Management, 34*(3), 375–409.

Rosing, K., & Frese, M. (2009). *Leadership in the Innovation Process: The Importance of Ambidexterity*. Manuscript submitted for publication.

Stewart, T. A., & Raman, A. P. (2007). Lessons from Toyota's long drive. *Harvard Business Review, 85*(7/8), 74–83.

Takeuchi, H., Osono, E., & Shimizu, N. (2008). The contradictions that drive Toyota's success. *Harvard Business Review, 86*(6), 96–104.

Van de Ven, A. H., Polley, D. E., Garud, R., & Venkataraman, S. (1999). *The innovation journey*. New York: Oxford University Press.

Van Dyck, C., Frese, M., Baer, M., & Sonnentag, S. (2005). Organizational error management culture and its impact on performance: A two-study replication. *Journal of Applied Psychology, 90*(6), 1228–1240.

West, M. A., & Farr, J. L. (1990). Innovation at work. In M. A. West & J. L. Farr (Eds.), *Innovation and creativity at work: Psychological and organizational strategies* (pp. 3–13). Chichester: John Wiley & Sons.

Zhou, J., & George, J. M. (2003). Awakening employee creativity: The role of leader emotional intelligence. *Leadership Quarterly, 14*(4–5), 545–568.

Innovation Through Market Pull and Technology Push in the Heavy Equipment Business: The Voith Case

Bertram Staudenmaier and Michael Schürle

Contents

1 Introduction 205
2 Over 140 Years of Innovation at Voith 206
3 Push or Pull: Choosing the Right Strategy for Innovation 208
4 Voith Paper: Partner and Pioneer for the Paper Industry 209
5 Innovation at Voith Paper 210
   5.1 The Innovation Process 210
   5.2 The Idea Generation Process 212
   5.3 A Climate for Innovation 214
   5.4 Market Launch: Investment Security Through the Paper Technology Center (PTC) 215
   5.5 A Practical Example: How Customer Orientation and Technology Orientation Can Work Together 216
6 Conclusion 216

1 Introduction

Today more than ever before, the international competitiveness of a company depends on its ability to launch superior, innovative products in the global marketplace. For that reason, investments in research and development are not a luxury, but an absolute necessity. This is especially true in the B2B sector, where the customer is another company which must also maintain its competitiveness. As a result, the demands and importance placed on new developments in the B2B sector are high: The cost of investments must pay off, and a faster return on investments (ROI) is also required. In addition, through investments in new products and production processes, B2B customers hope to differentiate themselves from competitors and

---

B. Staudenmaier (✉)
Voith AG, Heidenheim, Germany
e-mail: Bertram.Staudenmaier@Voith.com

strengthen their overall market position as well as their products' unique selling point (USP).

To keep a step ahead of the international competition, today's companies must make carefully planned investments in research and development. Over the last few years, European companies have recognized this need, as shown in the recent study "The 2008 EU Industrial R&D Investment Scoreboard" published by the European Commission.[1] This study shows that during the last five years, the pace of investment in research and development has steadily increased in 1000 of the EU companies surveyed. And for the first time in 2008, EU companies invested more in R&D than their U.S. counterparts. Japanese companies have remained behind the Europeans in terms of investments for the last three years.

This shows that we are moving in the right direction. And it will be exciting to see if this trend continues in the coming years. Then even in difficult financial times, research and development can't be allowed to stand still, because innovative products and services will always be decisive if a company is to remain competitive. On a day-to-day basis, every company should ask itself these questions:

> How can employees be motivated and supported to develop innovative technologies and products?
> Where do fresh ideas come from?
> How much basic research is needed?
> Which developments make economic sense?
> What good are research centers?
> And how much customer interaction is profitable?

Voith has asked itself these questions. And as one of the largest family-owned companies in Europe, Voith continuously promotes new product developments. That's also why Voith was listed as number 83 out of the 1,000 EU companies in the above mentioned study: Currently, over 5% of Voith's annual revenues are reinvested in research and development.

2 Over 140 Years of Innovation at Voith

The history of the Voith company began in 1867. And after only two years Voith was awarded its first patent for a wood-pulp processing machine, named the "Refiner," which was a technological breakthrough. Today, the Voith Group has

---

[1] Héctor Hernández Guevara; Alexander Tübke; Andries Brandsma: "The 2008 EU Industrial R&D Investment Scoreboard". Luxembourg: Office for Official Publications of the European Communities, 2008.

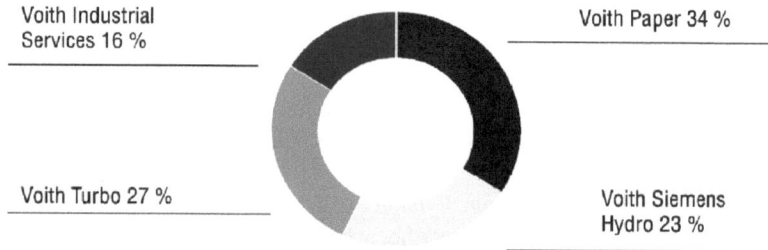

Fig. 1 Orders received per corporate division 2007/2008. Total: € 6092 million

43,000 employees, an order volume of 6 billion euros and operations at 270 facilities worldwide. It remains one of Europe's largest family-owned companies being active in the paper, energy, mobility and services sectors.[2]

Because Voith invests heavily in research and development, the company also has over 10,000 registered patents and every year 400 new patents are added to that total.[3] One reason for this wealth of new patents is Voith's worldwide network of modern research and development facilities.

The latest example of this innovative power is the first Voith locomotive, named "Maxima". In only 15 months, Voith engineers completed the locomotive's design concept. And today, the Maxima is the most powerful diesel-hydraulic locomotive available, with a rating of 3,600 kW (or 5,000 PS).

In the field of renewable energy, Voith is also forging ahead to develop technologies which can harness the unused forces of the oceans to deliver emission-free electricity in the near future. Here, renewable energy derived from the oceans' waves, tides, and currents represents a potential source of 1,800 gigawatts, which is approximately the output of 2,000 large coal-fired power plants. Gaining access

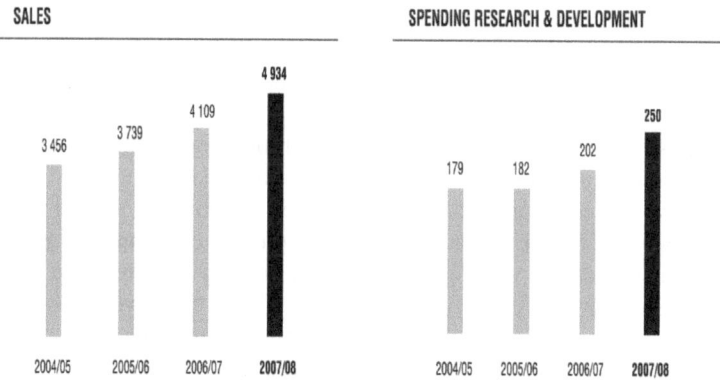

Fig. 2 Revenue growth and research & development spending from 2004 to 2008

---

[2] Voith Annual Report 2007/2008.

[3] Voith Report Special Edition "140 Years at Voith" and "Figures. Data. Facts. 07/08".

to this incredible reservoir of energy is certainly one of the major challenges for the future. In Scotland, the world's first wave-driven power plant has been producing electricity for the local power grid since 2006. In South Korea, a tidal current power plant is currently under construction. And Voith has also been active in the field of wind-power, where it is developing both new propulsion concepts and maintenance services.[4]

Voith Paper is one of the four divisions within the Voith Group and is a leading manufacturer of paper machines. Seen from a global perspective, one out of every three sheets of paper (i.e. one third of all paper) is produced on Voith Paper machines. Voith Paper is also the industry's technology leader and the only supplier with a comprehensive spectrum of products as well as services for the entire range of paper manufacturing – from graphic to special papers, and to board and packaging papers. Voith Paper consists of four functional divisions: Fiber & Environmental Solutions, Paper Machines, Automation and Fabric & Roll Systems.

In the field of paper technology, Voith concentrates on developing innovative products which significantly reduce the need for water, energy and raw materials – with the combined goal of cutting manufacturing costs as well as preserving natural resources. For example, Voith recently introduced a process for producing premium hygienic paper that can reduce energy consumption by 35%.

3 Push or Pull: Choosing the Right Strategy for Innovation

Every year, dozens of innovations are introduced to the paper industry which deliver tangible advantages to customers, through improved processing technologies, better output, and reduced operating expenses. Such market-ready innovations can be contrasted with more basic innovations, which can fundamentally change current production processes or entire business sectors. What's decisive, however, is choosing the right strategic direction for such new developments. Here, new developments are often categorized as either customer-oriented (pull) innovations or technology-oriented (push) innovations and these terms represent two completely different business strategies.

While push innovations are usually created through research and development efforts, pull innovations are created in response to a market demand, i.e. a tangible need for a new product or process. In this second case, a potential market for the new product already exists and this can represent a profitable opportunity for a technology leader in competition with other companies. Strategy experts at the worldwide business consultancy McKinsey[5] have long anticipated the current business trend away from push strategies and toward more promising pull strategies.

Today, the Internet is a stunning example of a pull strategy. Here, the Internet user has a need for information and chooses the content that he or she wants from the

---

[4] Voith Annual Report 2006/2007.
[5] The McKinsey Quarterly 05 No. 3, "The Next Frontier of Innovation".

World Wide Web. This type of interaction is what makes the Internet a classic pull medium and it has started a trend which is also affecting our traditional industries, particularly in terms of innovations which address an existing market demand. Then pull innovations are customer or market-oriented and as a result they have a minimal risk of becoming a flop. Especially in mature or saturated markets, the presence of an unsatisfied demand on the part of the customer is an absolute prerequisite for targeted offers and effective advertising.

Before our customers can decide which product or company is most interesting, they usually make competitive comparisons of a company's potential, internal processes and offers, according to the economist Professor Wulff Plinke.[6] That's why, in addition to customer contacts, our company's standing with respect to our competitors and suppliers plays an important role. Equally important is a recognition of the company's competitiveness and its efforts to improve its relative positioning, such as through better cost/benefit ratios. From the customer's standpoint, the following questions need to be answered: Does the new Voith Paper product or process offer me a benefit? Does it solve an existing problem? Or does it help cut costs?

Unfortunately, in the paper industry, real innovations don't usually come from cooperative efforts with customers, because of their high investment costs. The paper industry is therefore extremely reluctant to take risks and cooperative projects with our customers tend to focus on improving only our current line of products.

To implement new trends at an early stage in the paper industry, fundamental or basic research is critically important. At Voith Paper this research and development work is conducted worldwide, as well as in close cooperation with universities and research institutes. These efforts even include the creation and financing of professorships. In addition, a special program also exists for post-doctorate students in the natural sciences and engineering, which allows them to conduct research on selected topics for three years. In this way, Voith wants to draw on the knowledge and experience of outside experts from a wide range of fields. Here, Voith's intention is to sponsor key specialties and promote research projects which might otherwise be canceled due to a lack of personnel or funding. In this case, all of the program costs are financed by the Voith Holding.

In summary, you could say that at Voith Paper both technology-oriented innovations as well as customer-oriented innovations are being created. And quite often, innovations are indeed both technology-oriented and customer-oriented at the same time. How can that be?

4 Voith Paper: Partner and Pioneer for the Paper Industry

Successful product innovations depend upon the ability to put oneself in the customer's position, or to see the world from a customer's perspective – while at the same time being able to take advantage of technical advances coming out of the

---

[6] Professor Wulff Plinke, 1995.

research and development labs for new or existing products. A good example of this can be seen in carbon fiber reinforced plastic (CFRP), whichs offer significant advantages over steel rolls. Because of their light weight, CFRP rolls consume far less energy during operation and, being corrosion-free, they also have a significantly longer lifespan.

Placing yourself in the customer's position also helps you to be open to new ideas and able to cooperate with customers when developing new solutions. Eventually, the customer's needs become your own goals. And through the use of advanced technologies and materials, new products can be developed which create new demands.

From Voith's standpoint, being a partner and pioneer for the paper industry reflects our dual strategy of being both customer-oriented and technology-oriented. As a partner, we always listen to our customer and try to discover "hidden needs". There tailor-made solutions play an important role which can reduce the total cost of ownership, as well as customer services which cover the entire lifespan of a paper machine. As a result, tangible added-value can be generated for customers.

As a pioneer, we constantly search for new innovations that can shape the future of the paper industry. In addition to increased efficiency and quality improvements, we also see the responsible use of our natural resources as an important factor in the development of new products and technologies. Through standardization we can cut costs even further, because despite our technology orientation we always keep an eye on cost efficiency, too. In fact, high-tech and cost efficiency go hand-in-hand.[7]

5 Innovation at Voith Paper

5.1 The Innovation Process

Voith depends on innovations to secure its own future. Because it's only through technological innovations that Voith can remain a long-term preferred partner for its customers. A partner, who fully appreciates the customer's business and needs– while at the same time developing and offering customers new solutions. In the end, this approach creates both the foundation of our success and the basis for future growth. Innovation is of course essential for all growth. And our business' organic growth, i.e. growth without acquisitions, over the last ten years has averaged 4.1% per year.[8]

In principle, the innovation process at Voith is pretty simple. One approach is to recognize a problem that exists in the paper industry, such as increasing energy costs, and then to define a challenge statement which targets that problem – in this case to reduce energy costs. A second approach considers the potential of new technologies discovered through basic research, which are then also defined as a challenge statement. In both approaches, the challenge statement is converted

---

[7] Voith Report 02/05.
[8] Voith Report 02/05.

into a solution which can then lead to the development of a prototype. One of the most important aspects of this entire process however, is a continuous review of the project feasibility. Because the more time and effort we put into an innovation, the higher its investment cost and its flop risk become.

For this reason, Voith experts from selected technical fields and business sectors continuously monitor the development of all new products and ensure that these developments also support Voith's long-term business strategy. Here, it is particularly important for experts to be aware of any long-term risks, because in the case of major innovations it may take up to 10 years before a new product is established in a market.

Even a perfectly organized innovation process however, cannot guarantee the success of innovations. The key factors are of course the men and women who bring the innovation process to life. These Voith employees need to have: an ability to see developments from the customer's perspective, plus enough experience to determine whether or not an idea can be implemented within a reasonable time-frame and at acceptable costs. Added to that, they must also understand the technologies used in a new project, the market potential of a new product, and the strength of the competition that such new products will have to face. Clearly, these are tough requirements that no individual employee could fulfill and therefore Voith relies on a dedicated team of innovation specialists.

Paper machines contain extremely complex technologies. In fact, they have five times as much circuitry as an Airbus. That's why expertise from many different departments is so important when developing new paper products and processes. Here, experts in technology, research, development, product management, marketing, and sales all need to work closely together as a team – with the marketing expert as a coordinator.[9]

Voith makes every effort to support the open exchange of ideas between these groups of experts: for example, through its "New Technologies" forum with lectures from outside specialists. This gives Voith employees involved in the innovation process an ideal opportunity for open informal discussions.

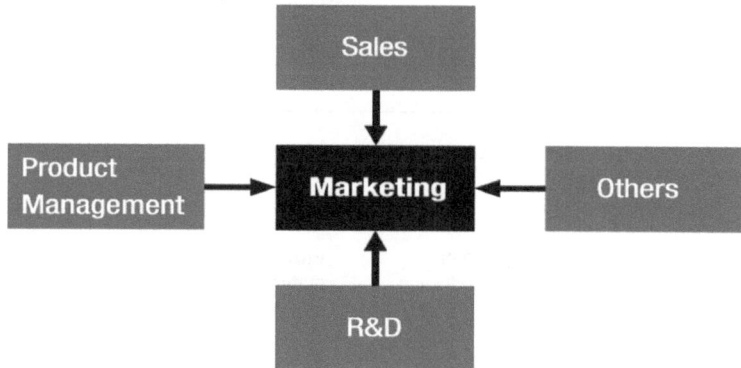

Fig. 3 Marketing ideas guide R&D efforts

---

[9] Voith Paper Marketing Toolbox: "Marketing in the Innovation Process" p. 3.

Another example was the Voith Group Conference 2005,[10] which was held under the motto "Working On Our Innovations". This conference is held every three years and deals with major issues affecting the Voith Group. Over 700 managers and experts from all group divisions attended the conference, with the majority being active in research and development. During the conference, the Voith Board of Management discussed ways of securing the company's future organic growth with conference participants, as well as how the innovation process can help achieve that goal. One key issue was the funding of innovations. To promote new ideas that are threatened by financial constraints, the Voith Board of Management established a special R&D budget. However, before you start thinking about investing in new ideas, you need to have those ideas. So, how does Voith Paper plan to create new ideas?

5.2 The Idea Generation Process

During the idea generation, the issues of competitiveness and business growth should be considered. For example, which product ideas can guarantee our company's competitiveness and profit margins? Which ideas can foster organic growth?

Before attempting to answer these questions, let's look at the so-called Three-Wave-Model together, which illustrates a product's lifecycle.

A typical product lifecycle starts with a development phase, being followed by a market launch phase, as well as a growth and maturity phase. Finally, the product lifecycle ends in the saturation and decline phase. The entrepreneurial task is to ensure a full product pipeline.

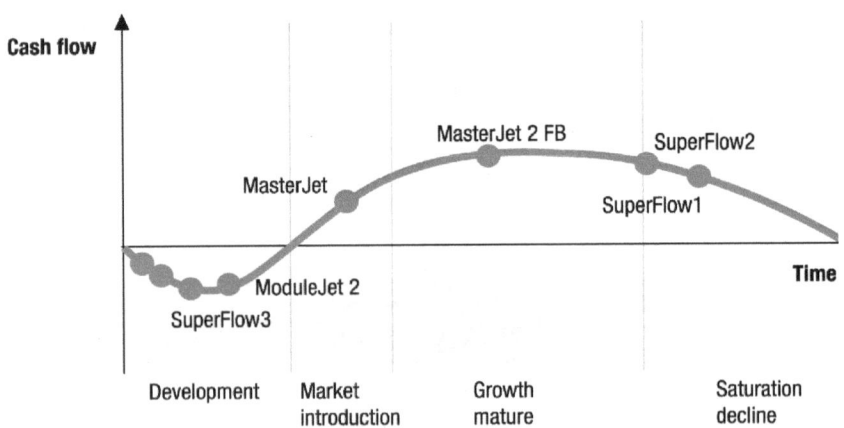

Fig. 4 Product lifecycle

---

[10] Voith Report 02/05, p. 13.

The goal of Voith's innovation strategy is to add enough new product ideas to the development phase, to replace its existing products which are nearing the end of their lifecycles. Most of the innovation effort concentrates on this kind of new product development. In this way, Voith ensures the generation of sales and profits generated by existing products. This is called the "secure level", which means securing our core business.

However if you want long-term growth, a second factor comes into play: the "build up level". At this level, entirely new products are developed that don't replace an existing product. As a result, these new products also generate new revenues, such as new services that allow customers to operate their equipment more efficiently. In fact, for most industrial manufacturers the entire service sector is a high growth business.

At the third level, completely new markets are opened up which require new expertise and applications. At Voith, good examples of product ideas at the so-called "create level" are power generators using wave action and Voith Paper systems which recycle the waste water in paper mills.

At Voith Paper, new ideas come from a wide range of sources. The knowledge about customer's needs exists in different internal and external sources which should be systematically examined and shared. This is the classic arena for marketing research, with its customer surveys and detailed databases. But often, these research practices simply don't go deep enough to reveal a customer's actual needs and expectations. Here, far more information and communication is needed, to build strong customer-supplier relationships, as well as to discover the latent needs of (potential) customers.

One possible method for doing this is to sponsor group discussions on specific topics for selected customers. During such discussions, the different opinions and viewpoints of the participants can often provide unique insights and fresh ideas.

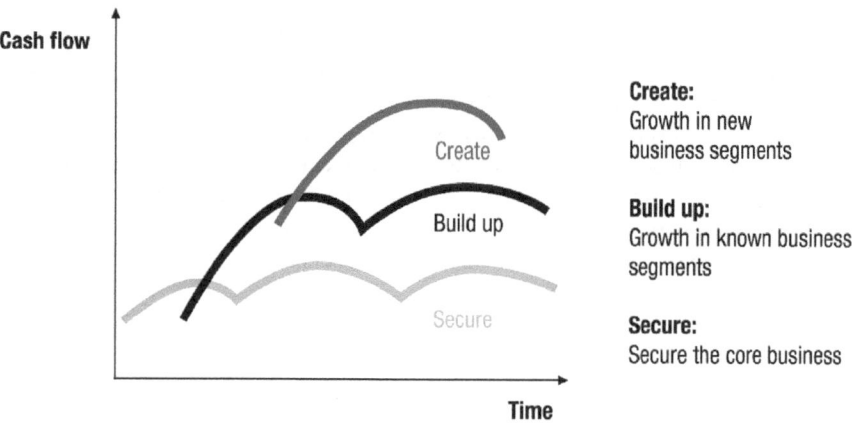

Fig. 5 Three wave model

The "Lead User"[11] concept is another method which takes this process one step farther. The term "Lead User" refers to an exceptionally well qualified customer, who is also willing to participate in the development of new products and services. The advantage of the "Lead User" concept is that customers can influence the innovation process at its early stages. The concept is implemented through a workshop with an outside moderator, where open discussions are conducted without the presence of Voith representatives.

It would be wrong, however, to limit the analysis of customer needs only to external sources. Many of Voith Paper's employees have extensive experience with customers from various projects and possess detailed knowledge concerning customer requirements. What's needed is a multifunctional approach to make the best possible use of this knowledge, by collecting key information and passing it along to the R&D experts. As a result new ideas can be created that are both market-relevant and customer-oriented.[12]

In 2006 for example, seven multi-divisional workshops were conducted. There, 450 ideas were created through brainstorming and 17 of those ideas entered the product development process.

In addition, Voith Paper offers an online idea-box to its intranet users. Via the idea-box, potential development ideas are registered and automatically transferred to R&D departments for evaluation.

5.3 A Climate for Innovation

Despite all of the organizational tools and processes which are designed to simplify the handling of innovative ideas, we should never forget one thing: The creativity of our 40,000 Voith employees is our most precious resource!

In order to turn creative ideas into innovative products that customers can use – a business' organization must have the right climate. Primarily, it is a manager's leadership responsibility: to promote creative thinking, to provide direction, to qualify employees and to share knowledge throughout an organization.

But the most important responsibility of all is to promote an open professional atmosphere, which accepts setbacks and allows employees to admit mistakes to superiors. When every innovation means crossing a technological frontier and along an uncharted course, some form of mistake is inevitable. But the effects of such mistakes can be minimized: success estimates for a project should be realistic, over-optimism should be avoided, and managers must accept negative developments. Here, the motto should be: "Tell me early, tell me the truth."

---

[11] The term "Lead User" was introduced in 1986 by Eric von Hippel.
[12] Voith Paper Marketing Toolbox: "Marketing in the Innovation Process", p. 3.

5.4 Market Launch: Investment Security Through the Paper Technology Center (PTC)

At the end of the innovation process is the market launch. Especially in the case of real innovations which are entirely new to both customers and us, the product must be fully tested to ensure its reliable performance in a working environment. Customers must never be allowed to suffer from untested or unproven technologies which could cause a serious financial loss. To minimize these types of risks, in 2006 Voith opened the Paper Technology Center (PTC) at its headquarters in Heidenheim, Germany – the world's most advanced paper research center. The Paper Technology Center represents an investment of 75 million euros and was the largest single investment in Voith's history.

The PTC offers the paper industry a unique range of capabilities found nowhere else in the world. Here, for the first time the entire paper production process can be tested and optimized under realistic working conditions. And thanks to the PTC's modular systems, different production concepts can be quickly tested and compared. For customers planning to invest in a new paper machine this is a major advantage – because at the PTC customers can test various production concepts in advance. In the end, customers can even see and feel the final results in their own hands, namely paper, as well as testing its printability.

From the customer's point of view, the PTC therefore also provides maximal investment security. And for investments of several 100 million euros, that's a decisive factor: New paper machines must be capable of producing the required paper quality and paper volumes within the agreed time-frame. Expensive experiments are the last thing a customer needs. That's why they prefer fully tested concepts and remain skeptical of many so-called improvements. Here, the PTC contributes decisively to minimizing risks and helps customers to see the real advantages of innovation.

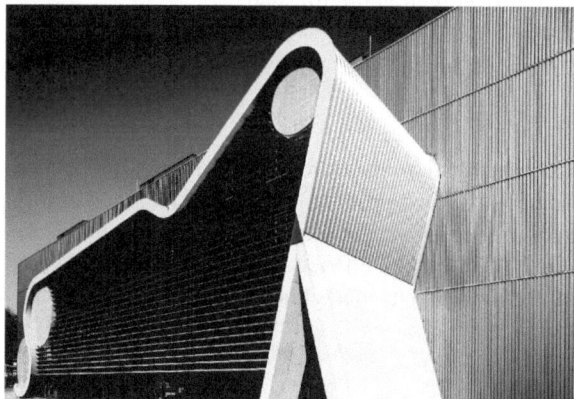

In 2006, the Paper Technology Center (PTC) was opened in Heidenheim, Germany, and is the world's most advanced paper research institute.

5.5 A Practical Example: How Customer Orientation and Technology Orientation Can Work Together

In this example, the "FeltView" demonstrates how a problem in the paper industry eventually leads to the development of a new product. This is a particularly good way to show how a customer's need was fulfilled by a new technology. Inside a paper machine, there are a large number of wires and felts (the generic term is machine clothing). The function of the machine clothing is to dewater and transport the paper web. The FeltView is an automated, online measurement system which monitors the machine clothing in the paper mill's press section. Here, the press felts are continuously checked and the customer receives an instant diagnosis of their condition, as well as the paper's quality – so that if necessary, adjustments can be made immediately.

Before the FeltView was developed, the condition of a paper machine's felts were checked only sporadically, felt measurements were seldom reproducible – and it was a dangerous job, usually done by hand, in positions which were difficult to reach. In fact, an operator's decisions were often based on a "gut feeling", given the lack of an adequate database. Today, however, the FeltView provides operators with the hard facts needed to make good decisions.

In the case of the FeltView product, initially a customer needed a better way of checking the condition of the press section's felts and optimizing felt washing. In response, Voith Paper then developed a product to monitor these felts with the aid of new sensors and software programs. Thanks to this new information, productivity, process efficiency and product quality can be improved.

The following product goals were defined during the development phase:

- Shortening of the warm-up period for new felts
- Reliable, uniform felt running times
- Comparison of different felts over their entire lifespan
- Improved job safety during felt monitoring
- Efficiency improvements in the press section through greater process transparency

Various experts from the press section, measurements, information technology, and machine clothing developed the new product. Three divisions of Voith Paper were involved which, thanks to their close working relationship, made this development project a success. In 2003, the first FeltView prototype was installed by the customer. And in 2005, the FeltView had its market launch and has now established itself on the market.

6 Conclusion

At Voith Paper, the market pull and the technology push strategies are of almost equal importance. Being customer-oriented is always important for a B2B company;

however most cooperative projects with customers tend to concentrate only on improving our existing products.

To create real innovations, basic research is essential. Here new technologies or materials can be transferred into the paper industry. During the innovation process, the challenges faced by the paper industry must always be remembered, such as the need for lower investment and operating costs. This avoids a product development which ignores market demands.

Voith Paper offers customers integrated solutions that enhance the entire production process, innovative processes that cut delivery times, and ecological technologies that improve the environmental footprint of paper manufacturing. Here, the use of standardized components and modular systems plays a key role in minimizing investment costs,[13] which is an important benchmark for all innovations. And in the end, customers receive a significantly higher added-value.

The Voith Group is not interested in quick profits. Long-term thinking is our strength, and our customer's long-term success is the basis of our own success. At Voith our motto is: "Never let the customer down". That gives us the stamina to pursue new product developments and see them mature to become a market success.

Good ideas haven't been abandoned at Voith just because of money. This strong commitment to innovation reflects the company's culture. Because, R&D investments that produce real benefits for our customers, are the foundation of the Voith Group's success.

---

[13] Georg Küffner "On the Role of Paper", p. 316.

Principles of Collaborative Innovation: Implementation of Open and Peer-to-Peer Innovation Approaches

Gerhard Satzger and Andreas Neus

Contents

1 Introduction: Applying ICT for Collaborative Innovation 219
2 Dimensions of Innovation: Openness and Peer-to-Peer Structure 222
   2.1 Information Flow and Ownership: Closed vs. Open 222
   2.2 Coordination Mechanism: Hierarchy vs. Peer-to-Peer 224
3 The "C^4" Framework for Innovation 225
   3.1 Alternative Definitions of Collaborative Innovation 226
   3.2 Attributes of Different Innovation Models 226
4 Collaborative Innovation in Services 228
   4.1 Characteristics of Service Innovation 228
   4.2 Empowerment of Service Users 229
5 Deriving and Implementing Principles of Collaborative Innovation 230
   5.1 Principles for Openness and Peer-to-Peer Structures 231
   5.2 Managing Change for Implementing New Principles 233
6 Conclusion and Outlook 235
References 235

1 Introduction: Applying ICT for Collaborative Innovation

More than three decades ago, innovations in information and communication technology (ICT), such as the Internet, have started an unprecedented success story slashing traditionally high costs of connecting and coordinating people and information. Along with this, we have seen some radical challenges to established concepts of how to organize innovation and value creation.

G. Satzger (✉)
Karlsruhe Service Research Institute, Karlsruhe Institute of Technology (KIT) Karlsruhe, Germany
e-mail: gerhard.satzger@kit.edu

For illustration, let us first look at three examples in different industries: What do Encyclopedia Britannica, proprietary UNIX variants and early online services such as CompuServe or AOL have in common? First, they were all incumbent and formerly dominant offerings in their respective markets – based on an expert-centric, closed, hierarchical development model for innovation and value creation. Second, they all lost significant market share due to the success of competing offerings – Wikipedia, Linux and the Internet – which are based on open, peer-to-peer approaches to providing value. These approaches entail a much lower, decentralized cost structure, and allow much faster adaptation to changing user requirements due to their openness and lack of hierarchical gatekeepers (cf. Table 1).

In all three cases, technology has served as an enabler, but the main competitive advantage of these new offerings was not introduced by purely technological, but rather structural differences in the organization of the information flow and ownership – *open vs. closed* – and in the coordination mechanism – *peer-to-peer vs. hierarchical*. All three new entrants were able to leverage a dramatic reduction in transaction costs to coordinate, innovate and provide value in *smarter* – not just *faster* or *cheaper* – ways. They applied features of what commonly is called "collaborative innovation". We will discuss this in more detail later.

These open, peer-based challengers were able to expand the reach of contributors to the innovation and value creation process far beyond the confines of any single organization. As they were using such radically different approaches (leading to cost structures that beat the incumbents' ones by orders of magnitude), these are prime examples of disruptive innovation leading to "asymmetric competition" (Hecker, 2005; O'Reilly, 2006): competition between rivals following very different motivations based on different assets and business models.

Table 1 Incumbents vs. new entrants in information-based value creation over 10 years[1]

Offering type	Market situation around 1998	Market situation around 2008
Online Encyclopedia	Wikipedia did not exist. US Market was dominated by proprietary offerings: Encarta.com was launched in 1993, Britannica.com was launched in 1994.	Wikipedia wins 96.7% of all US web traffic to encyclopedias, 1.3% goes to Encarta.com, 0.6% to Britannica.com.
Server Operating System (top 500 supercomputers)	Proprietary UNIX variants: 99.4% Linux: 0.2% Others: 0.4%	Proprietary UNIX variants: 4.6% Linux: 87.8% Others: 7.6%
Online Services	AOL 13.5m US subscribers (about 25% share of 56m Internet users).	AOL has ca. 9.3m US subscribers (<5% share of 221m Internet users).

[1] Data based on: Hopkins (2009), Top500.org (2008), Time Warner (2008, p. 4), Nielsen (2008, p. 2), Newburger (1999, pp. 3–6).

Interestingly enough, the radicality of the approaches enabled by ICT advances in communication and coordination is fueled by two additional concepts built into the approaches above:

- The idea of *common "ownership"* of the jointly created artifacts encourages re-use and building upon the best solution – opposed to a situation in which every organization spends time and effort to re-invent the wheel, just to have exclusive ownership: Liberal "open" licensing schemes for information-based products or services typically allow for commercial exploitation, but not monopolization.
- *Non-monetary incentive systems* are increasingly being applied based on intrinsic motivation, topical interest, philanthropy, passion or reputation – motivators whose advantages have been recognized already long before the age of the Internet: "[H]e was sure that one volunteer was worth five hirelings", as expressed by one of Kipling's protagonists (1896, Chap. 10, par. 1).

Both concepts contribute to significant innovation and co-creation of value – of a kind which prior to the enabling ICT capabilities used to be available only from closed, hierarchically structured, centrally managed, and commercially funded entities. In fact, quantitative studies of defects have found that information products developed in an open, peer-to-peer fashion can offer quality that is as good (or in many cases even significantly better) than their commercial counter-parts developed in a closed, hierarchical fashion (Reasoning, 2004, pp. 5–7).

As a consequence, the collaborative co-creation of value has proven so efficient in some areas that we can now observe that many commercial incumbents quickly change (or better: have to change) their hierarchical approach to value creation: in 2005, Encyclopædia Britannica still adamantly rejected a "Nature" journal study concluding that Wikipedia's collaborative model was approaching a level of quality comparable to its own (Giles, 2005, p. 900; Encyclopædia Britannica, 2006, p. 2). However, it has since then announced a departure of its hierarchical editorial model that had been in place for the last 240 years in favor of a more open, peer-based, Wikipedia-like system that allows for more people to contribute and edit content (Encyclopædia Britannica, 2008, 2009). Another main encyclopedic publisher in France, Larousse, has opened up to contributions in a similar move (Murphy, 2008). In March 2009, Microsoft announced the sunsetting of its encyclopedia offering "Encarta", both online and offline version, by the end of 2009 (Ahmed, 2009).

Even organizations that traditionally use closed and centralized approaches for confidentiality or security reasons inherent in their missions, such as the US Patent and Trademark Office (USPTO) or the National Aeronautic and Space Administration (NASA), are now leveraging the power of an open approach to improve quality without increasing cost: in the "Peer-to-Patent" project, the USPTO opens the patent examination process to public participation, inviting community-based reviews of patent applications and prior art (Noveck, 2006, pp. 152–154). Likewise, in its "CosmosCode" Project, NASA is soliciting public peer participation in a project aimed at jointly creating open source software to be used in future space exploration (Wooster, Simmons, & Hofstetter, 2007, pp. 2–3).

Obviously, we can observe a disruptive type of "collaborative" innovation and co-creation of value in real-world examples. The two main dimensions of this collaborative innovation are sketched in Part 2, before we use these dimensions to propose a framework for defining and understanding different innovation models and their implications. In Part 4, we will argue that new collaborative approaches are particularly relevant to *service* innovation. Part 5 analyzes guiding principles and implementation obstacles for organizations aiming to leverage the advantages of collaborative innovation. Finally, Part 6 concludes with an outlook on trends and suggestions for further research.

2 Dimensions of Innovation: Openness and Peer-to-Peer Structure

While "closed" and "collaborative" innovation may sometimes be considered two extremes along a single dimension (Gloor & Cooper, 2007), we find that there are actually two distinct dimensions at work, which we will consider in more detail in this section. First, we look at the free flow of information moving across a firm's boundary (*closed vs. open*), before we move beyond the traditional, hierarchical coordination mechanism for innovation (*hierarchical vs. peer-to-peer*).

2.1 Information Flow and Ownership: Closed vs. Open

From traditional, in-house R&D on the "closed" end to more "open" approaches that freely cross company boundaries, this dimension centers on the constrained vs. the unconstrained flow of information. As a prototypical definition let us consider Chesbrough (2003, p. 177):

> In the old model of closed innovation [...] successful innovation requires control. [...] Today, though, the internally oriented, centralized approach to R&D is becoming obsolete in many industries. Useful knowledge is widely disseminated, and ideas must be used with alacrity. If not, they will be lost. Such factors create a new logic of open innovation, in which the role of R&D extends far beyond the boundaries of the enterprise. Specifically, companies must now harness outside ideas to advance their own businesses.

This definition focuses on opening information flows across company boundaries and enabling them to harvest ideas from the outside. However, the real test of any proclaimed "openness" in innovation is how the "open" information flow is reflected in how the ownership of an innovation developed in such an "open" setting is organized: Unless there is some kind of reciprocal agreement, under which the participants in the "open innovation" do not only share the effort, but also the benefits of the joint innovation or value creation, the "openness" may really be only an attempt to free-ride at another party's expense. Over the last few years, the question of whether the monopolization of innovation benefits – e.g. through patent systems – actually help or hinder innovation has seen a renewed[2] interest. An argument can be made that legal uncertainty, created by an overly broad patent system, may be impeding innovation, even though the patent system's declared goal and intended outcome is to support innovation by creating incentives for innovators (e.g. Bessen & Meurer, 2007, pp. 4–5). The traditional view claims that making innovations proprietary by temporarily granting exclusive rights ("monopolization") is a necessary prerequisite for innovation activity: The concern is that without exclusive rights, organizations would decide not to innovate as they do not see a return on investment for the effort spent innovating, and thus innovation and value creation might cease as competitors might free-ride on the innovator's investments. However, dramatically reduced transaction costs and explicit open source or "Creative Commons" licensing models provide a workable alternative to monopolization, as we will illustrate below.

In today's world, where the cost of sharing information is practically zero, such joint innovation, development, and ownership of information products and services have proven their value and utility in many areas – from Open Source (e.g. GPL licensed software like Linux) to Open Content (e.g. Creative Commons licensed works and derived services). Commercial businesses have found that a profit can be made by using business models built not on monopolizing ownership of ideas, but rather by building on top of common assets and providing a differentiating value on top of open standards, open source or open content, such as convenience or excellence in execution (Behlendorf, 1999). While this may sound counter-intuitive at first, we are well familiar with this concept in other contexts: All kinds of profitable businesses operate on the shared space of publicly funded and maintained street networks – from taxi companies to logistics – without any of them needing to exclusively own the commodity of the street. They simply compete on a higher level, using the same basic infrastructure. Likewise, they may use the GPS satellite service to optimize their routing – again without any exclusivity with regards to using the infrastructure.

But while these infrastructure examples are often funded through taxpayer money as "public goods", the realm of information products and services does hold some interesting models of the private provision of public goods (Kelsey & Schneier, 1999, p. 4): an open license regulating ownership of a collaborative innovation can provide an efficient, low-cost means for the participating partners to ensure they receive a benefit from their investment – by preventing any participant from monopolizing the collaborative effort and by ensuring access also to any subsequently derived version, i.e. access to the value injected by other parties. As Newton (1676)

---

[2]In the 18th century, when many of the modern patent systems were introduced, there had already been a discussion on this. The result was the belief that the temporal monopoly granted through a patent was necessary to incentivize innovative activity.

had declared: "If I have been able to see further, it is only because I am standing on the shoulders of giants" – meaning all those researchers who had come before him, on whose work he was able to build. Finally, with *innovation speed* gaining in importance (Kessler & Chakrabarti, 1996, p. 1145), the low transaction costs of an explicit open ownership license has a clear speed and transaction cost advantage over individually negotiated, customized contracts or elaborate non-disclosure agreements.

2.2 Coordination Mechanism: Hierarchy vs. Peer-to-Peer

The coordination mechanisms used in Open Source or Open Content projects have attracted a considerable amount of attention for two reasons: first, for the fact that they obviously work (judging by the output), and secondly, for the surprising observation that most of them had not been consciously designed using traditional means of coordination (hierarchy or markets) and incentives (money in form of employment contracts or performance pay).

Basically, they represent a *different* option to implement an organization: Jensen and Meckling (1992, p. 251) note the objective of an organization is to solve two fundamental problems: the "rights assignment problem" (who should exercise a decision right) and the "control problem" (making self-interested agents exercise their rights in alignment with the overall goal). The traditional solution to these problems has been to *move the knowledge* necessary for the decision upwards in the hierarchy to those who have the decision rights. Another option would be to *distribute the decision rights* to those who have the knowledge. This latter option is exactly what happens in the peer-to-peer collaboration structures found in open source: "Open-source collaboration is an organizational form that permits the exchange of effort rather than the exchange of products, and it does so under a regime in which suppliers of effort self-identify like suppliers of products in a market rather than accepting assignment like employees in a firm" (Langlois & Garzarelli, 2006, p. 5).

Table 2 Coordination and property systems (based on Benkler, 2002, p. 394)

	Property system more valuable than implementation costs	Implementation costs of property system higher than opportunity cost
Market exchange more efficient than organizing or peering	Markets (Farmers markets)	Commons (Ideas & facts; roads)
Organizing more efficient than market exchange or peering	Firms (Automobiles; shoes)	Common property regimes (Swiss pastures)
Peering of more efficient than organizing or market exchange	Proprietary "open source" efforts (Xerox's Eureka)	Peer production processes (NASA Clickworkers)

Principles of Collaborative Innovation

The hierarchy vs. peer-to-peer dimension may stretch our traditional concept of how to coordinate innovation efforts. Based on Coase's (1988, p. 43) explanation of the emergence of firms and Demsetz's (1967) explanation of property rights, Benkler (2002) proposes an expanded model to explain the occurrence of peer production, as shown in Table 2.

Having laid the conceptual groundwork for the dimensions of openness and peer-to-peer structure, we will now present a conceptual framework that integrates these two dimensions.

3 The "C^4" Framework for Innovation

Based on the two orthogonal dimensions within the collaborative innovation concepts presented in the previous section, we can now formulate a framework for mapping different approaches to joint innovation or joint value creation – and provide a mental "map" for moving from the traditional, closed and hierarchical innovation approach into new territory:

- The dimension "information flow and ownership" separates closed from open innovation approaches
- The dimension "coordination structure" differentiates hierarchy and peer-to-peer organization.

Figure 1 depicts the resulting framework and the four "C"-models of innovation arising from it: closed shop, conglomerate, crowdpicking, and collaboration.

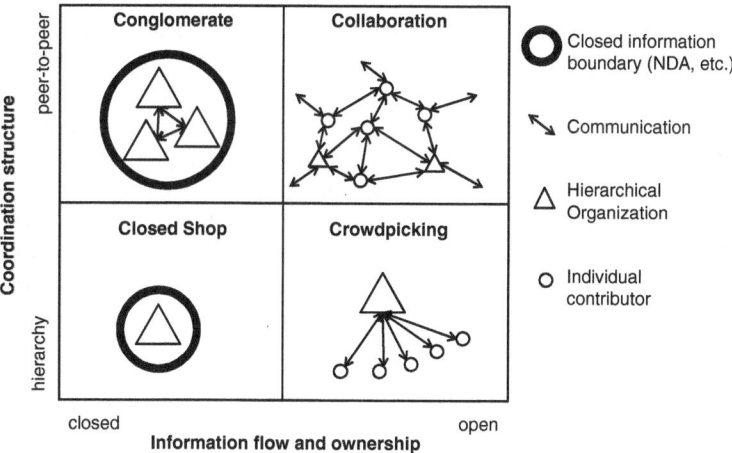

Fig. 1 The "C^4" – Framework for Innovation

3.1 Alternative Definitions of Collaborative Innovation

The map in Fig. 1 is also suited to illustrate and to delimit the two types of definitions of "collaborative innovation" that can be found in the literature: some authors call every step beyond traditional in-house R&D "collaborative" – that is all quadrants *except for the lower left*. More closely defined though, collaborative innovation only takes place across company boundaries with a peer-to-peer co-ordination mechanism. While the participating organizations may have hierarchies internally, these do not serve to coordinate the collaboration. This is identical to the *top-right* quadrant in our frame-work labeled "collaboration".[3] We believe the latter, more narrow definition is more helpful, as it highlights a new area where neither of the traditional strengths of incumbent organizations (hierarchy and enforced boundaries) yields an advantage, and where new ways of organizing, incenting, and distributing value are being developed.

3.2 Attributes of Different Innovation Models

While the four models set out in our framework are of course a simplification – companies or projects can fall anywhere on the two axes yielding a unique set of coordinates – it may still be useful to characterize the four models along some of their attributes to better understand how they operate, and where they differ.

The *closed shop* model is characterized by an internal, expert-driven, hierarchical approach that aims at monopolizing and then exploiting in-house innovation. In the *conglomerate* model, innovation is pursued across company boundaries, but within strict, closed legal borders and the intent to partner with a few selected peers to oligopolize an innovation. The *crowdpicking* model recognizes users as sources of value in the innovation process, but retains the hierarchical relationship, like in the USTPO example mentioned earlier. The *collaborative* model is both open in terms of information flow and ownership, and driven by self-organizing, peer-to-peer structures. Companies as well as individuals can be part of a collaborative innovation effort – but the companies are acting as peers rather than owners. Due to the open ownership approach, competitive advantage is sought by "out-innovating" the competition. Figure 2 positions generally known examples within this framework as well as the examples mentioned previously, while Table 3 summarizes the characteristics of each model. Having developed a framework for understanding different innovation models, we now consider why collaborative innovation is especially relevant in services – enabling service users to become co-innovators.

---

[3] Gloor and Cooper (2007) also use this peer-to-peer property in their definition of collaborative innovation networks, but this definition is limited by its proximity to traditional definitions of *Communities of Practice* (Wenger, McDermott, & Snyder, 2002).

Principles of Collaborative Innovation

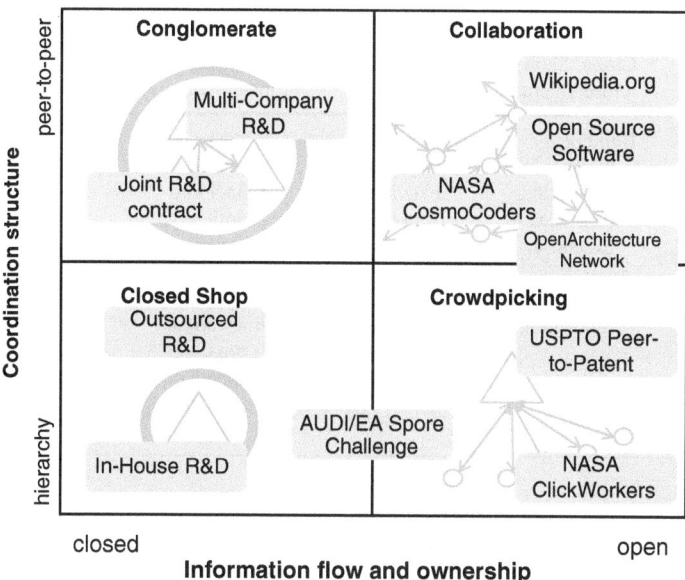

Fig. 2 Mapping of examples of collaborative innovation and value creation

Table 3 Overview and comparison of the 4 "C" – innovation models

Typical attributes	Closed Shop (closed/hierarchy)	Conglomerate (closed/peer)	Crowdpicking (open/hierarchy)	Collaborative (open/peer)
Coordination mechanism	Hierarchy	Network of hierarchical project teams	Communication with users organized by company	Network of "amateurs", companies self-organizes
Constitution mechanism	Nomination & mandate	Nomination & mandate	Provider decides to open a communication channel. Users may choose to use it	Self-selection & common vision
Intellectual property	Internal to company by employment contracts	Internal to consortium by special contract	Discussion content typically "open" for both users and company to use	Open, "commons"-type license
Incentive systems	Company's incentive systems	Company's incentive systems	Company's incentive systems for company side; no- or low-cost incentive system for participating users	No- or low-cost incentive system

Table 3 (continued)

Typical attributes	Closed Shop (closed/hierarchy)	Conglomerate (closed/peer)	Crowdpicking (open/hierarchy)	Collaborative (open/peer)
Innovation approach	Build better proprietary innovation from idea to execution	Build better joint proprietary innovation from idea to execution	Get ideas and feedback to funnel into proprietary innovation process	Stand on "the shoulders of giants", and compete on execution
Innovation strategy	Monopolize innovation	Oligopolize innovation	Monopolize innovation if possible; compete on best understanding of user needs	Best/fastest execution/ exploitation of invention

4 Collaborative Innovation in Services

Collaborative approaches to innovation seem to be particularly related to and relevant for services. We will argue this along two lines: First, there is evidence that due to the *specific characteristics of service innovations* new ICT-enabled collaboration setups are playing a more important role for service providers compared to manufacturing industries. Second, the ICT-driven *empowerment of the user* as a co-creator of value in services gives rise to peer-to-peer models of innovation.

4.1 Characteristics of Service Innovation

For innovation in services, *organization* has been found to play a far greater role than it does for manufacturing: Analyzing a survey of over 3000 EU firms on product, process and organizational innovation, Howells and Tether (2004, p. 16) found that only 8% of manufacturers classified their main innovative activities as *organizational innovation*, while 35% of the services companies did so. As we have already discussed, ICT as an enabler of new forms of coordination and communication can play a very disruptive role.

Innovation in services has also been found to pose different challenges compared to innovation in physical goods. Using an innovation lifecycle perspective, Barras (1986) found that with service companies the introduction of innovation follows a "reverse cycle" when compared to manufacturing. In the latter, the innovation moves from major disruption to incremental improvement, while in services he finds this order reversed (Table 4).

Using a dimension perspective, when comparing innovation in manufacturing and services, Salter and Tether (2006) have found that service innovations are particularly characterized by open or distributed organization, "soft" assets and benefits (reputation), as well as significant organizational change driven by technological change (see Table 5).

Table 4 Reverse product cycle of innovation in service industries (based on Barras, 1986, p. 163)

Phase of Development	Type of innovation	Characteristics
1. Maturity	Increase efficiency	Application of new technology designed to increase the efficiency of delivery of existing services
2. Growth	Improve quality	Technology is applied to improving the quality of services
3. Introduction	New services	Technology assists in generating wholly transformed or new services

Table 5 Characteristics of manufacturing vs. service innovation (based on Salter & Tether, 2006, pp. 3–5)

Dimension	Manufacturing innovation	Service Innovation
Organization of Innovation	Closed and internal	Open or distributed; multiple actors
Protection of Innovation	Patents	Copyright, Trademark, complementary assets (e.g. reputation)
Impact of technological change on organization	Impact of technological change often limited (e.g. steps in the production process)	Strong element of organizational change coinciding with technological change

Janner, Schroth, and Schmid (2008, p. 145) find that "especially in the services sector, we observe a shift from closed and company-internal innovation processes to more open and collaborative forms of innovation (e.g. via various types of communities that emerge in enterprises' surrounding ecosystems)."

In conclusion, the research on service innovation indicates that (a) service innovation tends to follow a different path from innovation in manufacturing, (b) service innovation already is more open, distributed and collaborative, involving multiple actors and (c) due to the much higher focus on *organizational* innovation, ICT can be expected to have a strong impact as an enabler of service innovation along the disruptive path introduced in Part 1: Open and peer-to-peer innovation.

4.2 Empowerment of Service Users

However, in addition to being a source of innovation that can be leveraged for competitive advantage by the service *providers* – e.g. by innovating the delivery process (electronic banking) or creating new service offerings (such as VoIP) – ICT also has a second vector of impact as an enabler for the service users (both commercial and private) by providing low-cost tools for collaborating on service innovations – effectively offering innovation users similar (or perhaps even lower?) transaction costs for these tasks than usually found in incumbent service provider organizations.

Transaction cost theory states that the decision to organize a transaction within an organization (and thus to grow the organization) depends on the internal transaction costs being lower than the cost associated with organizing the same transaction across an organization's boundaries, i.e. via the market (Coase, 1988, pp. 43–45). Over the last 15 years, ICT has lowered market transaction costs significantly in many countries, giving rise to business models that would not have been profitable before (e.g. end-user auctioning on commodity goods).

We can therefore ask if the most efficient place for service innovation may today lie *outside* of service provider organizations, i.e. within peer-networks of users who are intrinsically motivated to support innovation (and thus do not have to receive a dedicated pay-check) of a service important to them. Von Hippel (2005, pp. 1–2) has called this phenomenon "democratizing innovation" meaning that "...users of products and services – both firms and individual consumers – are increasingly able to innovate for themselves". This type of "user-driven" innovation is especially fitting for the characteristics of services (Miles, 2008, p. 117; Salter & Tether, 2006, pp. 3–4)

This is becoming true even for highly specialized services used in innovation processes – such as 3D designing and modeling of future automobiles – which used to be the exclusive realm of closed, centrally organized structures hosting experts employed full-time. Even this area has now been opened to average users with average skills due to combined advances in communication infrastructure, increased computing power on the user's end, advances in the power and ease-of-use of 3D modeling software, and increased computer literacy of users. For example, Audi has partnered with computer-game publisher Electronic Arts to use the interface of the latter's evolutionary game "Spore" to ask users to design their dream car of the future (Audi, 2009). The *Open Architecture Network* provides a platform for collaboration on open architectural innovation – with the resulting innovations placed under creative commons licenses (Open Architecture Network, 2009).

We have seen that a collaborative approach, leveraging transaction costs lowered by ICT, is particularly applicable to the services field. While the *technology* enablers for collaborative innovation are available to all companies, there are major non-technical factors that determine whether the benefits of collaborative innovation can actually be captured by an organization.

5 Deriving and Implementing Principles of Collaborative Innovation

The written and unwritten assumptions based on which actors in a given system or environment evaluate their options and choose their path of action have been given different names: mental models (Johnson-Laird, 2004, pp. 180–182), heuristics (Tversky & Kahnemann, 1974, p. 1124), theory-in-use (Argyris & Schön, 1978, p. 12), paradigms (Kuhn, 1962, p. 10) and guiding principles (McTaggart, 1997, p. 25). The common element among these denominations is that they describe basic principles and beliefs by actors about how a (complex) system works, which they use to successfully interact with their particular environment – their colleagues, the formal organization and the external world. If these principles and mental models are not compatible with the way their organizational context works, frustration will occur (Mathieu, Goodwin, Heffner, Salas, & Cannon-Bowers, 2000).

5.1 Principles for Openness and Peer-to-Peer Structures

We therefore examine the main principles for projects and organizations working on collaborative innovation and value creation, such as the development of open source software and open content (for open information flow and ownership), and for self-organized communities of practice (for peer-to-peer organization structures).

In Table 1 we have gathered the principles that are quoted in a variety of sources in the literature and that may pose significant barriers for application of open and peer-to-peer collaboration – given the current predominance of the closed-shop innovation model.

We can group these principles into 3 clusters:

- *Challenging the experts*: The value of someone's contributions are not measured by their formal expert status or previous glory, but only by the value they add in the current project.
- *Pragmatic self-organization*: Coordination is done bottom-up in a pragmatic fashion – contributors volunteer for both management and contributor roles.
- *Experimentation and speed*: There is a bias for experimentation and execution speed, for taking risks rather than playing it safe.

An overview of the clustered principles is provided in Table 6.

Table 6 Overview of collaborative innovation principles found in literature

Principle cluster	Collaborative innovation principle	Explanation	Sources
Challenging the Experts	Given enough eyeballs, all bugs are shallow	Somebody finds the problem, and somebody *else* understands it. And I'll go on record as saying that finding it is the bigger challenge.	Torvalds, quoted in Raymond (1999, p. 30)
	There are no secrets	There are no secrets. The networked market knows more than companies about their own products.	Levine et al. (1999, p. xiii)

Table 6 (continued)

Principle cluster	Collaborative innovation principle	Explanation	Sources
	Treat your users as your most valuable resource	If you treat your beta-testers as if they're your most valuable resource, they will respond by becoming your most valuable resource.	Raymond (1999, p. 34)
	Not all the smart people work for us	Not all the smart people work for us	Levine et al. (1999, p. xiii)
Pragmatic Self-Organization	Freedom of choice in selecting work (vs. tasks assigned by the manager)	Open source developers are attracted to the occupation for its freedom of choice in assignments. Both paid and unpaid [...] participants to some degree can select the work which they prefer.	Elliott and Scacchi (2002, p. 8)
	Benevolent Dictator (vs. project or line management)	The ability to fork the code keeps the benevolent dictator benevolent.	Raymond (1999, p. 68)
	Rough Consensus and Running Code	We reject: kings, presidents and voting. We believe in: rough consensus and running code.	Borsook (1995)
	Org charts are not useful for knowledge work	Org charts worked in an older economy where [...] detailed work orders could be handed down from on high.	Levine et al. (1999, p. xv)
	Empower users to innovate	If we empower users to participate in innovation and let them sort out the most relevant ideas, we will win.	Viitamäki (2007)
Experimentation and speed	Be prepared to be wrong	The most striking and innovative solutions come from realizing that your concept of the problem was wrong.	Raymond (1999, p. 39)

Table 6 (continued)

Principle cluster	Collaborative innovation principle	Explanation	Sources
	Be prepared to throw the first implementation away	When a new system concept or technology is used, one has to build a system to throw away, for even the best planning is not so omniscient as to get it right the first time.	Brooks (1995, p. 116)
	Release early & often	Release early and release often. Treating your users as co-developers is your least-hassle route to rapid code improvement.	Raymond (1999, p. 34)

As we can see, many of the principles do not match well with traditional means of coordinating work or managing the information flow and ownership. While some of the principles may be thought of as extensions of existing models, others are diametrically opposed to the principles found in most organizations today.

5.2 Managing Change for Implementing New Principles

With the obvious advantages of using collaborative innovation, what keeps organizations from embracing it wholeheartedly? While the enabling ICT can easily be obtained and set up, the new principles that are needed to make the collaboration work in practice may conflict with the established organizational culture and power structure.

As Machiavelli (1532, Chap. 6) stated: "It ought to be remembered that there is nothing more difficult to take in hand, more perilous to conduct, or more uncertain in its success, than to take the lead in the introduction of a new order of things. Because the innovator has for enemies all those who have done well under the old conditions, and lukewarm defenders in those who may do well under the new."

Unfortunately, although approximately 500 years have passed since Machiavelli described this problem, it appears that affecting the change associated with innovation is still an acute problem. In a study involving over 1500 project professionals worldwide (Jørgensen, Owen, & Neus, 2008) the top change challenges have been identified as "Changing Mindsets and Attitudes" and "Corporate Culture" (Fig. 3).

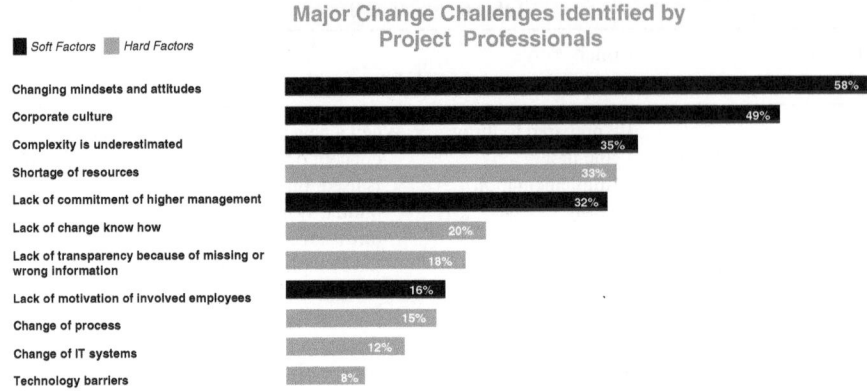

Fig. 3 Major change challenges identified by project professionals

Therefore, we can expect introducing new principles and mental models for a collaborative innovation approach into an existing organization will not be a simple task, especially when existing power structures are threatened. As Upton Sinclair phrased it: "It is difficult to get a man to understand something when his salary depends upon his not understanding it" (Ariely, 2008, p. 227).

Table 7 summarizes the key implementation issues and approaches recommended in the literature to overcome them. For each of the three clusters, while

Table 7 Implementation hurdles and approaches to overcome them

Principle cluster	Clash with typical corporate principles	Possible implementation path
Challenging the Experts	The traditional approach to innovation is expert-driven – users are only a source of data, not of independent insight.	Extend existing ways of including users in innovation: Panels, market research, etc. Support challenging the expert – e.g. using the "Emperor's Clothes" metaphor.
Pragmatic Self-Organization	Tasks are assigned by the manager to steer effort, a top-down organization structure is designed to increase efficiency.	Dedicating part of employee's time budget (10–20%) to drive their "pet" projects allows to leverage this kind of self-selection (Fuchs, Blachfellner, & Bichler, 2007, p. 300)
Experimentation and speed	Efficiency paradigm (e.g. Six Sigma etc.) tries to eliminate experimentation and variation. Risk-aversion is built into most organizations' control and audit processes.	Celebrating failure can shift the culture away from risk-aversion. (Watson, 2008)

they clearly clash with established corporate principles, we can nevertheless see approaches how to implement them in a limited fashion. In order to gain experience with collaborative innovation, at least a small group should be either created or identified – e.g. perhaps there is already a group which has close contact with an active community of potential contributors – in which these principles can be introduced and piloted.

6 Conclusion and Outlook

In this paper, we have shown that ICT – as an enabler for communication and coordination – has given rise to new ways of innovating and creating value. We have analyzed how collaborative innovation combines disruptive changes towards openness of information flow and ownership, as well as towards peer-to-peer structures, and that these dimensions come with their own set of principles which are markedly different from those of traditional "closed-shop" innovation.

Of course, there is much still to be understood about collaborative innovation: Non-monetary incentive models for the participants, (new?) legal frameworks for sharing and distributing the collaborative innovation value, new organization models and their benefits, new ways to "design" and implement the principles as cultural cornerstones of collaborative innovation, as well as the impact of business model innovation through the use of collaborative innovation and delivery in services.

In order to leverage the advantages and efficiencies of collaborative innovation, some widely held traditional beliefs about information flow and ownership, as well as coordination, may need to be questioned.

References

Ahmed, M. (2009). Microsoft accepts defeat to Wikipedia and kills off Encarta. Times Online, March 31, 2009. http://technology.timesonline.co.uk/tol/news/tech_and_web/article6008575.ece. Retrieved 2010-03-12.

Argyris, C., & Schön, D. (1978). *Organizational learning: A theory of action perspective*. Reading, MA: Addison Wesley.

Ariely, D. (2008). *Predictably irrational: The hidden forces that shape our decisions*. Harper Collins.

Audi (2009). Audi/Spore Challenge 2025, http://microsites.audi.com/ea_spore_onlinespecial/. Retrieved 2009-03-16.

Barras, R. (1986). Towards a theory of innovation in services. *Research Policy, 15*, 161–173.

Benkler, Y. (2002). Coase's penguin, or, linux and the nature of the firm. *The Yale Law Journal, 112*, 369–446.

Bessen, J. E., & Meurer, M. J. (2007). The Private Costs of Patent Litigation, Law and Economics Working Paper No. 07-08, Boston University School of Law, http://www.bu.edu/law/faculty/scholarship/workingpapers/documents/BessenJ-MeurerM050107REV.pdf. Retrieved 2009-02-27.

Behlendorf, B. (1999). Open source as a business strategy. In: *Open sources: Voices from the open source revolution*. O'Reilly.

Borsook, P. (1995). How Anarchy works: On location with the masters of the metaverse. The Internet Engineering Task Force, Wired 3.10.

Brooks, F. P. (1995). *The mythical man month: Essays on software engineering*. Anniversary Edition 1995/1975, Boston: Addison Wesley.

Chesbrough, H. W. (2003). *Open innovation. The new imperative for creating and profiting from technology*. Boston: Harvard Business School Press.

Coase, R. H. (1988). *The firm, the market and the law*. University of Chicago Press.

Demsetz, H. (1967). Towards a theory of property rights. *The American Economic Review, 57*(2), 347–359.

Elliott, M. S., & Scacchi, W. (2002). Communicating and Mitigating Conflict in Open Source Software Development Projects, http://www.ics.uci.edu/~melliott/commossd.htm. Retrieved 2009-03-16.

Encyclopædia Britannica (2006). Fatally Flawed: Refuting the Recent Study on Encyclopedic Accuracy by the Journal Nature, http://www.corporate.britannica.com/britannica_nature_response.pdf. Retrieved 2009-02-28.

Encyclopædia Britannica (2008). Britannica's New Site: More Participation, Collaboration From Experts and Readers. June 3rd, 2008, http://www.britannica.com/blogs/2008/06/britannicas-new-site-more-participation-collaboration-from-experts-and-readers/. Retrieved 2009-02-27.

Encyclopædia Britannica (2009). History of Encyclopædia Britannica and Britannica. Taking a Great Legacy into the 21st Century, http://www.corporate.britannica.com/company_info.html. Retrieved 2009-02-16.

Fuchs, C., Blachfellner, S., & Bichler, R. (2007). The Urgent Need For Change: Rethinking Knowledge Management, in: Knowledge Management: Innovation, Technology and Cultures, http://www.icts.sbg.ac.at/media/pdf/pdf1487.pdf. Retrieved 2009-03-16.

Giles, J. (2005). Internet encyclopaedias go head to head. *Nature, 438*(7070), 900–901.

Gloor, P., & Cooper, S. (2007). The New Principles of a Swarm Business, MIT Sloan Management Review, April 1, http://www.sloanreview.mit.edu/the-magazine/articles/2007/spring/48312/the-new-principles-of-a-swarm-business/. Retrieved 2009-03-16.

Hecker, F. (2005). Asymmetric Competition, http://www.blog.hecker.org/2005/09/09/asymmetric-competition/. Retrieved 2009-03-07.

von Hippel, E. (2005). *Democratizing innovation*. Cambridge, MA: The MIT Press.

Howells, J., & Tether, B. S. (2004). Innovation in Services: Issues at Stake and Trends, http://www.europe-innova.org/servlet/Doc?cid=6372&lg=EN. Retrieved 2009-02-27.

Hopkins, H. (2009), Britannica 2.0: Wikipedia Gets 97% of Encyclopedia Visits, http://weblogs.hitwise.com/us-heather-hopkins/2009/01/britannica_20_wikipedia_gets_9.html. Retrieved 2009-02-16.

Janner, T., Schroth, C., & Schmid, B. (2008). Modelling service systems for collaborative innovation in the enterprise software industry. In: *IEEE International Conference on Service Computing, 2*, 145–152.

Jensen, M., & Meckling, W. (1992). Specific and general knowledge, and organizational structure. In: Werin, L., & Hijkander, H. (Eds.), *Contrast economics*. Cambridge, MA: Basil Blackwell.

Johnson-Laird, P. N. (2004). The history of mental models. In: Manktelow, K., & Chung, M. C. (Eds.), *Psychology of reasoning: Theoretical and historical perspectives* (pp. 179–212). New York: Psychology Press.

Jørgensen, H. H., Owen, L., & Neus, A. (2008). Making Change Work – Continuing the Enterprise of the Future Conversation. IBM Institute for Business Value.

Kelsey, J., & Schneier, B. (1999). The Street Performer Protocol and Digital Copyrights, First Monday, 4, 6–7. http://www.firstmonday.org/htbin/cgiwrap/bin/ojs/index.php/fm/article/view/673/583. Retrieved 2009-02-27.

Kessler, E. H., & Chakrabarti, A. K. (1996). Innovation speed: A conceptual model of context, antecedents, and outcomes. *Academy of Management Review, 21*(4), 1143–1191.

Kipling, R. (1896). Captains Courageous: A Story of the Grand Banks. http://www.gutenberg.org/files/2186/2186-h/2186-h.htm. Retrieved: 2009-03-03.

Kuhn, T. S. (1962). *The structure of scientific revolutions*. University of Chicago Press.

Langlois, R. N., & Garzarelli, G. (2006). Of Hackers and Hairdressers: Modularity and the Organizational Economics of Open-Source Collaboration, DRUID Summer Conference 2006 on Knowledge, Innovation and Competitiveness. http://www.2.druid.dk/conferences/viewpaper.php?id=101&cf=8. Retrieved: 2009-03-03.

Levine, R., Locke, C., Searls, D., & Weinberger, D. (1999). *The Cluetrain manifesto: The end of business as usual.* Cambridge, MA.

Machiavelli, N. (1532). The Prince, http://www.gutenberg.org/files/1232/1232.txt. Retrieved 2009-03-16.

Mathieu, J. E., Goodwin, G. F., Heffner, T. S., Salas, E., & Cannon-Bowers, J. A. (2000). The influence of shared mental models on team process and performance. *Journal of Applied Psychology, 85*(2), 273–283.

McTaggart, R. (1997). Guiding principles for participatory action research. In: McTaggart, R. (Ed.), *Participatory action research: International contexts and consequences.* SUNY Press.

Miles, I. (2008). Patterns of innovation in service industries. *IBM Systems Journal, 47*(1), 115–128.

Murphy, E. (2008), French Publishing Group Sets up Rival to Wikipedia. In: The Independent, 14 May 2008. http://www.independent.co.uk/news/world/europe/french-publishing-group-sets-up-rival-to-wikipedia-827705.html. Retrieved 2009-02-27.

Newburger, E. C. (1999). Computer Use in the United States: Population Characteristics, US Census Bureau, http://www.census.gov/prod/99pubs/p20-522.pdf. Retrieved 2009-03-07.

Newton, I. (1676). Letter to Robert Hooke, Feb. 5th, 1676.

Nielsen (2008). Nielsen Online Reports Topline U.S. Data for March 2008, http://www.nielsen-online.com/pr/pr_080414.pdf. Retrieved 2009-03-07.

Noveck, B. S. (2006). "Peer to patent": Collective intelligence, open review and patent reform. *Harvard Journal of Law & Technology, 20*(1), 123–162.

O'Reilly, T. (2006). Purpose-Driven Media, http://www.radar.oreilly.com/archives/2006/04/purpose driven-media.html. Retrieved 2009-03-07.

Open Architecture Network (2009). About the Open Architecture Network, http://www.openarchitecturenetwork.org/about. Retrieved 2009-03-16.

Raymond, E. (1999). *The cathedral and the bazaar.* O'Reilly.

Reasoning LLC (2004). How Open Source and Commercial Software Compare: A Quantitative Analysis of Database Implementations in Commercial and in MySQL 4.0.16. http://www.reasoning.com/pdf/MySQL_White_Paper.pdf. Retrieved 2009-03-16.

Salter, A., & Tether, B. S. (2006). *Innovation in services: Through the looking glass of innovation studies.*

Time Warner (2008). Time Warner Annual Report 2007, http://files.shareholder.com/downloads/TWX/574216884x0x172551/CA55EB21-7F45-47BA-BE51-BFF899548A24/2007AR.pdf. Retrieved 2009-03-07.

Top500.org (2008). Top 500 Supercomputers – Operating System Share, http://www.top500.org/charts/list/32/os. Retrieved 2009-02-16.

Tversky, A., & Kahnemann, D. (1974). Judgement under uncertainty: Heuristics and biases. *Science, 185*(4157), 1124–1131.

Viitamäki, S. (2007). Crowdsourcing Innovation Principles, http://www.samiviitamaki.com/2007/03/15/crowdsourcing-innovation-principles/. Retrieved 2009-03-08.

Watson, R. (2008). Celebrate Failure, Fast Company Magazine, July 2008, http://www.fastcompany.com/resources/innovation/watson/112105.html. Retrieved 2009-03-15.

Wenger, E., McDermott, R., & Snyder, W. (2002). *Cultivating communities of practice.* Harvard Business School Press.

Wooster, P. D., Simmons, W. L., & Hofstetter, W. K. (2007). Opening Space for Humanity: Applying Open Source Concepts to Human Space Activities, AIAA SPACE 2007 Conference & Exposition, http://wiki.developspace.net/w/images/6/68/AIAA2007-Opening SpaceforHumanity.pdf. Retrieved 2009-02-27.

Managing Open Innovation Networks in the Agriculture Business: The K+S Case

Alexa Hergenröther and Johannes Siemes

Contents

1 Introduction 240
2 Innovation as a Key Economic and Social Necessity in Agriculture 240
   2.1 Initial Considerations in Agriculture Yesterday, Today and Tomorrow 240
   2.2 Importance of Innovation in Agriculture 242
   2.3 Innovative Developments in Plant Nutrition to Date 243
3 Value Chains as the Starting Point of Networks 245
   3.1 The Porter Model 245
   3.2 Agricultural Value Chain and the Agri-Food Business 246
   3.3 Specific Value Chain of Industrial Inputs 247
4 Creating and Maintaining Networks 248
   4.1 Significance of Networks in the Economy 248
   4.2 Innovation Networks 250
   4.3 Networks in Agriculture 251
   4.4 Sample Projects at Individual Levels of the Specific Value Chain 252
   4.5 Promotion of Open Networks 254
5 Opportunities for Innovation in Agriculture 255
   5.1 Fields of Opportunity 255
   5.2 Adoption and Diffusion Processes in Agriculture 257
   5.3 Challenges in Network Management 258
6 Summary 259
References 259

A. Hergenröther (✉)
K+S Aktiengesellschaft, Kassel, Germany
e-mail: alexa.hergenroether@k-plus-s.com

1 Introduction

In its 150-year history, the German potash industry has successfully implemented a long series of innovative steps both on a corporate level as well as in the market segment of agricultural plant nutrition and fertilization that is relevant to its products. In this respect it is possible to identify two lines of development in the mid-19th century as forming the starting point of this story.

Human and animal nutrition is predominantly based on the crop yields and harvests, so that plant growth is of decisive significance. The increasing food needs are the first line of development. In 1840, the agricultural chemist Justus von Liebig formulated the theory of mineral nutrition, according to which plants do not obtain their nutrition from humus, but from mineral substances such as phosphate, potassium, nitrogen and other inorganic minerals (Knittel & Albert, 2003). The promotion of plant growth results from supplying these mineral nutrients, previously removed by earlier harvesting (law of substitution), as farmers would otherwise engage in "predatory" cultivation. The simultaneously formulated "law of the minimum" states that each plant requires a certain quantity of these minerals, together with water, light, air and heat, in order to achieve plant growth and also food for human nutrition. It is the minimum amount of nutrient available that determines the growth of the respective plants. The other line of development results from the prevailing sense of a new beginning prompted by industrialization that characterized the mid-19th century. This period saw the larger scale extraction of rock salt, for which there was a greater need as a result of higher demand on the part of a growing population and its increased use for industrial purposes. One by-product of this was the discovery of those salts that contain potash and which could be processed to provide mineral nutrition to plants (K+S, 2006).

The combination of Liebig's discoveries with the higher availability of potash resulted in the establishment of more potash mines and corresponding factories for the processing of the crude salt into potash fertilizers, as we can read in the history of K+S AG. These potash fertilizers together with nitrogen, phosphate and other minerals were used to enhance plant growth and thus secure and increase farmers' crop yields (K+S, 2006). At the same time, a constantly growing global population is demanding a further increase in the volumes of agricultural products and food. The use of these mineral substances to provide better nutrition for plants is referred to colloquially as fertilization and the products used for it as fertilizers or mineral fertilizers (Mengel & Kirkby, 1982).

2 Innovation as a Key Economic and Social Necessity in Agriculture

2.1 Initial Considerations in Agriculture Yesterday, Today and Tomorrow

Following an introductory overview of the origins of the need for food and of the potash industry, to put the innovative steps into context, we will now examine the

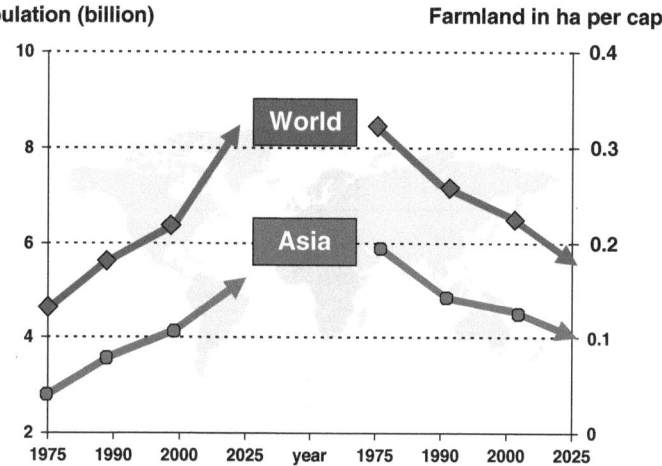

Fig. 1 Population growth and available farmland. From FAO, IFDC, IPI, Siemes (2002)

driving forces in more detail. The cornerstones are, on the one hand, the available resources on offer and, on the other, the different technical production phases in the food input and product chain, together with the need and demand for food on the part of the consumer. On the demand side, the needs of a growing world population are the main driver for innovations in the food production chain. (Abel, 1978) The extreme increase in the global population since the 19th century, which became even more rapid during the 20th century and is expected to accelerate further over the coming decades, is striking in this regard (Fig. 1).

In addition, changed eating habits are a further factor, this entailing the greater consumption of high-quality food as well as increased consumption of fast food, together with the tendency for more of the rural population to move to the highly populated urban centers (Siemes, 2002). Increased demand for meat-based food products has, of course, also affected plant cultivation, because it takes 2 kg of cereals to produce 1 kg of chicken, and 3–4 kg of cereals to produce 1 kg of pork. In recent years, a further long-term driver of demand has increasingly been the need for bioenergy and its use as a material by industry.

However, as there is only limited availability of soil or agriculturally cultivable area, and this falls per capita as the population grows, growing demand for food is forcing farmers to increase their production per unit of surface area, i.e., to intensify their exploitation of the land. Additionally, the risks of climate change may put a cap on potential crop yields (Fig. 2). It is therefore assumed that the creation of the necessary agricultural products over the coming decades will have to be based on an increase of only 0.8% p.a. in usable arable land. On the other hand, experts expect a necessary increase in yields of 2.8% p.a. through the additional use of improved inputs and their more intensive use (Marcinowski, 2009).

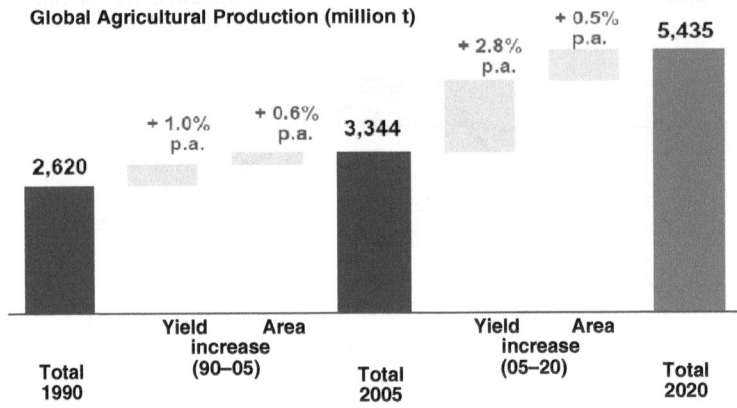

Fig. 2 Agricultural production to 2020 and food needs. Modified from Marcinowski (2009), BASF based on USDA/FAPRI

2.2 Importance of Innovation in Agriculture

The food supply is determined at several stages by the rising demand for nutrition and restricted by the available production factors. The dependencies can be illustrated using the example of the food chain (Fig. 3).

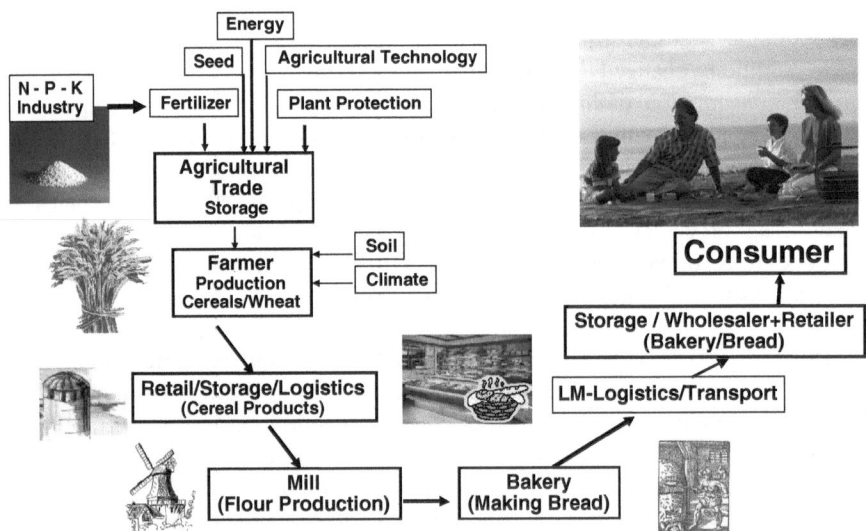

Fig. 3 Agri-food chain (value chain example: cereals/bread)

The food chain starts with the preliminary products and inputs (fertilizers, plant protection, seed, energy, machinery etc.) of the agricultural sector, which are mainly used to improve plant growth and to thus increase and secure the quantity and quality of crop yields. The agricultural products (e.g., wheat) form the basis for the manufacture of intermediate products (e.g., flour) and these, in turn, are used to manufacture finished products for the consumer (e.g., baked goods) (Theuvsen, Spiller, Peupert, & Jahn, 2007). The increases in yields in agriculture result from natural circumstances and operating conditions such as the soil, climate, workforce, technology and capital and from an extremely wide variety of input innovations.

The relevant fields of innovation in agriculture include, in addition to land and water, improved mechanization, the targeted application of fertilizers and plant protection agents, the increased use of higher-yield seed and green biotechnology. The effects of innovations in German agriculture over the past century can be seen in the changed performance figures of the sector. On the one hand, the average cereal harvest has quadrupled; on the other hand, the number of people employed to work on each unit of surface area has fallen to 10% of the original value 100 years ago. While in 1900 the agricultural production of a farmer in Germany was able to feed 4 people, this century the figure is over 120 people, which would not have been possible without appropriate innovations and the resultant increases in yields (DBV, 2009).

The need for sustainability and a certain degree of risk aversion on the part of farmers often result in a hesitance to adapt and diffuse innovations. Such short-term effects may also have a positive impact on, for example, the conservation of resources and help to keep production efficient over the long term. (Christen, 2006) They thus contribute both to a reconciliation of economic and ecological factors and to the securing of mankind's food supply in a sustainable manner. The tension between economic and ecological factors is assessed very differently, depending on the standpoint of the observer (specialist orientation, political attitude and remote or close-up view of the farmer). This can be seen from the discussion regarding the use of modern means of production like plant protection agents, green gene technology and organic farming (Welthungerhilfe, 2000). Thanks to decades of better education for farmers and their comprehensive knowledge and skills, as well as state-funded and private consulting activities, the balance between economic and ecological factors is now an important part of their target system.

2.3 Innovative Developments in Plant Nutrition to Date

The requirements made of the agricultural sector with regard to the demand for food, and the resultant improvement and increase in production and productivity, are the driving forces behind innovation. Naturally, they not only include the production relations of the agricultural sector, but also the upstream and downstream sectors. On the input side the activities of the farmer are the main points addressed by innovations in the corresponding companies in the upstream sector, like the fertilizer industry, plant protection industry, seed growers and agricultural technology industry as well as the corresponding trade sector dealers.

A wide range of innovations have been developed in the fields of plant nutrition and fertilizers (Fig. 4). In ancient times and during the Roman period, it was already known that plant growth could be improved by the use of organic waste (dung etc.) (Columella, 1981). In the Middle Ages and the early modern period, many references can be found to the use of dung, ash, compost, leaves, bone meal and pond mud etc. for enhancing plant growth (Rösener, 1993). The decisive breakthrough came in the mid-19th century, as a result of, on the one hand, Liebig's theory of mineral nutrition, and on the other as a result of the industrial development of companies processing potassium (1860) and phosphate (1879) (K+S, 2006). Further innovations included the Haber-Bosch process for the synthesis of ammonia as the basis for nitrogen fertilizers (1913), the manufacture of complex fertilizers (Nitrophoska), the use of liquid fertilizers, and the production of improved spreading and application technology (BASF, 1990). Recent times have also seen the development of inhibitors and micronutrient fertilizers (Kummer & Zerulla, 2006) as well as the use of sensors and GPS technology to achieve the optimal effectiveness of fertilizers. The other sectors, like plant protection and seed, have also been characterized by a long series of innovative developments. A good example of the interaction of improved plant nutrition and plant protection together with seed, as a result of innovations in the corresponding industries, is the increased wheat harvest over the past 50 years (Fig. 5) The combination of these input factors, together with more efficient technology, quadrupled the wheat harvest during this period (Knittel & Albert, 2003).

Early and Roman period	Use of organic waste / manure etc.
Middle Ages	Use of manure, waste, compost, leaves, marl, etc.
from 1800	Liberation of the serfs and end of three-field crop rotation system
	Thaer / Thünen: Conservation of soil fertility
1840	Sprengel / Liebig: Theory of substitution minimum in agricultural chemistry
ca. 1840	Import of guano from Chile as fertilizer
1860	Availability of potassium from mines
1979	Discovery fo phosphate deposits in Florida
ca. 1885	Use of Thomas slag from the steel industry as a phosphate fertilizer
189	Fertilizer Advisory Service through potash syndicate for farmers and dealers
1913	Haber-Bosch Synthesis of nitrogen (ammonia) at BASF
ca. 1925	Increased production of mixed and compound fertilizers (1927: "Nitrophoska")
from 1950	Growing importance of soil survey for phosphate, potash, calcium, etc.
ca. 1960	Increasing development of centrifugal fertilizer spreaders
ca. 1970	Growing importance of magnesium as plant nutrient
ca. 1975	Use of liquid fertilizers via field spraying (AHL/NP)
ca. 1990	Use of micronutrients ("Microtop"), N-inhibitors, N-stabilizers ("ENTEC")
ca. 1990	Growing importance of sulfur as plant nutrient
ca. 1995	Development of precision farming and N-sensors

Fig. 4 Innovation in plant nutrition and fertilizers

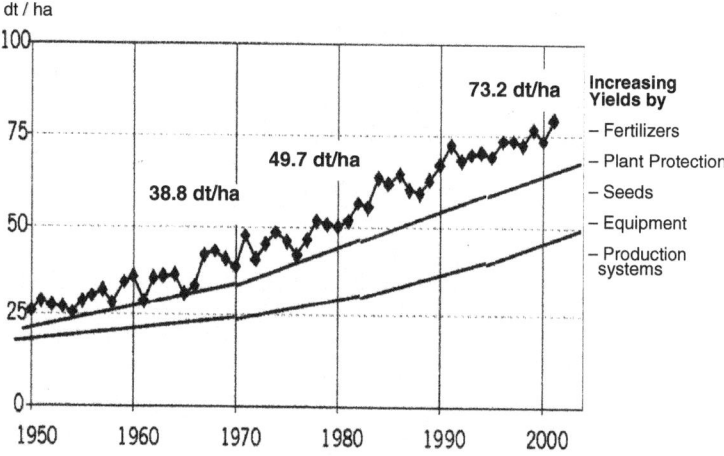

Fig. 5 Yield development of winter wheat 1950–2000 (average yearly yield in Germany). Modified from Knittel and Albert (2003)

3 Value Chains as the Starting Point of Networks

3.1 The Porter Model

On the sectoral level, the agricultural sector, and thus agriculture including the upstream sector as well as agricultural trading, the food industry and the food retail and wholesale trade all the way through to the consumer are bound together by a complex nutrition and value chain. Of equal and central importance for each individual company is its own internal value chain. The goal of each company in a market economy is to achieve appropriate earnings and profits by supplying services or products. The relevance of these many and varied activities to corporate results can be systematically determined on the basis of Porter's model (Fig. 6) (Porter, 1985).

This model for analysing the industry's structure offers a wide variety of benefits for each individual company throughout the entire agri-food chain; however, the inevitable dependencies within this chain also have to be taken into consideration. When examining the industry, individual activities are analyzed with regard to their benefits for the respective customer and supplier as well as the possible risks from potential competitors and substitute products. On the one hand, this is used to build up competitive advantages, and on the other, to identify a better strategic direction for the company (Nestlé, 2008). The advantage of sectoral and value analysis is that the individual operating activities focus on their relevance to the customer and market and on their differentiation from the competition. In companies within the fertilizer industry (e.g., the potash industry), such comparisons of individual cost items and their relevance to production have been made by external consulting firms over the past 20 years, achieving a stronger competitive position. Similar benchmark activities also exist for other sectors of the upstream and downstream industry as

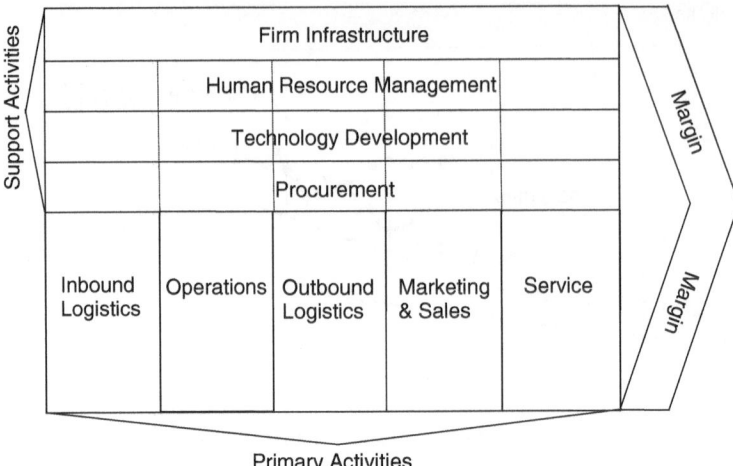

Fig. 6 Activities of the value chain. From Porter (1985)

well as for agricultural production systems - so-called "agri benchmarks" (Zimmer et al., 2007).

3.2 Agricultural Value Chain and the Agri-Food Business

Companies in the fertilizer industry and the plant protection industry, as upstream industries within the agricultural sector, are located at the start of the agri-food chain. They play a role in the complex chain that is created by mankind's need for food. As these product groups play a direct role in food production in the form of preliminary products, the success of these companies depends decisively on the use of their products in this chain (Fig. 3). The agricultural sector creates plant and animal products which are in turn processed by agricultural trading companies as agricultural raw materials in the food industry and food trades to create food products ready for the consumer to purchase and eat or drink. These finished products are then marketed directly to the food-service industry and, to a greater extent, to the consumer by corresponding logistics and trading companies like supermarkets (Schmitz, 2008).

The material dimensions of the agri-food chain can be seen from the revenues figures at the individual stages of the agri-food business (Fig. 7). Consumer spending on food in Germany totaled € 245 billion in 2006/2007 (Schmitz, 2008). Revenues for the food manufacturing industry in the domestic food chain totaled € 138 billion. The corresponding value of German agriculture was € 45 billion. The inputs amounted to approximately € 34 billion. This included about € 4.1 billion for fertilizers, plant protection and seed, of which € 2.2 billion can be attributed to fertilizers and about € 1.2 billion to plant protection. If we turn to those employed within the

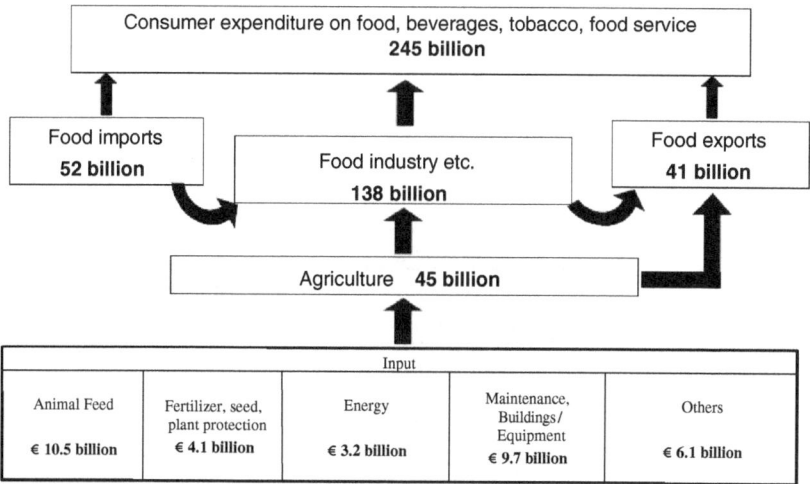

Fig. 7 Agri-food business in Germany (revenues in €). Modified from Schmitz (2008)

agri-food business, this sector provides employment in Germany for about 4 million people, which is equivalent to 10% of the total working population. The agri-food business's share in Germany's gross value added reached 6.8% in 2007.

3.3 Specific Value Chain of Industrial Inputs

The value chains within companies providing industrial inputs for agricultural production involve different manufacturing processes, and thus also different product and value chains, due to the specific products involved. Taking the example of the production of potash and phosphate fertilizers, we obtain a simplified model of the value chain (Fig. 8). On the one hand, to produce fertilizers these companies themselves require a series of inputs such as energy, machinery, auxiliary materials and services. On the other hand, mining operations have many specific features for obtaining raw materials and their treatment in special factories involving a wide variety of different process for concentrating the substances. By means of distribution and logistics functions, the different varieties of fertilizer are marketed via private companies and cooperatives to farmers who use them to improve the growth of their crops and to attain optimal crop yields (K+S, 2006). A further example of a subsection of a value chain for preliminary products and services is the development of a plant protection agent (IVA, 2005). Very different units of a company have to work together, in areas ranging from chemistry, biology, toxicology and ecotoxicology through process technology, field testing, production technology and packaging to the marketing of the finished product.

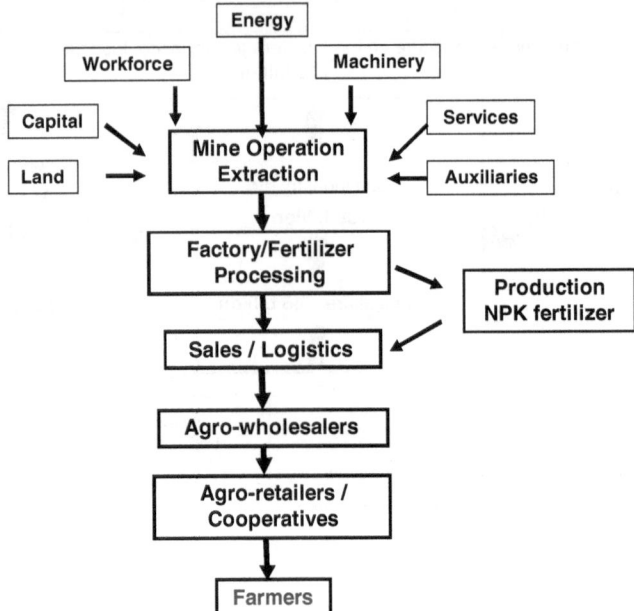

Fig. 8 Value chain of industrial upstream services (for example, fertilizer production (potash/phosphate))

The contribution of an individual company to the creation of value can be illustrated using the example of K+S AG (K+S, 2008). In 2007, revenues reached € 3.34 billion. After deducting expenses and depreciation, the company achieved added value of about € 960 million, which corresponds to about 28% of its revenues.

4 Creating and Maintaining Networks

4.1 Significance of Networks in the Economy

A network is defined as technical systems which have specific structures and can be allocated to the IT sector. The term is also understood to refer to cooperative relationships between persons or groups of persons. As a result of the increasing complexity and specialization of our knowledge society, which is constantly evolving, such networks between people offer great opportunities for improving the transfer of knowledge and thus for innovations (Küster, Schuhmacher, & Werner, 2008). In companies and also in other institutions, in addition to the existing documents and publications, a significant part of the intelligence and knowledge can be found in the heads of individual persons. This employees' knowledge is exploited by many companies, which set up a knowledge management system in order to

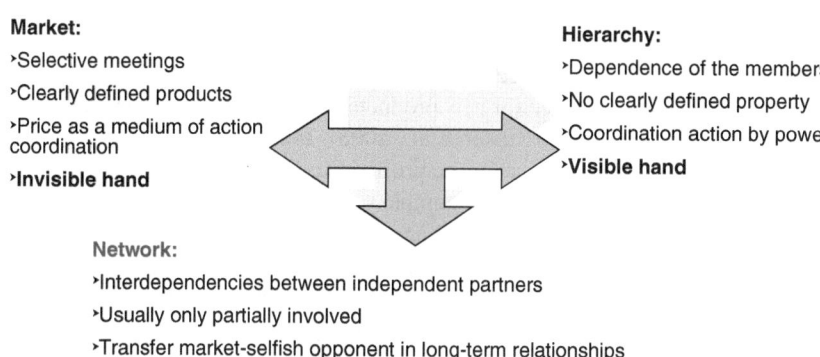

Fig. 9 Network diagram and network relations. Modified from Wagner (2005)

improve efficiency and gain competitive advantages on the market. These network relationships, even in the everyday practice of companies, are often very complex and, in some cases, not directly identifiable (Fig. 9).

There exists a certain tension between the market activities, the existing hierarchy and the networks, so that some form of coordination of the activities is normally necessary (Wagner, 2005). The basis of networks is a certain degree of trust between those involved. As regard innovations, networks are therefore of enormous importance, because it is precisely in informal networks that participants normally contribute their implicit knowledge and, on occasion, their speculation and hypotheses. The coordination of such forms of knowledge then feeds a wide variety of creative solutions, because innovations themselves arise from creative ideas, which individuals have worked out. These thoughts and ideas are often exchanged between people of very different types and backgrounds, who themselves exist within a varied web of relationships resulting in innovations (Hauschildt, 1997; Wahren, 2003).

Networks can be organized as a closed network within a company or as an open network in a combination of company members and participants from outside. The classical restriction on a research and development department with other operating units in companies is an old-fashioned style of networks on innovation. But for decades, a lot of companies have additionally used external expertises in a more informal cooperation. Such interactions with various groups in a company as well as with external experts including customers result nowadays in a more formal kind of an open innovation network and offer a lot of advantages and opportunities for a company in their market field. These will also help to avoid flop, failure rates, costly product ideas, marketing mistakes etc. The interactive relationship and modern society show a growing relevance of such open economic networks and also open innovation networks for all partners on the marketplace.

4.2 Innovation Networks

Innovation is a central part of the success for social and economic development and also for companies. The market economy is based on competition between companies and this is the driving force for new products, new technologies, new marketing strategies etc. (Mensch, 1977; Küster et al., 2008). Innovation in the various areas of a company is the central factor for the profitability and success as well as for the further development and growth of a company. Nowadays with the growing complexity, extent and variety of knowledge and the broader information network e.g., via the internet, companies establish a kind of innovation network and have opened such networks to various external partners (e.g., suppliers, customers, universities, research centers, laboratories, governmental organizations, private organizations, pressure groups etc.). They incorporate the external creativity in their own innovation process. As regards customers and markets needs, the customer integration in the innovation process has increased in the last decade. There are a lot of examples for customer integration in innovation networks like the automotive industry, consumer goods industry, food industry etc. (Küster et al., 2008, Reinwald & Piller, 2006). The cooperation of various partners in a product chain results in a higher innovation intensity and a changed life cycle of products resulting in competitive advantages.

The above mentioned value chains and product chains within companies described in the previous section are supplemented by parallel chains of comminication (Fig. 10).

The diagram shows the various dimensions of the chains (Ermann, 2001). These three elements, product, value and communication, form the basis further innovations in the form of different networks at different levels of companies. Internal or closed and external or open networks can both offer considerable business opportunities, but also result in fields of conflict in the market and with the respective

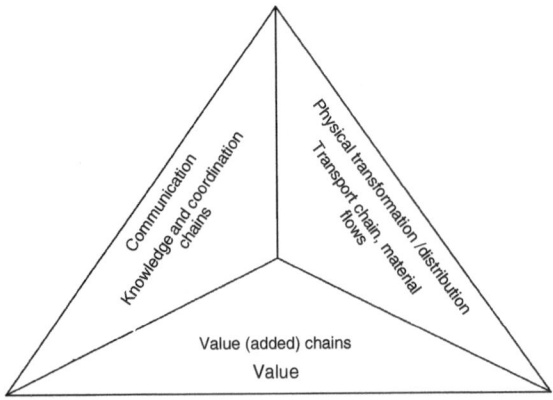

Fig. 10 Dimensions of product, communication and value chain in and between companies. Modified from Ermann (2001)

hierarchy. There will also be tension between trust and mistrust, in particular as far as external and informal networks are concerned. Thus, memberships in national and international associations and comparable organizations are necessary levels of information and institutions for the coordination of interests within a particular sector. Moreover, they lead to a series of informal networks of employees of those companies involved. They can also have a positive or negative impact on their own interests and innovations.

4.3 Networks in Agriculture

A structure of small and medium-sized companies can predominantly be found in the agricultural sector. In addition to this, the production relationships result in different production conditions, as a result of such natural resources as the soil and climate or the distance to the main markets. These natural and historical facts explain the early creation of agricultural associations and interest groups in the 19th century, following peasant emancipation, and also explains the activities of the founder of agricultural science, Albrecht Thaer (Rösener, 1993). Such associations functioned like networks and their goal was the transfer of knowledge and the introduction of innovations in agriculture. Examples of this are the function of local and regional peasant associations and the agricultural itinerant teachers of the 19th century, the creation of chambers of agriculture, the farmers' association and the Raiffeisen organization, together with the foundation of the German Agricultural Society (DLG).

Nowadays, there are a large number of regional, national, European and international associations and organizations in the agricultural sector. To this can be added agri-food business associations and federations, ranging from the sector's upstream industry (e.g., Industrieverband Agrar, European Fertilizer Manufacturers Association, International Fertilizer Industry Association), the Farmer's Association and Raiffeisen organization, the German Agricultural Society and agriculturally-oriented academic societies, the associations of the food manufacturing industry and food retail and wholesale trades (e.g., the Federation of German Food Industries, Confédération des Industries Agro-Alimentaires, Federal Association of German Food Products) through to the consumer organizations (e.g., Federation of German Consumer Organization, Food Watch). These are supplemented by a large number of sector-oriented lower-level industrial and craft organisations. The specific bodies of these associations are often used as innovation networks for the individual members and their companies. As a source of information and level of communication in the agricultural and food sectors, the various networks cover a very wide, heterogeneous and complex field. They include the manufacturers of inputs for agriculture such as K+S, BASF, KWS, Claas, wholesalers like Agravis, Baywa and related service providers such as banks, special consultants, subcontractors, machinery syndicates, private and cooperative agricultural dealers etc. On the agricultural sales side, there are food industry companies such as Nestlé, Humana,

Westfleisch, etc. and the related crafts (bakeries and butchers) as well as food retailers like Aldi, Edeka, Rewe etc. A wide-ranging network in the agri-food business in Germany is the German Agricultural Society (DLG), with around 20,000 members, an independent, non-political umbrella organization. Its objective was to implement scientific knowledge in agricultural practice. Nowadays, the DLG sees its function in providing a "basis for innovation and progress" and attempts to influence the entire agricultural and food sector by means of a wide variety of activities. The main area of activity of the DLG is personal contact for creating and using networks as well as the exchange of opinions between its members and representatives of the industry of all kinds (DLG, 2009). At various levels, K+S is also active in the DLG by bringing its specific knowledge to the table in a mutually beneficial manner.

4.4 Sample Projects at Individual Levels of the Specific Value Chain

Cooperation between the different stakeholders in the specific sectors of the agrifood chain occurs in very heterogeneous fields and involves a high degree of specialization and complexity. As a result of the fact that this sector has already existed for a long time, it is often the case today that the innovations which gradually find their way into agricultural practice are relatively small. This form of relatively small innovations can also be found, in part, in a series of industrial production processes, such as those seen in the potash industry.

An example of the cooperation between the different specialists, when it comes to innovations in agriculture, is the use of GPS (global positioning systems). Such systems are used to determine one's position anywhere on earth (Fig. 11) (PrecisionAg, 2009, Knittel & Albert, 2003). Within the framework of "precision farming", GPS can be used to determine the position of machines in fields and permits the more exact working of arable fields.

Together with recorded information about the plant nutrients in the soil, the type of soil and the current nutrient needs of the cultivated crop, a more precise and appropriate application of fertilizers is achieved. In the development of these new technologies, it was necessary for farmers, plant cultivation and soil specialists, agricultural equipment technologists for fertilizer spreaders, fertilizer manufacturers (like K+S), and last but not least, electronic engineers and programmers to work together. The outcome of such innovation is that fertilizers are applied in a manner that better suits the soil and plants, and this has both economic and ecological benefits. The use of an N-sensor results from similar cooperation between experts, in this case, sensor technicians and optoelectronic engineers (Fig. 12) (YARA, 2009).

By optically sampling the leaves of plants, the sensor determines their need for the plant nutrient nitrogen. On the basis of a pre-setting for plant needs, the electronics control how much fertilizer is spread as the spreader crosses the respective field area. A further example of a specific product development is the offer of a stabilized

Fig. 11 Potash application with a GPS map: manual control of the quantity. From Knittel et al. (2003), AGRO-SAT Consulting

Fig. 12 N-tester and N-sensor: Time- and crop-specific nitrogen fertilization. From Yara/Agricom

nitrogen fertilizer, e.g., ENTEC (Fig. 13) This product from BASF/Compo/K+S releases the nitrogen contained slowly, which, if applied at the right time, corresponds approximately to the growth needs of the plants over time (K+S, 2006). This brings particular benefits in horticulture and landscape gardening, for example, where repeated applications of fertilizer can be avoided and, simultaneously, nitrogen prevented from entering the subsoil. Also, with the development of the

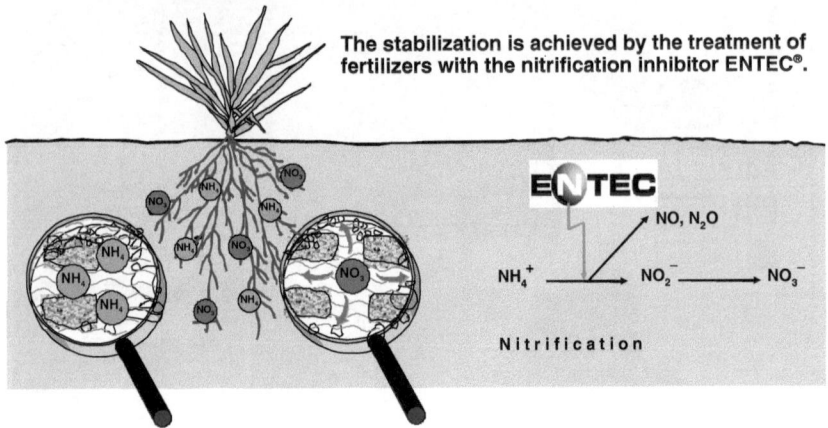

Fig. 13 Efficiency benefits of balanced fertilization: ENTEC – Principle system. From BASF/COMPO/K+S

EPSO products, K+S has entered the specific market of micronutrients for plants following the needs for farmers with intensive crop production.

A completely different project is Fraunhofer-Allianz's "Food Chain Management" (Fraunhofer Gesellschaft, 2005). Against the backdrop of food scandals, food safety, food quality and the ability to retrace raw materials is becoming ever more important. The research project undertaken by the Fraunhofer-Allianz and its institutes hopes to use microsystems technology to render the individual flow of goods in the food chain, from the very first point through to the consumer, trackable and controllable, so that it will be possible to identify food risks at an early stage and eliminate them. This work is intended to boost consumer confidence in food.

4.5 Promotion of Open Networks

Economic networks are instruments of value creation in companies. Especially open networks on the field of innovation result in new creativities, new ideas and later on in new products, technologies, concepts and also in new market opportunities. With the higher speed of market changes, companies to adopt a broader view and quicker insight for customers needs, which results in the customers' integration in the company's innovation process (Küster et al., 2008, Hauschildt, 1997). Therefore the promotion of open networks especially of open innovation networks produces a lot of competitive advantage and also value creation for companies. On the other hand, these positive effects, which are closely connected with trust in a company's employees, will also increase the risk of a negative influence via early outside information on new business ideas. Looking at the future development of markets, customers' behavior and competition, companies have to take the opportunities to organize their own open innovation network but also to try to place their own employees in third parties' open networks (like DLG). Such a concept could be time consuming and also partly against the classical hierarchical organization system, especially when it comes to disruptive innovation which could create new market opportunities. Generally, however, companies have to stimulate a lot of key employees in the research, technical and marketing area to join such external, open networks by offering various kinds of personnel and group incentives and also by using the image of the company as a specific leader in the market field with a concept of sustainable value creation (Satzger & Neus, 2009).

5 Opportunities for Innovation in Agriculture

5.1 Fields of Opportunity

The look at fields of innovation in agriculture, on the basis of a case study on K+S, a manufacturer of fertilizers and downstream supplier for the agricultural sector, will deliberately and appropriately be limited to the more narrow field of plant nutrition and the use of fertilizers. A thorough examination of the opportunities for innovation in agriculture would go beyond the scope of this publication. The plant nutrition and fertilizer application sector is a subsector of plant production, which in turn is part of the production systems of an agricultural business. Agriculture itself is part of the value chain of foodstuffs, which are ultimately necessary to feed mankind. Agriculture uses a series of inputs and services such as fertilizers, plant protection agents, agricultural machinery, energy, consulting services etc. for the production of agricultural products. Although the field of plant nutrition and fertilization has been researched for more than 100 years, there are constantly new requirements to be met in the further development of agriculture and the agri-food chain. Nowadays, for example, farmers in Western Europe already exploit many specific research findings relating to plant cultivation extremely successfully, thanks to their good education and the high level of production intensity (DLG, 2009; K+S, 2006; Rentenbank, 2006; KWS, 2007; Christen, 2006; Asenso-Okyere & Davis, 2009). The opportunities for innovation in plant nutrition and fertilization can be divided into four areas, namely biological, technical, organizational and marketing opportunities. This division does not claim to be exhaustive, but is intended to facilitate an overview. The further, future development of the sector constantly reveals new issues and interdependencies, which in turn stimulate innovations.

Opportunities for biological innovation concern plant biotechnology and plant breeding, which hope to optimize plant performance through the modification of plants (Fig. 14). Via its own agricultural extension service, K+S still has over 100

- Plant biotechnology: modification of the physiological process and genome chains related to plants, cultivation of plants for nutrient efficiency, salt tolerance, specific ingredients for human / animal nutrition

- development of nutrient inhibitors regarding soil and air (e.g., nitrogen)

- Secondary substances to improve nutrient uptake and use of synergies between fertilizers, plant protection and seeds

- Increasing awareness of sustainability of water and energy efficiency as well as the dynamics of nutrition in agricultural production systems and crop rotations

- Environmental impacts of plant nutrients, soil fertility and protection

- Monitoring and further development of new and existing agro-production lines regarding nutritional needs and dynamics (bioenergy crops, etc.)

- Better use of nutrients in organic waste materials (manure, secondary raw materials, etc.)

Fig. 14 Biological innovation opportunities in plant nutrition and fertilization

years' experience with field trials on fertilizer use as a knowledge base for farmers and their optimal crop production. Furthermore, nutrient inhibitors, new knowledge about nutrient dynamics, improved water and nutrient efficiency together with synergies between fertilizers, plant protection and seed are all gaining in importance. Additionally, environmental effects and the sustainability of the entire plant production system, together with the more effective use of organic residues, are attracting an increasing amount of attention.

Opportunities for technical innovation are directed towards improving spreading processes (Fig. 15). However, they also concern the manufacturing processes e.g.,

- Development of specific electronics, integrated sensors / GPS - based processes in spreading technology integrated with farmers' production systems

- Improvement in spreading technology / applications, electronic identification of fertilizer inside the spreader and automatic spreader setting by volume, speed, width and other spreading characteristics, etc.

- Plant-specific and area - specific registration and data collection

- Targeted, demand-oriented supply of nutrients such as fertigation, subsoil fertilization, crop - and soil - specific straight, complex and speciality fertilizers

- Utilization of the Interaction between fertilization, crop protection and seed, as well as the miscibility of fertilizers with pesticides

- Further development of the production process of fertilizers, improvement of products, product quality, quality management, optimization of logistics as well as storage capacity and capability

Fig. 15 Technological innovation opportunities in plant nutrition and fertilization

of fertilizers, and their quality management. The further development of spreading technology for mineral and organic fertilizers, including electronics and the use of data, along with a more selective, needs-based supply of nutrients for plants, are receiving more and more attention, as are interactions with other input factors. For the companies in the fertilizer industry, it still remains necessary to optimize production processes and work on the supply of improved, tailored products with the corresponding product quality, as is the case with K+S.

Opportunities for organizational innovation can be found in the sphere of electronics and networking with other information systems within the agricultural sector, both over time and in relation to the soil, plants and crop yields. The more targeted use of organic residues, while taking account of critical substances, is of increasing importance. Changed scales and structures in the agricultural sector are resulting in a more differentiated demand for spreading equipment, e.g., for large or specialized farms. As a result of the further growth in knowledge, the need for knowledge transfer is becoming a key focus, and in particular the need for special advice and the necessary acceptance of new developments. Opportunities on the marketing and sales sides are resulting from globalization in general and thus leading to a higher level of corporate risk being involved in the fertilizer market. In addition to this, agriculture's upstream sector is undergoing structural change and the trade has to expect changes in sales and distribution possibilities as a result of the greater use of electronic media. The specific services provided by the sector's market partners in the service area (e.g., in relation to customer loyalty) and modern logistics are having a considerable impact on market activity. This is complemented by tailored products for specific fields of application. Media and personal consulting act as a supporting element in the market, so that e.g., the flow of goods can be tracked and the application of a K+S's own products monitored. Last but not least, as a result of the retraceability of flows of materials, the integration of fertilizers and fertilization strategies in the agri-food chain will become increasingly necessary, so that risks to consumers and manufacturers can be detected and remedied at an early stage.

5.2 Adoption and Diffusion Processes in Agriculture

Opportunities for innovation are offered by a number of new development possibilities in a sector. However, in order to derive a benefit from an innovation, it is becoming increasingly important not only to examine the economic viability of an innovation, but also the necessary distribution of the innovation, for example, in agriculture (Fig. 16) (Rogers, 1995). Adoption, i.e., the taking over, of an innovation depends to a considerable extent on the people as well as on the complexity of the innovation, along with its compatibility with the existing production system of the farmer. Communication, the quality of the information, the use of opinion-leaders and the trustworthiness of consultants and the sales force determine whether an innovation will be adopted. Ultimately, social and political acceptance may hinder adoption, as can currently be seen in the case of green gene technology. On the other hand, innovations are adopted more quickly if they expand

Adaptation depends on:

- Complexity of innovation for the farm and the farmer
- Compatibility in the production systems, testing and observation
- Relative advantage, technical benefits and economic risk
- Communication and available information as well as their qualities
- Use of opinion leaders and trust in consultants and traders
- Social and political acceptability and social significance

Diffusion depends on:

- Communication through mass media, information channels and exhibitions, as well as practical demonstrations
- Experience with previous innovations and their success
- Success and reputation of the lead users / innovators
- Economic weight in farm operations and the risk involved
- Risk aversion and reversibility of such a decision
- Social change in the industry and the society

Fig. 16 Adaptation and diffusion of innovation in agriculture

or add to an existing production system, as the development of bioenergy shows (Langert, 2007). The diffusion of an innovation, i.e., its market penetration, and the adoption of an innovation over time precisely follow the classic maturity curve. This is shaped by the first steps in adoption and implementation. It results initially from the early experience of the first users and from the information channels. Due to risk aversion on the part of farmers to taking on innovations, the economic effects of a possible failure and the possible reversibility of decisions are of relatively great importance. At the same time, the political climate and a faster rate of social change influence the time needed for an innovation to penetrate the market in any sector.

According to agricultural literature, it was previously assumed that, for example, in the case of special crops, the diffusion period totaled about 30 years and that this has now diminished due to better communication (Bohnemeyer, 1996). Comparable periods can be identified in relation to the diffusion of the use of mineral fertilizers during the past century. From early on, the fertilizer industry (as in the case of K+S AG's predecessor companies, for example) therefore already undertook its own field tests and used its own agricultural extension service since about 1890, but still took decades to convince farmers of the benefits and necessity of using mineral fertilizers, as the history of K+S shows (K+S, 2006).

5.3 Challenges in Network Management

If innovations are to be utilized quickly, it is necessary to turn our attention to those who "dig their heels in" and resist the adoption and diffusion process in the agricultural sector. Cooperation between different partners in the agri-food chain and the related information chains in the form of networks offer a series of opportunities to improve the acceptance of innovations in agriculture and accelerate their diffusion and adoption. As outlined earlier, there is a complex network in the agri-food business, which today is already being utilized intensively by many participants, such as the DLG (DLG, 2009). At this level, farmers, consulting institutions, representatives of the upstream industry (e.g., K+S), representatives of the food trade and retail and wholesale trade are collaborating. In cooperation with professional associations and the companies along the agri-food chain, in future too there will be a series of opportunities to intensify this sector-oriented network management and thus implement many specific innovations more quickly and successfully. But there are also opportunities for specific open networks of companies of the fertilizer industry and other groups in the agri-food chain.

6 Summary

Innovations in the agri-food chain are gaining greatly in importance in today's knowledge society. They form the basis for the sustainable success in value creation of companies and thus secure their future on the market. Taking the example of the fertilizer product segment of K+S AG, we have examined the importance of plant nutrition as an input in agriculture in depth and, in relation to it, explored the value chain in the agri-food business. The complexity of this chain and the mutual dependencies in the form of product chains reveal the necessity to cooperate in the sector if innovations are to be identified and implemented successfully. Although the example of the fertilizer industry in Germany as a sector has existed since the second half of the 19th century, and thus for around 150 years, a close examination of the plant nutrition and fertilization sector in agriculture shows further biological, technical, organizational and marketing-side opportunities for innovation, which make up a broad field of possibilities that are yet to be tapped. Even if, from the historical point of view, agriculture is, to a certain extent, torn between tradition and progress, the activities of the DLG and cooperation with partners in the agri-food chain show, for example, that the networks that exist here have, over recent decades, had a very positive impact on the adoption and diffusion of innovations in agriculture. The participation of various interest groups, including employees of the fertilizers companies (such as K+S) for example, has increased the transfer of knowledge and the use of innovations by farmers and offers many opportunities for the sustainable positioning of companies in their markets.

References

Abel, W. (1978). *Agrarkrisen und Agrarkonjunktur*. Hamburg.
Asenso-Okyere, K., & Davis, K. (2009). *Knowledge and innovation for agricultural development*. Washington: IFPRI.
BASF. (1990). *Chemie der Zukunft*. Ludwigshafen.
Bohnemeyer, A. (1996). *Innovationsadaption und –diffusion in der Landwirtschaft*. Münster.

Christen, O. (2006). *Entwicklungen, Probleme und Konzepte für den Pflanzenbau von morgen*, ILU-Heft 12, Bonn.
Columella, L. I. M. (1981). Zwölf Bücher über Landwirtschaft (De Re Rustica Libri Duodecim), Reprint, Vol. 1, München.
COMPO. (2005). *Wachstum braucht Sicherheit*. Münster.
DBV. (2009). *Situationsbericht 2009*. Berlin.
DLG. (2009). *Landwirtschaft 2020*. Frankfurt.
Ermann, U. (2001). Regional essen? Wert und Authentizität der Regionalität von Nahrungsmitteln, AGEV Tagung.
Fraunhofer Gesellschaft. (2005). *Food Chain Management*.
IVA. (2005). Pflanzenschutz heute, "Journalist", April 2005.
Hauschildt, J. (1997). *Innovationsmanagement*. München.
Knittel, H., & Albert, E. (2003). *Dünger und Düngung*. Bergen.
K+S. (2006). *Wachstum Erleben*. Kassel.
K+S. (2008). *Unternehmens- und Nachhaltigkeitsbericht 2007*. Kassel.
Küster, S., Schuhmacher, M., & Werner, B. (2008). *Open innovation in innovation networks*. Mannheim.
KWS. (2007). *Erfolg kann man säen*. Göttingen.
Kummer, K-F., & Zerulla, W. (2006). *Ablauf und Eigenarten der Wirkstoffforschung in der Pflanzenernährung*. Limburgerhof.
Langert, M. (2007). *Der Anbau Nachwachsender Rohstoffe in der Landwirtschaft Sachsen-Anhalts und Thüringens*. Halle-Wittenberg.
Marcinowski, St. (2009). *Welternährung 2020*. Berlin: Agrarforum.
Mengel, K., & Kirkby, E. A. (1982). *Principles of plant nutrition*. Bern.
Mensch, G. (1977). *Das technologische Patt*. Frankfurt.
Nestlé. (2008). *Bericht zur gemeinsamen Wertschöpfung*. Vevey.
Porter, M. (1985). *Competetive advantage*. London.
PrecisionAg. (2009). PrecisionAg in the UK 2009, http://www.precisionag.com.
Reinwald, R., & Piller, F. (2006). *Interaktive Wertschöpfung*. Wiesbaden.
Rentenbank. (2006). *Landwirtschaftliche Rentenbank, Organisatorische und technische Innovationen in der Landwirtschaft, 21*, Frankfurt.
Rogers, E. (1995). *Diffusion of innovations*. New York.
Rösener, W. (1993). *Die Bauern in der Europ*. München: Geschichte.
Satzger, G., & Neus, A. (2009). *Principles of collaborative innovation*.
Schmitz, P. M. (2008). *Bedeutung des AgriFoodBusiness für den Standort Deutschland*. Gießen.
Siemes, J. (2002). Outlook for Potash, "Fertiliser Round Table" Conference, Charleston, SC.
Theuvsen, L., Spiller, A., Peupert, M., & Jahn, G. (2007). *Quality management in food chains*. Wageningen.
YARA. (2009). N-Sensor for site-specific variable application of nitrogen, http://www.yara.com.
Wagner, J. (2005). *Vertrauen in Netzwerkbeziehungen*. München.
Wahren, K.-H. (2003). *Erfolgsfaktor Innovation*. Berlin.
Welthungerhilfe. (2000). *Jahrbuch Welternährung*. Frankfurt.
Zimmer, Y., Nehring, K., Moellmann, T., & Witte, T. (2007). Agri Benchmark, Cash Crop Report 2007, FAL, Braunschweig.

Part III
Capital Markets, Finance and Innovation Performance

Part III

Modern Valuation Approaches for Corporate Innovation Activities

Andreas Krostewitz and Martin Scholich

Contents

1 Foreword 263
2 The Valuation of Corporate Innovation Activities 264
3 Traditional Valuation Approaches and Corporate Innovation Activities 266
   3.1 Introduction 266
   3.2 Cost Approach and Costs Related to Corporate Innovations 267
   3.3 Market Approach and the Marketability of Innovation Activities 268
   3.4 Income Approach and the Valuation of Cash Flows of Innovation Activities . . 268
   3.5 The Certainty Equivalent Method 270
4 Modern Valuation Approaches and Corporate Innovation Activities 271
   4.1 Real Option Approach and Innovation Opportunities 271
   4.2 Risk Compound Valuation Approach for the Valuation of Corporate
       Innovation Activities 274
5 Conclusion 277
References 278

1 Foreword

Due to shortened technology and product life cycles, companies need to introduce new products and services in shorter periods. Companies that best meet changing demands or take markets in new directions are more likely to have long-term success.[1] Furthermore, companies need to attract external capital to finance operations and projects. More and more, external financiers assess companies on their innovativeness due to the fact that these companies promise the highest return on their investments in the long run. Therefore, companies need to be innovative to keep up

---

A. Krostewitz (✉)
PricewaterhouseCoopers AG, Marie-Curie-Str. 24–28, 60439, Frankfurt am Main, Germany
e-mail: andreas.krostewitz@de.pwc.com

[1] See Schumpeter (1942).

with changing market conditions and to attract external capital. Both challenges demand an evaluation of the companies' innovation activities to allocate corporate resources to the most promising innovation projects and to communicate the value impact on the company. But the valuation task faces two problems. First, the complexity of corporate innovation activities makes it difficult to evaluate them as they apply to a wide variety of technical, product, process or administrative aspects. Besides, innovations like process innovations are mostly bound up with the company so that it is hard to separate them from the company. Second, it depends on the valuation context which kind of valuation method is most appropriate. For internal management purposes a subjective valuation with respect to internal compound effects to other activities and processes is adequate. External communication purposes require an objective valuation approach. Before starting a valuation of corporate innovation activities, the management needs to identify the relevant innovation activity and the valuation purpose. The valuation purpose determines the method to be used and the aspects to be considered.

The following article discusses the valuation of corporate innovation activities in the light of the aspects described above. Traditional and modern valuation approaches are analyzed as to how they comply with the requirements of the valuation task. Therefore, Section 2 focuses on the key aspects and facts, which determine corporate innovations. Section 3 gives an overview over traditional valuation methods and shows what problems might come up applying these methods and how they are applicable to a certain valuation task. Section 4 introduces new valuation approaches, which might better capture the specifics of corporate innovations in a more appropriate way. The paper closes with a summary of the findings.

2 The Valuation of Corporate Innovation Activities

Being innovative means understanding, evaluating and being able to manage innovation activities within a company. Corporate innovation activities can, therefore, be defined as development and implementation of new ideas by people who engage in transactions with others within an institutional order.[2] Innovation can be understood as the creation, implementation and introduction of new products, services or processes on the market.[3] Product innovations mean completely new products or new features, which differentiate a product from competitive products. Service innovations relate to stand-alone or product-related intangible activities. Process innovations correspond to refinements or development of work flows to enhance efficiency and effectiveness.[4] An activity is called a radical innovation when it has never been produced or sold by anyone before; hence a market does not exist yet. An innovation is called incremental when it changes or enhances existing products, services or processes to meet changing demand.

---

[2] See Ven (1986), p. 591.
[3] See Scholich and Robers (2007).
[4] See Cooper (2002).

Corporate innovation activities can be considered as processes, which can be divided into stages and phases. Innovation activities are not something static but permanent within the company. In fact, they can be considered as processes bound up with the company, which follow similar patterns. Several process models describe the phases and stages of innovation activities.[5] Following the Stage-Gate Process of Cooper and Kleinschmidt (1990), the innovation activity can be divided into distinct stages and should be separated by management decision gates. Every innovation activity starts with an idea and ends – when the idea is followed up until the end – with the launch of a new product, service or company process.[6] Preceding each stage there should be a decision point or gate, which should serve as a Go/No-Go decision point. At the gates, mediocre projects should be stopped and resources should be allocated to other promising projects.

The main characteristic of corporate innovation is its uniqueness. A project can hardly be compared to other existing projects as every innovation project has its own complexity. For radical innovations, markets or comparable products or services normally do not exist yet. The non-marketability means that there is no active market, on which homogeneous products are traded between informed buyers and sellers.[7] For incremental innovations of existing products or services, comparable markets generally exist. Depending on the stage of the innovation process the information complexity of an innovation can differ significantly. In the early stages of innovation activities there is a high degree of incomplete information. The scope of the innovation and its influence on corporate success can hardly be estimated. In later stages, when the innovation gets more concrete, the information basis may improve that it becomes easier to estimate the impact of an innovation on corporate success. Furthermore, innovations can be tangible or intangible in nature. Intangible innovations have no physical substance, e.g. a process or service innovation. If the innovation project results in a product or patent, innovation becomes tangible.

A corporate innovation project needs to be identified and separated from the rest of the corporate innovation activities to be evaluated. Identifiability means that it is possible to estimate future cash flow distributions for an innovation project. Principally, this is possible in each stage of the innovation process. Due to an improved information basis as the innovation process proceeds, the probability distribution for the related cash flows may change. Furthermore, it is irrelevant for the valuation purpose whether the result of the project is tangible or intangible.[8] The only specific aspect which is relevant for the valuation of the innovation project is

---

[5] For example the Phase-Review Process by Hughes and Chafin (1996), p. 92; the Stage-Gate Process by Cooper and Kleinschmidt (1990), p. 46; the Phase Models by Ulrich and Eppinger (1995), Thom (1992), Brockhoff (1999) or Witt (1996); the Value Proposition Cycle by Hughes and Chafin (1996), p. 93.

[6] The way to this point can be divided into a series of activities (stages) and decision points (gates). Stages on the way to the launch are preliminary investigation (stage 1), detailed investigation (stage 2), development (stage 3), testing a validation (stage 4) and launch (stage 5).

[7] See Pellens, Fülbier, Gassen, and Sellhorn (2008), p. 287; Longstaff (2001, 1995); Kahl, Liu, and Longstaff (2003).

[8] However, this is relevant for the estimation of the cash flows.

the marketability of the cash flows.[9] It depends on this aspect whether a market value can be determined or not. The non-marketability of the innovation activity urges the employment of valuation approaches, which can handle this characteristic. Before starting a valuation it is necessary to define what is meant by risk or uncertainty of the innovation project or the related cash flows and how they should be measured.[10]

For the valuation of corporate innovation projects, the related cash flow, its probability distribution and the valuation context are relevant. The valuation context determines what kind of risk needs to be considered and which value concept (subjective or objective) has to be assumed. For internal management purposes, the risk compound to other projects and assets is relevant. During the valuation process these compound effects and their impact on the value of the project and the company have to be identified. Clarification is needed as to how the corporate innovation changes the risk structure of the company. The determined value is subjective and depends on entity-specific aspects and on the individual risk attitude of the investor. For external communication purposes, a market value (e.g. fair value) has to be determined. Therefore, only the systematic risk of the project is relevant. Entity-specific compound effects or unsystematic, project-specific risks do not matter. An objective value concept has to be applied, which does not depend on individual risk attitudes. In general, the determination of a market value will be difficult when no liquid market exists. This is the normal situation for innovation projects, especially for radical innovations.

Finally, the decision which valuation approach to apply depends on the valuation context and the marketability of the innovation activity. Every other innovation-specific aspect (e.g. its tangibility) can be reflected in the respective cash flow distribution, which makes the innovation project – theoretically – assessable like any other project or asset. Even if these approaches do not seem to be appropriate to evaluate radical innovation projects, they might be applicable for incremental innovations.

3 Traditional Valuation Approaches and Corporate Innovation Activities

3.1 Introduction

The valuation of corporate innovation activities and the decision for an appropriate approach is difficult in the following sense. First, some corporate innovation projects are so intimately bound up within the company, e.g. process innovations, that it is difficult to identify and estimate the relevant cash flows precisely. Second, especially in the early stages of the innovation process, the weak information basis complicates the estimation of cash flow distributions. The impact of the project on the company

---

[9] See Mayers (1973); Brito (1977).
[10] See Kürsten (2007); Kürsten and Straßberger (2004).

and other assets can hardly be foreseen. Third, radical innovation projects are non-marketable so that an objective market valuation is not possible without specific assumptions. The valuation of cash flows of a single innovation project, however, does not differ from the valuation of a cash flow of a tangible or (other) intangible asset in principal. Therefore, the knowledge and experience gained in the valuation of tangible and intangible assets can be applied to the valuation of innovation projects.

Traditional valuation approaches, which are used to evaluate intangible assets, can be a starting point. A comparable complexity of intangible assets makes it reasonable to use these special valuation approaches for the valuation of corporate innovations projects. However, they prove to be problematic in consideration of the non-marketability of corporate innovation projects. Whereas these approaches try to determine a market value of an asset, other methods like the certainty equivalent approach, aim to determine a subjective value. The following approaches will be reviewed for their applicability:[11]

- Cost Approach
- Market Approach
- Income Approach
- Certainty Equivalent Approach.[12]

The aim of the first three approaches is to ascertain an objective fair value. The last approach is a representative, but theoretically doubtful method for subjective valuation. The following sections explain these approaches and discuss their applicability to innovation projects.

3.2 Cost Approach and Costs Related to Corporate Innovations

The Cost Approach determines the value of an asset according to the costs needed to reproduce it with a utility equivalent asset (reproduction method) or duplicating it (replacement method).[13] The cost approach is based on the assumption that a buyer is not willing to pay more for an asset than the amount needed to reproduce or replace it. The approach is applicable when the costs involved can be reliably estimated and if these costs adequately represent the future benefits of the respective asset.

If applied to corporate innovation activities, the cost approach could determine the costs of reproducing or replacing the innovation. However, the main characteristic of an innovation is its uniqueness regarding its future economic benefits. In

---

[11] See Jäger and Himmel (2003).
[12] See for example Schwetzler (2000).
[13] See Scholich, Mackenstedt, and Greinert (2004), p. 497; Mackenstedt, Fladung, and Himmel (2006).

general, it is not possible to reproduce or replace the innovation with existing, comparable products, services or processes on the market. This is especially true for radical innovations. If it could be reproduced or replaced, it would not be an innovation in the sense described. Furthermore, the aim of corporate innovation is to generate additional utility for the customer or the company. So, it might be hard to find an equivalent asset or project, which generates an equivalent utility. The Cost Approach is not applicable for corporate innovation activities as long as comparable projects to replace or reproduce it cannot be found.

3.3 Market Approach and the Marketability of Innovation Activities

With the Market Approach, the value of an asset can be derived from prices obtained for comparable assets in an open, liquid market. The comparison with market transactions presumes that prices paid for comparable assets in prior transactions are good indicators of the value of the respective asset. This approach is based on observable market prices and comparable market transactions.[14] Often adjustments are made to reflect individual facts and circumstances of the asset in question.

The definition of a corporate innovation states that it is the first of its kind in the market. Therefore, the main problem of the Market Approach is to find comparable market transactions. Corporate innovation activities are unique and have their own specific complexity and risk. Due to non-marketability of radical corporate innovation, it is hard to find comparable transactions in the market and a derivation of reasonable market multiples is extremely difficult. Valuations based on multiples may be inappropriate or impossible. In the case of incremental innovations, comparable assets or transactions might be easier to find and the Market Approach might be better applicable in these cases. Nevertheless, the theoretical basis of the Market Approach using multiples is weak.[15]

3.4 Income Approach and the Valuation of Cash Flows of Innovation Activities

The Income Approach estimates the value of an asset by calculating the present value of future cash flows, which are expected to be generated by the asset.[16] An appraiser has to perform two main tasks. First, he has to estimate future cash flows

---

[14] See Matschke and Brösel (2007); Mackenstedt, Fladung, and Himmel (2006); Löhnert and Böckmann (2005); Scholich, Mackenstedt, and Greinert (2004); Jäger and Himmel (2003); Buchner and Englert (1994).

[15] See Krostewitz (2008) with further references.

[16] See Mackenstedt, Fladung, and Himmel (2006); Scholich, Mackenstedt, and Greinert (2004), p. 498; Jäger and Himmel (2003); Reilly and Schweihs (1999).

attributable to the asset over its estimated remaining useful life time. Second, the corresponding discount rate has to be determined. When using the income approach, the determination of the asset-specific cash flow stream is the main challenge. Different methods for determining attributable cash flows have been established:

- Relief-from-Royalty Method
- Multi-Period Excess-Earnings Method
- Incremental Cash Flow Method.

With the *Relief-from-Royalty Method*, the appraiser determines the royalty savings attributable to the owner of the asset. The ownership of an asset makes it unnecessary to obtain a license for a comparable asset and, therefore, reliefs the owner from paying royalties. This method is often used to evaluate trade names, brands or production technology. The reliability of the royalty rates applied increases, if market based royalty rates between third parties are available. Ideally, a market is needed, on which the asset is traded or licensed out. The royalty savings are then discounted by an appropriate asset-specific discount rate.

The *Multi-Period Excess-Earnings Method* is a residual income model and starts with the cash flow which is somehow related with the asset to be valued. Generally, an asset is generating cash flows in combination with other assets and very seldom purely on its own. In order to separate the cash flows specific for the asset under consideration, other complementary assets have to be identified and the cash flows related to these assets, the so-called Contributory Asset Charges (CAC), e.g. leasing rentals for land, machinery, buildings, workforce, other intangible assets or net working capital, have to be subtracted from the combined cash flow stream. The resulting asset-specific cash flows will be discounted with a corresponding discount rate.

The *Incremental Cash Flow Method* determines the additional cash flow, which can be generated by using the asset in question. The value of the asset can be derived comparing the overall cash flows of the company with and without the asset. Cash flows resulting from cost savings or higher selling prices are determined by the valuation of process patents, formulas or production processes. Finally, the resulting incremental cash flows will be discounted with a risk-equivalent discount rate.

The transfer of the described Income Approaches to corporate innovation activities turns out to be difficult. Again, the non-marketability of radical corporate innovation activities is problematic. The Relief-from-Royalty Method assumes that the asset is marketable; otherwise market-based royalty rates do not exist. The application of capital market-related discount rates in all three methods theoretically implies the marketability of the innovation activity in question.[17] Especially, when discount rates are derived applying the Capital Asset Pricing Model (CAPM). This model assumes a perfect capital market where all assets are traded. Even if the

---

[17] In general, the weighted average cost of capital (WACC), adjusted by asset specific factors, is applied.

corporate innovation activity would be marketable, the application of the CAPM presumes that only systematic risk is relevant. Unsystematic, innovation-specific risk is not accounted for. This might be correct for a market valuation purpose but not for subjective valuation. In general, the non-marketability of corporate innovations prohibits a pure market valuation. Nevertheless, the described methods might be helpful in estimating the relevant cash flows of the innovation project in question. The Multi-Period Excess-Earnings Method and the Incremental Cash Flow Method disclose ways of separating the relevant cash flow stream of the innovation project from the company. As a result, the determination of market values according to the methods described has to be done with care and in regard to its limitations, e.g. the availability of relevant data.

3.5 The Certainty Equivalent Method

The Certainty Equivalent Method determines time-specific certainty equivalents $CE(X_t)$ of uncertain future cash flows for the periods $t=1, ..., n$ under consideration of the utility function u of the investor.[18] It is a representative for subjective valuation approaches. The certainty equivalent is the certain amount which has the same utility as the uncertain cash flow stream in time t.[19] The value of the cash flows V_0 in $t=0$ is derived by discounting the time-specific certainty equivalents with the risk free rate r_f:[20]

$$V_0 = \sum_{t=1}^{n} \frac{CE(X_t)}{(1+r_f)^t}.$$

Certainty equivalents are derived using Bernoulli-certainty equivalents and the approach assumes a risk-averse investor whose certainty equivalent is lower than the expected value of the uncertain future cash flows. The risk attitude of the investor is represented in the certainty equivalent for the relevant cash flows.[21]

Although it might be difficult for the investor to determine certainty equivalents for corporate innovation projects, the idea of subjective valuation, however, seems appropriate for valuing non-marketable assets, e.g. for internal management purposes. Despite these advantages the certainty equivalent method has deficits with respect to its theoretical basis. First of all, discounting of certainty equivalents with a risk-free discount rate is not covered by the expected utility theory, which forms the theoretical basis of the certainty equivalent method.[22] Second, the application

---

[18] See Drukarczyk and Schüler (2007), p. 68.
[19] See Drukarczyk and Schüler (2007), p. 51; Mandl and Rabel (2005), p. 62.
[20] See Kuhner and Maltry (2006), p. 135 f.; Schultze (2003), p. 263.
[21] See Spremann (2002), p. 316.
[22] See Kürsten (2002, 2003); Schwetzler (2002); Wiese (2003); Diedrich (2003).

of a risk-free rate implies the existence of a capital market, in which the cash flows are embedded. In that sense, the valuation cannot be seen as purely subjective but depends on possible portfolio transactions, which might be carried out by the investor in the assumed capital market.[23] Due to these theoretical weaknesses, the certainty equivalent method will not be discussed further in this paper.

4 Modern Valuation Approaches and Corporate Innovation Activities

4.1 Real Option Approach and Innovation Opportunities

4.1.1 Introduction to Real Option Valuation

Corporate innovation projects set up additional opportunities within the company. Therefore, it is reasonable to apply real option valuation methods to value innovation projects and the opportunities related to them. Real option valuation is a means to determine market values of real options representing action alternatives within a company. Applied to corporate innovation projects, the real option approach evaluates the opportunities related to innovation activities under certain assumptions.

In general, the real option approach is based on an analogy to financial option valuation.[24] A financial option is the right but not the obligation to buy or sell a certain underlying asset at a determined time or within a period at a prior fixed price. Many investment projects within the company have comparable characteristics, especially the existence of a decision situation. In that sense even corporate innovation projects can be seen as real options as they open up for new opportunities of the company, which can be realized, but do not have to be. The flexibility, which results from successful innovation projects, can be valued, because it sets up a new decision range for the company. The real option approach is especially practicable in cases, where decisions have to be taken in high risk environments. This is the case for companies which pursue intensive research and development. Comprising future opportunities related to innovation projects delivers more transparency to the valuation process.[25]

There are numerous types of real options in connection with corporate innovation projects. Growth and expansion options, on the one hand, cover all opportunities related to the future development of businesses. Termination and exit options, on the other hand, reflect the possibility of stopping a project in the case of an unfavorable

---

[23] See Kürsten (2003).

[24] See Hommel and Pritsch (1999); Rams (1999); Schwartz and Trigeorgis (2001); Spinler and Huchzermeier (2004); Trigeorgis (1996); Alvarez and Stenbacka (2006); Dixit and Pindyck (1994); Scholich and Wulff (2002); Bucher, Mondello, and Marbacher (2002); Baecker, Hommel, and Lehmann (2003); Hartmann (2006).

[25] See Bonduelle, Schmoldt, and Scholich (2003), p. 5.

development. Compound options are characterized as options on an option. When an option is exercised, a new option will arise.

4.1.2 Theoretical Background of Real Option Valuation

The starting point of the option pricing theory is the construction of a portfolio of risky assets and a risk-free asset, which exactly duplicates the cash flow profile of the option. In an arbitrage-free world, the cash flow of the option and the duplicating portfolio must have the same price. Based on this fundamental principle a risk-neutral valuation is possible, where individual risk attitudes are irrelevant. An advantage of the risk-neutral valuation is the possibility to by-pass the determination of risk-equivalent discount rates. This valuation principle is transferred to the valuation of real options.

Option pricing models can be divided into analytical and numerical approaches.[26] *Analytical approaches* deliver closed solutions for special valuation problems. Changes in the value of the underlying asset are modeled as continuous stochastic processes. The most popular approach is the Black-Scholes formula for the valuation of European call and put options.[27] However, real options are usually not "plain vanilla" European options, but far more complex. The advantages of analytical approaches are their easy application and their apparently low presumptions. In fact, the restrictive model assumptions of the analytical approach, e.g. a perfect and complete capital market and a liquid market for the underlying, confine the application of analytical methods for real option valuation.

Numerical approaches, e.g. Monte-Carlo simulations or binomial trees, are based on an approximation of the stochastic price processes of the underlying asset. They are intuitive and applicable to practical valuation situations. Complex valuation problems can be modeled as well.

During an option valuation based on the Monte-Carlo simulation, the expected value of an option cash flow process is determined, based on stochastic realizations of a risk-adjusted price process. Its present value represents the fair price of the option. This approach starts with the identification of value drivers. Based on empirical studies or subjective appraisals for each value driver, the probability distribution is specified. These data are the basis for the simulation of the cash flow process. In each simulation run, the parameters are determined by chance. After sufficient simulation runs, the frequency distribution of the results can be estimated and conclusions on the probability distribution of the cash flows and the expected present value can be drawn. An advantage of the Monte-Carlo simulation lies in the freedom of modeling and parameter estimation. It is possible to model the relationships between specific value drivers and the valuation output and it is possible to simulate complex situations. The approach is limited to the point where the number

---

[26] See Baecker et al. (2003), p. 26.
[27] See Black and Scholes (1973).

of dependencies increases and probability distributions of value drivers cannot be reliably estimated.

The real option valuation with binomial trees divides the time line in periods of similar length and estimates the price process of the underlying asset through discrete value variations. In the well-known binomial model of Cox, Ross, and Rubinstein (1979), the value can only be increased or decreased by an up or down factor. Cox, Ross, and Rubinstein assume increases and decreases of the same value on the knots of the tree. Based on the tree for the price development of the underlying asset, a decision tree can be modeled for the option values in each knot. The aim of a decision tree is to analyze whether it is advantageous to exercise the option before its duration ends. Expected option values are determined with risk-neutral probabilities and discounted with the risk-free rate. In its principles, the procedure requires limited methodological knowledge. The decision tree analysis demands the specifications of the underlying assets and the risks and chances related to the decision situation. A disadvantage of this numerical approach lies in the fact that the number of states relates to the number of time steps.

4.1.3 Application of Real Option Valuation to Innovation Activities

From a theoretical perspective the application of real option valuation approaches for the valuation of innovation projects seems appropriate. Corporate innovation projects set up new opportunities for corporate development, which can be realized, but do not have to be. The methods described above to determine future cash flows, e.g. the incremental cash flow method, can be used to approximate the cash flows of innovation projects. The representation of cash flows in decision trees discloses "up-side potentials" and "downside risks" of the innovation project. Despite of these advantages related to the real option approach on innovation projects, some limitations to the application exist, which will be discussed next.

In general, the application of the real option approach to innovation projects and the analogy to financial options has their limits. The valuation of financial options assumes the marketability of the underlying asset in a liquid market. It assumes that a market price for the underlying asset exists. However, especially radical corporate innovation projects are characterized by their novelty and incomparability with marketable assets. In this case an (traded and liquid) underlying asset may be hard to find. Moreover, risk-neutral valuation assumes that the cash flow of the option can be completely duplicated by marketable assets. In respect of the assumption of a complete market, individual risk attitudes are not accounted for and risk-neutral valuation is feasible. If the cash flow profile of the option cannot be entirely duplicated, risk-neutral valuation will not hold. A partial duplication of the cash flow profile can be due to incomplete information about the underlying innovation project or the related real options. Additionally, the investor's market access can be limited, corporate innovation projects are characterized by asymmetric information and the influence of the project on the company is hard to predict in early stages of the project. However, if it is not possible to duplicate the cash flow profile of an innovation real option in total, a true market value cannot be determined.

Thinking in alternatives or options and modeling the related risk profiles can be helpful in evaluating the advantages of innovation projects. However, the determination of the value of real options related to innovation projects must be applied with care. The Real Option Approach ignores the specific characteristics of innovation projects, which are their non-marketability (at least radical innovations) and the incomplete information about the related cash flows. Besides, it suggests a market valuation where it is actually not possible.[28]

4.2 Risk Compound Valuation Approach for the Valuation of Corporate Innovation Activities

4.2.1 Introduction to the Risk Compound Valuation Approach

The risk compound valuation approach allows the valuation of uncertain cash flows under the condition of an imperfect capital market. The imperfection of the capital market may have different sources. First, capital market transactions may not be sufficient in duplicating the cash flows. Second, there may be assets, which cannot be explained in full by capital market transactions or for which a market does simply not exist (non-marketable assets). As a result, a sound market value for a specific asset cannot be determined in practice. Instead, the evaluation has to be carried out with respect to individual preferences and risk attitudes. In case the investor has a portfolio of marketable and non-marketable assets, additional covariance risk between the cash flows of the considered asset has to be considered as well. The cash flows of the assets, which form the consumption base of the investor, are evaluated in consideration of the individual risk context and investor's individual preferences. The underlying idea of this approach goes back to Mayers (1972, 1973) and Brito (1977) who have examined the influence of non-marketable assets on the valuation of marketable assets. Wilhelm (2005) transferred the idea to the special situation of company valuation.

The valuation model assumes an investor who wants to maximize the expected utility of his consumption plan. The consumption plan can be realized by capital market transactions, non-marketable assets with a cash flow h or the respective valuation object with the cash flow X. It is assumed that the investor has only limited access to the capital market. The consumption associated with a capital market transaction results from risky, marketable assets and a risk-free investment opportunity with a risk-free rate r_f. Non-marketable assets are assets which are only duplicable with marketable assets to a certain extent. Trading restrictions exist for assets with special characteristics, e.g. social transfer payments, firm-specific know-how or lien-burdened assets. The investor who does not yet own the valuation object can acquire the object by paying a certain amount of cash. It is assumed that the investor

---

[28] An alternative approach can be found in Miao and Wang (2007) who expand the standard real option approach to an utility-based real option approach.

is willing to pay the amount V_X to get from the situation without the object to the situation with the object without reducing his utility. V_X can be understood as the value of the asset and considers the asset's cash flows and all portfolio effects between the cash flows of the marketable assets, the non-marketable assets and the object. The risk compound valuation approach allows the determination of the amount V_X taking into account limited access to the capital market and the existence of non-marketable assets. Under the assumption of a one-period valuation approach and normally distributed stochastic cash flows, the preferences of the investor refer to the expected values and variances of the stochastic cash flows. Under the additional assumption of an exponential utility function with constant risk aversion λ, the changes in utility caused by the asset to be valued can be quantified by the so-called Hybrid model.[29] The capital market transactions of the investor are limited to a static portfolio investment in risky and risk-free assets. It can be shown that the optimal portfolio of each investor consists of an investment in the so-called Tobin fonds and individual hedge portfolios. The Tobin fonds is identical for each investor, but the amount invested reflects the investor's utility function.[30] The individual hedge portfolios with cash flows w_h and w_X minimize the risk of the asset to be valued and the non-marketable asset (measured as the variance of the cash flows). It is not a complete hedge and some residual risk remains for the investor. However, with this so-called "covariance hedge" all covariance risks to the marketable assets are eliminated. In this situation, V_X is the certain amount the investor has to pay in $t=0$ for this asset to remain on the same level of utility as before the transaction:

$$V_X = \pi(w_X) + \frac{1}{(1+r_f)}\left\{E(X-w_X) - \frac{1}{2}\lambda\left[\text{Var}(X-w_X) + 2\text{Cov}(X-w_X;h-w_h)\right]\right\}.$$

The value of asset X depends on the residual risk $\text{Var}(X-w_X)$ of this asset and the correlation of the residual cash flows of this asset and the non-marketable asset. The value of the asset under consideration, V_X, the investor is willing to pay is composed of an objective and a subjective part. The objective part, $\pi(w_X)$, represents the market value of the duplication portfolio and can be interpreted as the market value of the respective cash flow of the valuation object. The individual part (swung brackets) depends on the investor's individual preferences symbolized by the expected value of the residual part of the cash flows, $E(X-w_X)$, reduced by a risk term. The risk term depends on the individual risk aversion of the investor, λ. The risk term expresses the relevant risk and the valuation context of the investor. Especially, the residual risk of the non-duplicable part of the cash flow, $\text{Var}(X-w_X)$, and the covariance risk between the non-duplicable parts of the non-marketable asset and the valuation object, $\text{Cov}(X-w_X;h-w_h)$, are of specific relevance for the valuation. This part can be considered to be the object-specific risk. The influence of this object-specific risk depends on the ability of the investor to hedge the risk of asset X by capital market transactions. If a perfect duplication would

---

[29] See Eisenführ and Weber (2003).
[30] See Tobin (1958).

be possible ($X=w_X$ and $h=w_h$), no residual risk (subjective part), $\text{Var}(X-w_X)$ and $\text{Cov}(X-w_X; h-w_h)$, will be induced.

4.2.2 Application to Corporate Innovation Activities

The risk compound valuation approach can be applied to corporate innovation projects by identifying the relevant cash flow and compound effects to other assets. The approach is applicable to innovation projects, for which a reliable market value based on traditional methods and the real options approach cannot be determined. As the valuation of the innovation project is solely based on its cash flow, a transfer of the valuation approach seems possible.

The central assumption of this transfer is the reduction of the innovation project to its uncertain cash flow X, which is embedded in the valuation context of the consumption plan of the investor. The sole criterion, which determines the value of the cash flow, is the marketability or the non-marketability and the induced risk. As explained above, radical and incremental innovations are distinguished by the degree of marketability. Cash flows of radical innovations might be hard to duplicate, as a market does not yet exist. The cash flows of incremental innovations might be easier to duplicate, since the probability of the existence of comparable assets traded in the market is considerably higher. Residual risk, which remains due to incomplete hedging, and covariance risk towards other assets reduces the value of the innovation project for a risk-averse investor. The residual, non-diversifiable variance risk represents the stand-alone risk of the innovation project, which remains after duplication. The covariance risk towards other corporate assets explains the influence of the innovation project on the risk structure of the company. The residual variance and covariance risk can be interpreted as project-specific risk. The better the cash flow of the innovation project can be duplicated, the less project-specific risk remains.

Consequently, the duplication depends on the information about the cash flow of the innovation project. If more details about the cash flow structure of the innovation project are available, e.g. in later stages of the project process, the duplication can be carried out in more detail and, therefore, to a more accurate valuation. In the early stages of the project, innovation projects can hardly be appraised. The impact on the value of the investor-specific risk term gains more importance, resulting in an increased subjective value component of the valuation.

To sum up, the risk compound valuation approach offers a combination of subjective and objective valuation under consideration of an imperfect capital market. If radical corporate innovation projects are understood as embedded assets within a company, the application of the approach in the valuation practice seems to be promising. Besides, the risk compound approach discloses the innovation project's influence on the risk structure of the company. Risk compound effects between the project and other assets within the company are taken into consideration. Moreover, the innovation project can be evaluated with respect to the risk attitudes of the investor. Despite the fact that this method is convincing from a theoretical standpoint, the practicability for valuing innovation projects has to be further evaluated.

5 Conclusion

The aim of this paper was to discuss the valuation of corporate innovation projects. In the course of the essay, traditional and modern approaches have been analyzed as to how they comply with the valuation task. The results can be summarized as follows.

Corporate innovation can be understood as the creation, implementation and introduction of new products, services and processes in the marketplace. They need to be evaluated to ascertain their value for the company to allocate resources to the most promising projects. The uniqueness, intangibility and non-marketability of corporate innovation projects need to be considered in the valuation process. Especially radical innovations are characterized by the lack of a market, whereas incremental innovations might refer to an existing market. The first task of the valuation process is the identification of the purpose of the valuation and the relevant value concepts (strategic value and market value).

The valuation context determines the valuation method to be used. The cost approach specifies the value of the innovation project by evaluating the costs of replacement or reproduction. Due to the uniqueness of the innovation project, the reproduction or replacement of the project is hardly possible. The market approach compares the innovation project with prior market transactions. As an innovation should be the first of its kind on the market, a comparable transaction might be hard to find. The income approach evaluates the cash flows created by the innovation projects. The determination of the market value of the relevant cash flows depends on comparable marketable assets. The certainty equivalent method aims to determine a subjective value of the innovation project, but has theoretical and especially practical limitations.

The real option approach evaluates opportunities related to innovation projects and is based on an analogy to the valuation of financial options traded in perfect capital markets. In general, the application to innovation projects is limited through the assumption of a marketable underlying on a liquid market. The risk compound valuation approach, however, allows for an imperfect capital market and does, therefore, better describe the reality. The non-marketability of assets and projects can be explicitly accounted for within this valuation approach and the valuation context of the investor is taken into account. The value of the innovation project comprises of an objective and a subjective part, taking into consideration the risk compound of the innovation project to other assets within the company. After all, the practicability of the risk compound valuation approach needs to be investigated.

In summary, the uniqueness, intangibility and non-marketability of corporate innovation projects makes a valuation a demanding task. In respect to the valuation context, only strategic values can be estimated. The determination of market values of radical innovation projects has its limitations due to the lack of a market. Market values of incremental innovations might be ascertainable depending on the existence of a market.

References

Alvarez, L. H., & Stenbacka, R. (2006). Takeover timing, implementation uncertainty and embedded divestment options. *Review of Finance, 10*, 417–441.
Baecker, P., Hommel, U., & Lehmann H. (2003). Marktorientierte Investitionsrechnung bei Unsicherheit, Flexibilität und Irreversibilität – Eine Systematik der Bewertungsverfahren. In: Hommel, U., Scholich, M., & Baecker, P. (Eds.), *Reale Optionen – Konzepte, Praxis und Perspektiven strategischer Unternehmensfinanzierung* (pp. 15–35). Berlin, Heidelberg, New York: Springer.
Black, F., & Scholes, M. (1973). The pricing of options and corporate liabilities. *Journal of Political Economy, 81*(3), 637–654.
Bonduelle, Y., Schmoldt, I., & Scholich, M. (2003). Anwendungsmöglichkeiten der Realoptionsbewertung. In: Hommel, U., Scholich, M., & Baecker, P. (Eds.), *Reale Optionen – Konzepte, Praxis und Perspektiven strategischer Unternehmensfinanzierung* (pp. 3–13). Berlin, Heidelberg, New York: Springer.
Brito, N. O. (1977). Marketability restrictions and the valuation of capital assets under uncertainty. *The Journal of Finance, 32*(4), 1109–1123.
Brockhoff, K. (1999). *Forschung und Entwicklung: Planung und Kontrolle*. Munich: Oldenbourg.
Bucher, M., Mondello, E., & Marbacher, S. (2002). Unternehmensbewertung mit Realoptionen. *Der Schweizer Treuhänder*, No. 9, 779–786.
Buchner, R., & Englert, J. (1994). Die Bewertung von Unternehmen auf der Basis des Unternehmensvergleichs. *Betriebs-Berater, 49*(23), 1573–1580.
Cooper, R .G. (2002). *Top oder Flop in der Produktentwicklung*. Weinheim.
Cooper, R. G., & Kleinschmidt, E. J. (1990). *New products: The key factors in success*. Chicago: American Marketing Association.
Cox, J. C., Ross, S. A., & Rubinstein, M. (1979). Option pricing: A simplified approach. *Journal of Financial Economics, 7*, 229–263.
Diedrich, R. (2003). Die Sicherheitsäquivalentmethode der Unternehmensbewertung: Ein (auch) entscheidungstheoretisches wohlbegründbares Verfahren. *Zeitschrift für betriebswirtschaftliche Forschung, 55*(3), 281–286.
Dixit, A., & Pindyck, R. (1994). *Investment under uncertainty*. Princeton: Princeton University Press.
Drukarczyk, J., & Schüler, A. (2007). *Unternehmensbewertung*. Munich: Vahlen.
Eisenführ, F., & Weber, M. (2003). *Rationales Entscheiden*. Berlin: Springer.
Hartmann, M. (2006). *Realoptionen als Bewertungsinstrument für frühe Phasen der Forschung und Entwicklung in der pharmazeutischen Industrie*. Dissertation, Technische Universität Berlin, Berlin.
Hommel, U., & Pritsch, G. (1999). Investitionsbewertung und Unternehmensführung mit dem Realoptionsansatz. In: Achleitner, A. K. & Thoma, G. F. (Eds), *Handbuch corporate finance*. Cologne: Verlag Deutscher Wirtschaftsdienst.
Hughes, G. D., & Chafin, D. C. (1996). Turning new product development into a continuous learning process. *Journal of Product Innovation Management, 13*, 89–104.
Jäger, R., & Himmel, H. (2003). Die Fair Value-Bewertung immaterieller Vermögenswerte vor dem Hintergrund der Umsetzung internationaler Rechnungslegungsstandards. *Betriebswirtschaftliche Forschung und Praxis, 55*(4), 417–440.
Kahl, M., Liu, J., & Longstaff, F. A. (2003). Paper millionaires: How valuable is stock to a stockholder who is restricted from selling it? *Journal of Financial Economics, 67*(3), 385–410.

Krostewitz, A. (2008). *Unternehmensbewertung im Risikoverbund – Neue Methoden der Unternehmensbewertung bei Mergers & Acquisitions.* Dissertation, Friedrich-Schiller-Universität Jena, Jena.

Kürsten, W. (2002). "Unternehmensbewertung unter Unsicherheit", oder: Theoriedefizit einer künstlichen Diskussion über Sicherheitsäquivalent- und Risikozuschlagsmethode. *Zeitschrift für betriebswirtschaftliche Forschung, 54*(2), 128–144.

Kürsten, W. (2003). Grenzen und Reformbedarfe der Sicherheitsäquivalentmethode in der (traditionellen) Unternehmensbewertung. *Zeitschrift für betriebswirtschaftliche Forschung, 55*(3), 306–314.

Kürsten, W. (2007). Kontextadäquates Risikomanagement: Eine komprimierte Einführung. *Jenaer Schriften zur Wirtschaftswissenschaft, 18*, 1–4.

Kürsten, W., & Straßberger, M. (2004). Risikomessung, Risikomaße und Value-at-Risk. *Das Wirtschaftsstudium, 33*(2), 202–207.

Kuhner, C., & Maltry, H. (2006). *Unternehmensbewertung.* Berlin, Heidelberg: Springer.

Löhnert, P., & Böckmann, U. J. (2005). Multiplikatorverfahren in der Unternehmensbewertung. In: Peemöller, V. (Eds), *Praxishandbuch der Unternehmensbewertung* (pp. 403–428). Herne & Berlin: Verlag Neue Wirtschaftsbriefe.

Longstaff, F. A. (1995). How much can marketability affect security values? *The Journal of Finance, 50*(5), 1767–1774.

Longstaff, F. A. (2001). Optimal portfolio choice and the valuation of illiquid securities. *The Review of Financial Studies, 14*(2), 407–431.

Mackenstedt, A., Fladung, H. D., & Himmel, H. (2006). Ausgewählte Aspekte bei der Bestimmung beizulegender Zeitwerte nach IFRS 3 – Anmerkungen zu IDW RS HFA 16. *Die Wirtschaftsprüfung, 16*, 1037–1048.

Mandl, G., & Rabel, K. (2005). Methoden der Unternehmensbewertung (Überblick). In: Peemöller, V. H. (Eds.), *Praxishandbuch der Unternehmensbewertung.* Herne: Verlag Neue Wirtschaftsbriefe.

Matschke, M. J., & Brösel, G. (2007). *Unternehmensbewertung: Funktionen – Methoden – Grundsätze.* Wiesbaden: Gabler.

Mayers, D. (1972). Nonmarketable assets and capital market equilibrium under uncertainty. In: Jensen, M. (Eds.), *Studies in the theory of capital markets: Papers of the conference on modern capital theory* (pp. 223–248). New York: Praeger Publishers.

Mayers, D. (1973). Nonmarketable assets and the determination of capital asset prices in the absence of a riskless asset. *The Journal of Business, 46*(2), 258–267.

Miao, J., & Wang, N. (2007). Investment, consumption, and hedging under incomplete markets. *Journal of Financial Economics, 86*(3), 608–642.

Pellens, B., Fülbier, R. U., Gassen, J., & Sellhorn, T. (2008). *Internationale Rechnungslegung.* Stuttgart: Schäffer-Poeschel.

Rams, A. (1999). Realoptionsbasierte Unternehmensbewertung. *Finanz Betrieb, 11*, 349–364.

Reilly, R., & Schweihs, R. P. (1999). *Valuing intangible assets.* New York: McGraw-Hill.

Scholich, M., Mackenstedt, A., & Greinert, M. (2004). Valuation of intangible assets for financial reporting. In: Fandel, G., Backes-Gellner, U., Schlüter, M., & Staufenbiel, J. E. (Eds.), *Modern concepts of the theory of the firm – managing enterprises of the new economy* (pp. 491–504). Berlin: Springer.

Scholich, M., & Robers, D. I. (2007). Vom Beginner zum Professional – Innovation bei PricewaterhouseCoopers. In: Schmidt, K., Gleich, R., & Richter, A. (Eds.), *Innovationsmanagement in der Serviceindustrie - Grundlagen, Praxisbeispiele und Perspektiven* (pp. 325–338). Freiburg, Berlin, Munich: Haufe.

Scholich, M., & Wulff, C. (2002). Ansätze und Methoden zur Bewertung von Wachstumsunternehmen. In: Hommel, U., & Knecht, T. (Eds.), *Wertorientiertes Start-Up-Management* (pp. 563–579). Munich: Vahlen.

Schultze, W. (2003). Kombinationsverfahren und Residualgewinnmethode in der Unternehmensbewertung: konzeptioneller Zusammenhang. *Kapitalmarktorientierte Rechnungslegung, 3*(10), 458–464.

Schumpeter, J. (1942). *Capitalism, socialism, and democracy.* New York: Harper and Row.
Schwartz, E. S., & Trigeorgis, L. (2001). *Real options and investment under uncertainty: Classical readings and recent contributions.* Cambridge, MA: MIT Press.
Schwetzler, B. (2000). Unternehmensbewertung unter Unsicherheit – Sicherheitsäquivalent- oder Risikozuschlagsmethode? *Zeitschrift für betriebswirtschaftliche Forschung, 52*(8), 469–486.
Schwetzler, B. (2002). Das Ende des Ertragswertverfahrens? – Replik zu den Anmerkungen von Wolfgang Kürsten zu meinem Beitrag in der zfbf. *Zeitschrift für betriebswirtschaftliche Forschung, 54*, 145–158.
Spinler, S., & Huchzermeier, A. (2004). Realoptionen: Eine marktbasierte Bewertungsmethodik für dynamische Investitionsentscheidungen unter Unsicherheit. *Zeitschrift für Controlling & Management, Sonderheft 1*, 66–71.
Spremann, K. (2002). *Finanzanalyse und Unternehmensbewertung, IMF – International Management and Finance.* Oldenbourg, Munich, Vienna.
Thom, N. (1992). *Innovationsmanagement.* Bern: Schweizerische Volksbank.
Tobin, J. (1958). Liquidity preferences as behavior towards risk. *Review of Economic Studies, 25*(2), 65–86.
Trigeorgis, L. (1996). *Real options, managerial flexibility and strategy in resource allocation.* Cambridge, MA: MIT Press.
Ulrich, K. T., & Eppinger, S. D. (1995). *Product design and development.* New York: McGraw-Hill.
Ven, A.v.d. (1986). Central problems in the management of innovation. *Management Science, 32*(5), 590–607.
Wiese, J. (2003). Zur theoretischen Fundierung der Sicherheitsäquivalentmethode und des Begriffs der Risikoauflösung bei der Unternehmensbewertung. *Zeitschrift für betriebswirtschaftliche Forschung, 55*, 287–305.
Wilhelm, J. E. (2005). Unternehmensbewertung – Eine finanzmarkttheoretische Untersuchung. *Zeitschrift für Betriebswirtschaft, 75*(6), 631–665.
Witt, J. (1996). Grundlagen für die Entwicklung und die Vermarktung neuer Produkte. In: Witt, J. (Eds.), *Produktinnovation.* Munich: Vahlen.

Value-Based Management of the Innovation Portfolio

Ulrich Pidun

Contents

1 Introduction 281
2 Requirements for Effective Innovation Portfolio Management 282
3 Instruments for Project Evaluation and Portfolio Selection 285
    3.1 Project Evaluation Methods 285
    3.2 Portfolio Selection Techniques 286
4 Practices of Innovation Portfolio Management 291
    4.1 Which Instruments for Project Evaluation and Portfolio Selection are Most Broadly Used in Practice? 291
    4.2 How Satisfied are Companies with Their Portfolio Management Approaches? ... 291
    4.3 What are the Observed Key Limitations of the Prevalent Approaches? 292
    4.4 What is the Performance of the Prevalent Innovation Portfolio Management Approaches? 292
    4.5 What Distinguishes the Best from the Rest? 293
    4.6 What Should a Company's Approach to Innovation Portfolio Management Depend Upon? 293
5 Summary and Call to Action 294
References 295

1 Introduction

Irrespective of the current economic turmoil, growing through innovation will remain at or near the top of most companies' agendas. In a recent survey among nearly 3,000 executives from around the world, two thirds of the respondents considered innovation one of their three most important strategic priorities and expected

---

U. Pidun (✉)
The Boston Consulting Group, Frankfurt, Germany
e-mail: PIDUN.ULRICH@bcg.com

rising spending on innovation over the medium term (Andrew, Haanaes, Michael, Sirkin, & Taylor, 2008).

At the same time, only a minority of executives are satisfied with their innovation performance (Cooper, Edgett, & Kleinschmidt, 1999, Andrew et al., 2008). When asked for the biggest obstacles to a higher return on innovation, managers rarely mention a lack of new ideas. The most prominent challenges are related to the management of the innovation portfolio: selecting the right ideas to commercialize, shortening development times by putting adequate resources behind the key projects, finding the right balance between return and risk of the innovation portfolio, and improving the internal coordination (Andrew et al., 2008).

Taking an integrated approach to the innovation portfolio thus seems to be one of the key success factors for effective innovation management. But can we draw a direct link between the portfolio approach and innovation performance? At least, there are some strong indications. For example, a recent global survey of innovation management practices showed that companies reporting the highest contribution to growth from their innovation projects tend to be more interested in pursuing and measuring their innovations as a portfolio (Chan, Musso, & Shankar, 2008). A study of innovation management at 205 US companies came to a similar conclusion: The importance that management accords innovation portfolio management is directly related to the end result – to management's perception and satisfaction with the outcome, and even to the performance of the portfolio itself (Cooper et al., 1999).

So, what can companies do in order to become more effective in managing their innovation portfolio? Building on a review of the relevant literature, interviews with corporate strategists and innovation managers, and personal experience of the author with innovation portfolio management in different industries, this article summarizes the current state of theory and practice and derives some specific recommendations for innovation practitioners.

In the next section, we will propose some key requirements for an effective value-based innovation portfolio management that can serve as a yardstick for identifying the key challenges and areas for improvement of existing systems. The third section will provide an overview of available instruments for evaluating and selecting innovation projects. We will see that the challenge is not a lack of methods or techniques, but how to combine the instruments for an effective innovation portfolio management. This will be addressed in the fourth section, where we will derive some learning points from best practice examples and discuss what an adequate approach will depend upon. In the final section, we will summarize the discussion and describe how companies can go about implementing value-based innovation portfolio management.

2 Requirements for Effective Innovation Portfolio Management

The portfolio approach to innovation management treats the selection of growth and R&D projects as a dynamic decision process. On a periodic basis, new projects are evaluated, selected and prioritized; existing projects may be accelerated, killed or reprioritized; and resources are allocated and reallocated to the active projects. The overall innovation portfolio should optimize the stated objectives of the company without exceeding available resources or violating other constraints (Archer & Ghasemzadeh, 1999).

The quality of an innovation portfolio can thus only be assessed in the context of the overall strategic goals of the company. The strategic direction of the firm must be clear before individual projects can be considered. This leads to the first proposition for effective innovation portfolio management:

Proposition 1: Decisions about the innovation portfolio should be made in the context of the overall corporate and business unit strategies

A lack of new ideas is rarely the bottleneck for innovation. In most companies, there are by far more proposals for new growth projects or new product development ideas than can be funded (Andrew et al., 2008). In order to keep the innovation management process efficient, companies must filter those projects that do not qualify for further consideration. This filtering should be done based on a set of carefully defined knock-out criteria. For example, some firms eliminate projects that do not match the strategic focus of the firm, do not yet have sufficient information upon which to base a decision, or do not meet a marginal requirement such as a minimum internal rate of return (Archer & Ghasemzadeh, 1999).

Proposition 2: A screening process should be used to eliminate projects from consideration before the detailed portfolio selection process is undertaken

Portfolio selection involves a comparison of alternative innovation and growth projects. There are a broad variety of methods for individual project evaluation that range from qualitative techniques to quantitative metrics (see the following section). Different types of projects may require different measures, but in the end, all projects compete for the same resources and must be compared equitably during portfolio selection.

Proposition 3: For an equitable comparison of different projects, some common metrics must be available which can be calculated separately for each project under consideration

However, an integrated approach to innovation management cannot stop at the level of the individual projects. It can be shown that the combination of individually good projects does not necessarily constitute the optimal portfolio (Chien, 2002). A mere ranking of projects will thus not be sufficient, portfolio effects have to be taken into account.

Portfolio effects result from interdependencies between the individual innovation projects (Floricel & Ibanescu, 2008; Loch & Kavadias, 2002; Verma & Sinha, 2002; Chien, 2002). Interdependencies can be classified into (1) resource interdependencies, (2) technology interdependencies, and (3) market interdependencies. Resource interdependencies result from sharing scarce resources between different projects and will typically lead to sub-additivity of project values because an increase in the resource level for one project would result in a decrease in the resource level of another project. Technology interdependencies result from leveraging common technology across multiple projects, frequently leading to super-additivity of project values (but also to increased cluster risks at the portfolio level). Market interdependencies can lead to super-additivity of project values (for example, if a new product utilizes a current product's market knowledge), but also to sub-additivity (for example, if a firm makes multiple redundant investments into a "winner-takes-all" type of market).

The relevance of interdependencies in innovation portfolios was confirmed by a number of studies. For example, Vassolo et al. investigated portfolios of biotech equity alliances of pharmaceutical companies (Vassolo, Anand, & Folta, 2004) and found evidence for sub-additivity (when a firm invested in multiple and competing projects) as well as super-additivity (in the case of fungibility of shared resources between different projects). Girotra et al. conducted an event study around the failure of phase III clinical trials and their effect on the market valuation of pharmaceutical companies (Girotra, Terwiesch, & Ulrich, 2007). They could explain the variance in the value of projects based on interactions with other projects in the firm's portfolio. In particular, they found that the presence of other projects targeting the same market and a build-up of projects that require the same development resources reduce the value of a development project.

The degree of interdependence between the projects in the innovation portfolio will vary from firm to firm, but in most cases it cannot be neglected. This leads to the fourth proposition for effective innovation portfolio management:

Proposition 4: Interdependencies between individual projects must be considered in innovation portfolio selection

In the next section, we will present a number of portfolio selection techniques that consider these interdependencies. However, even the most sophisticated instruments can only serve as support for the decision makers. Effective innovation portfolio management requires that the instruments be integrated into an overall decision making process. The system and process should be able to take into account multiple (even conflicting) objectives and also support group decision making because portfolio selection is a committee process in most firms (Archer & Ghasemzadeh, 1999). Moreover, practical experience has demonstrated that innovation management rarely works with the a priori definition of preferences and constraints and the subsequent derivation of the optimal portfolio (Stummer & Heidenberger, 2003). An effective innovation management system should support the iterative development of the innovation portfolio by providing interactive mechanisms and feedback on the consequences of portfolio changes. This leads to the final proposition:

Proposition 5: Decision makers should be provided with instruments that support them in crafting the optimal innovation portfolio in an interactive process

To summarize, the propositions developed in this section define the cornerstones of an effective innovation portfolio management process: The process starts with guidelines and objectives for innovation management that are derived from the overall corporate and business unit strategies. New project proposals must pass a screening process with carefully selected knock-out criteria in order to qualify for further consideration. Individual project analysis is based on pre-defined qualitative and quantitative measures, with some common metrics that can be used for an equitable comparison of different projects. However, portfolio selection must also consider interdependencies between projects and is done in an interactive process that usually involves group decision making.

What are the instruments that can support decision makers in this innovation portfolio management process?

3 Instruments for Project Evaluation and Portfolio Selection

Evaluation instruments for innovation portfolio management can be roughly grouped into two categories: (1) Project evaluation methods that measure each project's stand-alone contribution to the company's innovation objectives, and (2) portfolio selection techniques that support management in comparing the different projects in order to define the optimal innovation portfolio.

3.1 Project Evaluation Methods

Corporate decision makers use both qualitative and quantitative evaluation methods when considering new innovation projects. Qualitative methods are designed to make sure that management considers all critical success factors before starting a project. These methods vary depending upon an individual company's strategic priorities, but a good example is the R-W-W screen described below. Quantitative methods have a more common basis. They typically cover economic return metrics, but are increasingly supplemented by risk and optionality measures.

3.1.1 Qualitative Methods: R-W-W Screen

A good example for a stringent qualitative evaluation method is the R-W-W ("real, win, worth it") screen (Day, 2007). It is a simple tool to guide a development team to holistically assess each innovation project based on six fundamental questions:

- Real: Is the market real? Is the product real?
- Win: Can the product be competitive? Can our company be competitive?
- Worth it: Will the product be profitable at an acceptable risk? Does launching the product make strategic sense?

A definite no to any of these questions should lead to termination of the project, since failure is all but certain.

3.1.2 Economic Return

Economic return metrics are most widely applied for individual project evaluation (Remer, Stokdyk, & Van Driel, 1993). This includes Net Present Value (NPV), Internal Rate of Return (IRR), Return on Original Investment (ROI), Return on Average Investment (RAI), PayBack Period (PBP) and Benefit/Cost Techniques (BCT).

These methods include time dependency considerations of investment and income flows, but cannot explicitly account for the idiosyncratic risk of the project or the option character of staggered decision making. A 1991 survey of the use of the above techniques indicated a shift from the use of Internal Rate of Return (IRR) to Net Present Value (NPV) and a decrease in the use of the PayBack Period (PBP) (Remer et al., 1993).

3.1.3 Risk and Optionality

Companies are increasingly trying to quantify the specific risks associated with individual growth projects. A frequently used metric is the Value at Risk (VaR) that measures the potential loss from a project over a defined period for a given confidence interval (Damodaran, 2008). VaR can be calculated from probabilistic financial models with Monte Carlo Simulations that are supported by add-on programs to various spreadsheets, such as *At Risk* and *Crystal Ball*. VaR calculations for risk quantification have been successfully applied in such diverse industries as pharmaceuticals, chemicals, raw materials, telecommunication, electric utilities (Noor, Martin, & Bowman, 2005) and oil and gas (Balagopal & Gilliland, 2005).

VaR focuses on the negative effects of risk. However, uncertainty can sometimes also be a source of additional value, especially to those who are poised to take advantage of it. One method that tries to quantify the upside potential of risk is the real options approach (Faulkner, 1996; Damodaran, 2008). In essence, the additional value of a project that is treated as a real option stems from the fact that over the course of the project, managers can learn from observing what happens in the real world and adapt their behaviour to increase the potential upside and decrease the possible downside of the investment. Specifically, managers have the option to expand an attractive project, the option to scale down or even abandon an unattractive project, or the option to delay further investments if the future prospects are not clear. Real option techniques are well established in the energy, raw materials and pharmaceuticals industries, but have not found the widespread use that was predicted in the 1990s (Triantis & Borison, 2001).

3.2 *Portfolio Selection Techniques*

Portfolio selection involves the simultaneous comparison of a number of innovation projects with the objective to arrive at a ranking of projects as a basis for the definition of the optimal innovation portfolio. The task is complicated by the fact that frequently multiple conflicting objectives are associated with portfolio selection, and that projects may be highly interdependent. Classes of available portfolio selection techniques include ad-hoc approaches, comparative approaches, scoring models, optimization models and mapping approaches (Archer & Ghasemzadeh, 1999; Heidenberger & Stummer, 1999).

3.2.1 Ad-Hoc Approaches

Many companies do not use formal instruments for portfolio selection, but employ ad-hoc approaches to decide about innovation projects. In the worst case this will lead to a first-come-first-serve policy where seemingly attractive projects are funded in an opportunistic way until the budget limit is reached. In better cases, project selection involves an interactive process between project champions and responsible decision makers until a choice of the best projects is made. But without the adequate data support to ensure an equitable comparison of different projects, this approach can easily lead to a decibel-driven portfolio selection where the charisma or political talent of the project champion decides about project approval (Sanwal, 2007).

Another type of ad-hoc approach is the use of heuristics (Heidenberger & Stummer, 1999). These are simple decision rules that guide management in portfolio selection. For example, a firm that is focused on exploitation will fund line extension and product improvement projects with a higher priority than new product development projects. Heuristics may increase the efficiency of the portfolio selection process, but it is not clear whether they will lead to better decisions. In fact, simulation models suggest that the use of simple heuristics for resource allocation and project termination strategies in the pharmaceutical industry may cause unintended volatility of R&D performance (Gino & Pisano, 2006).

3.2.2 Comparative Approaches

Comparative approaches provide management with instruments for coming to a stepwise consensus about which projects to undertake. The process usually starts with the determination of the weights of the different objectives that should be considered. Alternatives are then compared on the basis of their contributions to these objectives. Once the projects have been arranged on a comparative scale, the decision makers can proceed from the top of the list, selecting projects until available resources are exhausted (Archer & Ghasemzadeh, 1999).

Examples of comparative approaches include analytical hierarchy methods, like pairwise comparison (Martino, 1995) and the Analytical Hierarchy Procedure (AHP) that divides a decision problem into smaller chunks and organizes the decision factors into a hierarchy so that complex decisions can be made through incremental judgments (Saaty, Rogers, & Pell, 1980). Another group comprises the behavioral and psychometric approaches, like Delphi and Q-Sort (Souder, 1984) that assist management in achieving group consensus. These methods can also be supported by voting software and hardware (for example, handheld voting machines) permitting the decision team to input their choices quickly and visually (Cooper et al., 1999).

Due to the large number of comparisons involved, comparative approaches are restricted to innovation portfolios with a limited number of alternative projects. Another disadvantage is that the ranking process must be repeated any time a project is added or deleted from the list. Moreover, comparative approaches cannot consider interdependencies between projects because they are only based on pairwise comparisons.

3.2.3 Scoring Models

Scoring models employ a relatively limited number of decision criteria to specify project attractiveness. The merit of each project is determined with respect to each criterion. Scores are then combined – typically based on different weights for each criterion – to yield an overall benefit measure for each project (Archer & Ghasemzadeh, 1999; Heidenberger & Stummer, 1999).

Scoring models are widely used instruments for portfolio selection (Bitman & Sharif, 2008). The criteria used often capture proven drivers of new product success, such as market attractiveness, product advantage, synergy with the base business, familiarity etc. An advantage of this approach is that criteria and their relative weighting can be easily customized to the specific needs of the company. As opposed to the comparative approaches, projects can be added or deleted without re-calculating the merit of other projects.

However, there are also some conceptual limitations to scoring models. A simple weighting of the different criteria may not be appropriate if there are certain knock-out criteria that cannot be compensated by other dimensions. The chosen set of criteria should be orthogonal to avoid redundancies and non-linear effects in averaging the dimensions. Moreover, like the comparative approaches, scoring models cannot consider interdependencies between projects because they only evaluate projects on a stand-alone basis.

3.2.4 Optimization Models

Optimization models generally use some sort of mathematical programming to select from the list of candidate projects a subset that maximizes a pre-defined objective function (for example, net present value). These models can capture project interactions such as resource dependencies and constraints, technical and market interactions, or program considerations (Martino, 1995). But they are usually not able to reflect uncertainties associated with the projects, which makes it impossible to attach risk measures to innovation portfolios (Gustafsson & Salo, 2005).

The first portfolio optimization models were already developed in the 1960s and 1970s. They employed mathematical techniques such as linear, dynamic, and integer programming (Jackson, 1983). However, despite their conceptual appeal and theoretical rigor, optimization techniques are not widely applied in practice. The main obstacles are that they are too complex and require too much input data, they fail to adequately treat uncertainty and interrelationships between projects, and they may just be too difficult to understand and use (Loch & Kavadias, 2002, Archer & Ghasemzadeh, 1999; Cooper, 1993).

Modern approaches employ optimization models as part of interactive portfolio analysis. For example, multiobjective integer linear programming can be used to determine the solution space of all efficient (i.e., Pareto-optimal) portfolios, which is then explored in an iterative process to find a portfolio that fits the decision maker's notion. However, the enormous computational effort to calculate all possible project combinations limits this particular approach to sets of some thirty projects (Stummer & Heidenberger, 2003).

Another recent trend is to link portfolio optimization models with decision trees in order to account for the interdependencies of projects as well as the uncertainty and managerial flexibility in decision making (Gustafsson & Salo, 2005).

3.2.5 Mapping Approaches

Among the different portfolio selection approaches discussed so far, optimization models are the first to explicitly consider interdependencies between projects, but they are limited by mathematical complexity and computational effort. Moreover, they are not able to appreciate the value of portfolio balance, which plays an important role in innovation portfolio management.

For this purpose, decision makers must resort to mapping approaches as strategic decision-making tools. They generally rely on graphical representations of the projects under consideration, using matrices that represent the critical dimensions for the decision. Various parameters can be plotted against each other in a bubble diagram format – plots such as economic benefit versus likelihood of success, or project attractiveness versus ease of undertaking.

In this way, mapping approaches can support not only the search for maximum benefit, but also for balance of an innovation portfolio. Important balance dimensions include growth versus profitability, risk versus return and short-term versus long-term value creation. To this end, companies should aim at innovation project portfolios with balanced distributions with respect to the project sizes, project life cycle stages and time to market, technologies and markets addressed, drivers of project specific risks etc.

For example, the R&D project portfolio matrix suggested by Mikkola builds on the original BCG growth-share matrix (Mikkola, 2001). Innovation projects are classified according to their benefits to customers and the size of their competitive advantage. This leads to four types of projects: STARs (high benefit, high advantage), FADs (high benefit, low advantage), SNOBs (low benefit, high advantage) and FLOPs (low benefit, low advantage). A balanced portfolio should contain STARs, FADs, SNOBs, and sometimes-even FLOPs. These projects are related to each other through the dynamics of innovation and imitation. FADs are important because they can be highly profitable at a reasonable risk since they are often based on imitation of existing products. SNOBs are equally important because they may possess the technological or product know-how that can provide breakthrough platforms for the firm. They typically require significant investments. However, when SNOBs do become STARs, they will bring revenues as well as strengthen core capabilities that are difficult to be matched by the competitors.

The objective of innovation to ensure constant renewal of the company can also be supported by the Life Cycle Portfolio (Sull & Houlder, 2006). Business opportunities typically progress through four stages: experiment, scale, mature and decline, which are labeled along the horizontal axis. The vertical axis denotes the age of an opportunity. The expected trajectory of an opportunity is from the lower left to the upper right. By mapping a company's growth opportunities – with the size of each ball denoting the amount of cash consumed (red) or generated (green) in a year – managers can identify imbalances and other trouble spots and surface important questions about opportunities that deviate from the expected trajectory.

A similar concept to balance the demands of the present with the promise of the future is the "horizons of growth" model (Baghai, Coley, & White, 2000). It distinguishes between growth opportunities in start-up, scaling and mature phases and recognizes that each growth horizon requires very different approaches to people, strategy, resource allocation and measurement. A company should have an innovation portfolio that is balanced across the three horizons, and executives should carefully spot opportunities that move into the next horizon, perhaps requiring a change in management or different forms of support and incentives (for a description of IBM's approach to the three horizons, see Garvin & Levesque, 2004).

Finally, an example for a tool that reveals the distribution of risk across a company's innovation portfolio is the Risk Matrix (Day, 2007; MacMillan & McGrath, 2002). Each innovation can be positioned on the matrix by determining its score on two dimensions – how familiar to the company the intended market is (x axis) and how familiar the product or technology is (y axis). Familiar products aimed at the company's current markets will fall in the bottom left of the matrix, indicating a low probability of failure. New products aimed at unfamiliar markets will fall in the upper right, revealing a high probability of failure. Most companies will find that the majority of their products cluster in the bottom left quadrant of the matrix, and only very few skew toward the upper right. This imbalance may be unhealthy: according to one study, only 14% of new-product launches were substantial innovations, but they accounted for 61% of all profit from innovations among the companies examined (Kim & Mauborgne, 1999).

To summarize, innovation managers can choose from a broad range of instruments for evaluating individual projects and selecting project portfolios. The challenge is to select an adequate and consistent set of instruments for a given organization – considering the specific needs and cognitive styles of decision makers – and to integrate these instruments into an overall framework. How can the proposed requirements for effective innovation portfolio management be fulfilled in practice? Which instruments work well and which don't? And what can we learn from the best?

4 Practices of Innovation Portfolio Management

This section will summarize the results of recent research studies on the practices of innovation portfolio management. We will discuss which instruments are most broadly used, how satisfied companies are with their existing approaches, which limitations can be observed, how these limitations translate into innovation performance, what sets apart the best innovators, and what should drive a company's specific approach to innovation portfolio management.

4.1 Which Instruments for Project Evaluation and Portfolio Selection are Most Broadly Used in Practice?

A recent industry survey of innovation portfolio management practices (Cooper, Edgett, & Kleinschmidt, 2001) confirmed the results of earlier surveys (Cooper et al., 1999; Liberatore & Titus, 1983): Financial return metrics are by far the most prevalent approach to project selection. They are used by 77% of the respondents, with 40% of businesses relying on them as their dominant portfolio method. The second most popular portfolio approach is to allocate the innovation budget to buckets that are defined based on corporate and business unit strategies, with projects being subsequently ranked or rated within those buckets (employed by 41% of the respondents). Mapping approaches are also employed by 41% of the respondents, but only 5% use them as their dominant method. Scoring models play a role in 38% of the companies (for 13% as the dominant method). On the other hand, comparative approaches and optimization models are not very relevant in practice. Interestingly, the average participating business uses 2.34 different techniques to select projects and manage its innovation portfolio.

4.2 How Satisfied are Companies with Their Portfolio Management Approaches?

On average, managers are moderately satisfied with their methods and instruments for innovation portfolio management (Chan et al., 2008; Cooper et al., 2001). However, a large spread in satisfaction responses exists between businesses. For example, one survey found only 10% of businesses to be very pleased with their established approach, while more than one-third of respondents would clearly not recommend their approach to others (Cooper et al., 1999).

Based on a factor analysis, one study identified two key drivers for satisfaction with the portfolio management method: (1) overall quality rating of the approach (for example, is it realistic, is it truly used to make go/kill decisions, is it user friendly, and would management recommend it to others?), and (2) management fit (for example, is it understood by management, does it fit with decision making style, is it seen as effective and efficient?) (Cooper et al., 1999). A cluster analysis of the participating companies revealed four types of businesses with respect to these two dimensions:

- Cowboy businesses (12% of respondents): low quality approach that, however, fits management very well
- Crossroads businesses (28%): high quality approach that is not (yet) well accepted by management
- Duds (18%): bad on all fronts – low quality approach and weak management fit
- Benchmarks (42%): highest scores on quality and management fit.

4.3 What are the Observed Key Limitations of the Prevalent Approaches?

When managers are asked what they perceive as the most important limitations of their current approaches to innovation portfolio management, typical answers include (Chan et al., 2008; Chien, 2002; Cooper et al., 2001):

- Inadequate treatment of multiple, often interrelated evaluation criteria
- Inadequate treatment of interrelationships among projects
- Inability to handle non-monetary aspects e.g., portfolio balance
- Models difficult to understand and use
- No explicit consideration of the experience and knowledge of decision makers.

We have mentioned most of these observed limitations as important requirements for an effective innovation portfolio management. But we have also shown that there are a number of well-established instruments for addressing these shortcomings. Before we turn to the best practice companies to see how they deal with the challenges, let us first look at how the observed limitations translate into innovation portfolio performance.

4.4 What is the Performance of the Prevalent Innovation Portfolio Management Approaches?

The performance of an innovation portfolio is not easy to measure because many other aspects besides good portfolio management, influence actual business performance in financial terms. Companies use a wide variety of input and output metrics for innovation management (Andrew et al., 2008; Chan et al., 2008). A set of perceptual measures that were found to be good indicators for the quality of an innovation portfolio (Cooper et al., 1999, 2001) consists of:

- Portfolio has the right number of projects for the resources available
- Projects are done on time (no pipeline gridlock)
- Portfolio contains high-value projects (profitable, high return)

- Projects are aligned with business's strategic objectives
- Spending breakdown mirrors the strategic priorities
- Portfolio has good a balance (e.g., long term vs. short term, high vs. low risk).

Based on these measures, the link between the dominant portfolio method used and the performance of the portfolio can be investigated (Cooper et al., 1999). The results are surprising: Financial models – despite being the most popular – yield the poorest performance results. In contrast, strategic approaches (letting the business's strategy decide upon resource allocation and even choice of projects) perform significantly best. Scoring models also produce positive performance, in particular in terms of yielding a portfolio containing high-value projects. It is ironic that financial return metrics – presumably chosen to select the highest return projects – perform worst on this dimension.

4.5 What Distinguishes the Best from the Rest?

The benchmark firms with the highest scores on the quality and management fit of their portfolio approach clearly outperform their peers across all six perceptual performance metrics described above (Cooper et al., 1999, 2001). But how do they achieve this, what are they doing differently? There are three main sources of differentiation between the best and the rest:

1. *Importance*: Senior management in benchmark firms views innovation portfolio management as very important.
2. *Formalization*: Benchmark firms have an explicit, well-established method for innovation portfolio management, with clear rules and procedures and broad management support. They consistently apply this method to all relevant projects, and consider all projects together and treat them as a portfolio.
3. *Multiple methods*: The best firms tend to use multiple portfolio methods. They rely much less on financial metrics and actively use business strategy related methods, scoring models and mapping approaches.

There is one more aspect that differentiates top performing innovators: They adapt their approach to innovation portfolio management to the specific situation of their company.

4.6 What Should a Company's Approach to Innovation Portfolio Management Depend Upon?

A company's approach to innovation management must be consistent with its other corporate processes and fit with the specific decision making style of the responsible senior managers. This will drive the acceptable degree of formality and complexity, and the mix of quantitative versus qualitative metrics. The key external contingency factor for an effective innovation portfolio management approach is the dynamics of the market and competitive environment.

When firms define their specific innovation processes, they can make choices along four dimensions (Floricel & Ibanescu, 2008):

- Structure: formality of the approach (formal versus ad-hoc)
- Emergence: direction of the decision process (top-down/planned versus bottom-up/emergent)
- Commitment: breadth of the portfolio (few projects with strong support versus many small options)
- Integration: consideration of interdependencies in project selection (strongly integrated versus loose/modular).

A recent survey study among 795 firms from the Americas, Europe and Asia found that these choices are significantly influenced by environmental dynamics (Floricel & Ibanescu, 2008). Firms in strongly growing markets with high and increasing resource requirements tend to use innovation portfolio management approaches with more structure and higher levels of resource commitment to each individual project. If high growth is accompanied by high velocity of change – such as fast advances in functionality, performance or cost – companies in addition are more likely to prefer integrated approaches that exploit synergies between the projects. Companies in turbulent environments with perceived strong discontinuities of past trends and anticipated directions acknowledge that a top-down approach to innovation management is less appropriate because it tends to reproduce past assumptions; these firms are consequently more open to emergent approaches and bottom-up processes.

5 Summary and Call to Action

Based on a review of the relevant literature and cumulative insights from industry interviews and personal project experience we have identified five key requirements for an effective innovation portfolio management:

1. Decisions about the innovation portfolio should be made in the context of the overall corporate and business unit strategies
2. A screening process should be used to eliminate projects from consideration before the detailed portfolio selection process is undertaken
3. For an equitable comparison of different projects, some common metrics must be available which can be calculated separately for each project under consideration
4. Interdependencies between individual projects must be considered in innovation portfolio selection
5. Decision makers should be provided with instruments that support them in crafting the optimal innovation portfolio in an interactive process.

For support in this process, managers can choose from a broad toolbox. Instruments for project evaluation include qualitative methods (like the R-W-W screen), economic return metrics, and techniques to capture risk and optionality. Project selection instruments include ad-hoc approaches and heuristics, comparative approaches, scoring models, optimization models, and mapping approaches.

Despite their dominant role for project evaluation in most companies, financial return metrics are not particularly effective in innovation portfolio management. The best companies tend to use multiple portfolio instruments, rely much less on financial metrics and actively use business strategy related methods, scoring models and mapping approaches. They have an explicit, well-established method for innovation portfolio management, with clear rules and procedures that they consistently apply to all relevant projects, considering all projects together and treating them as a portfolio. Moreover, innovation leaders adapt their portfolio management approach to the specific company situation.

How can your company get there? To begin with, you should perform a candid self-diagnosis of current innovation management processes: What stops your company from being a more effective innovator? The answer may be a lack of ideas or funding. But more probably, the true innovation bottleneck will be the non-strategic allocation of resources, the inability to say no and stop projects, or too strong a focus on exploitation rather than exploration. After diagnosis, the key is simply to start. You should pick what seems to be the right suite of instruments for your company – maybe focusing on the explicit inclusion of a risk perspective, or establishing a stronger link between innovation and business unit strategies, or using mapping approaches to manage the balance of the innovation portfolio. Once these instruments are put in place, you should use them consistently and do what's necessary to make them important to the right people internally. Over time, your company will develop its own effective method for innovation portfolio management.

This is the right time to care about your approach to innovation. Budget reductions are moments of truth for innovation management: a uniform cut for all projects would lead to much worse results than the termination of some projects that are selected from a portfolio perspective. In this way, value-based management of the innovation portfolio can support a company in times of financial crisis by freeing up cash without putting the long-term strategic position of the company at risk.

References

Andrew, J. P., Haanaes, K., Michael, D. C., Sirkin, H. L., & Taylor, A. (2008, August). Innovation 2008: Is the tide turning? *BCG Senior Management Survey*, 1–31.

Archer, N. P., & Ghasemzadeh, F. (1999). An integrated framework for project portfolio selection. *International Journal of Project Management, 17*(4), 207–216.

Baghai, M., Coley, S., & White, D. (2000). *The alchemy of growth: Practical insights for building the enduring enterprise*. New York: Perseus.

Balagopal, B., & Gilliland, G. (2005, July). Integrating value and risk in portfolio strategy. *BCG OfA*, 1–10.

Bitman, W. R., & Sharif, N. (2008). A conceptual framework for ranking R&D projects. *IEEE Transactions of Engineering Management, 55*(2), 267–278.

Chan, V., Musso, C., & Shankar, V. (2008, October). McKinsey global survey results: Assessing innovation metrics. *McKinsey Quarterly*, 1–11.

Chien, C. F. (2002). A portfolio-evaluation framework for selecting R&D projects, *R&D Management, 32*(4), 359–368.

Cooper, R. G. (1993). *Winning at new products*. Reading, MA: Addison-Wesley.

Cooper, R. G., Edgett, S. J., & Kleinschmidt, E. J. (1999). New product portfolio management: Practices and performance. *Journal of Product Innovation Management, 16*, 333–351.

Cooper, R. G., Edgett, S. J., & Kleinschmidt, E. J. (2001). Portfolio management for new product development: Results of an industry practices study. *R&D Management, 31*(4), 361–380.

Damodaran, A. (2008). *Strategic risk taking*. Upper Saddle River, NJ: Wharton School Publishing.

Day, G. S. (2007, December). Is it real? Can we win? Is it worth doing? *Harvard Business Review*, 110–120.

Faulkner, T. (1996). Applying "options thinking" to R&D valuation. *Research Technology Management, 39*, 50–57.

Floricel, S., & Ibanescu, M. (2008). Using R&D portfolio management to deal with dynamic risk. *R&D Management, 38*(5), 452–467.

Garvin, D., & Levesque, L. (2004). Emerging business opportunities at IBM, Harvard Business School case, No. 304-075.

Gino, F., & Pisano, G. (2006). Do Managers' Heuristics Affect R&D Performance Volatility?, Working Paper HBS Division of Research.

Girotra, K., Terwiesch, C., & Ulrich, K. T. (2007). Valuing R&D projects in a portfolio: Evidence from the pharmaceutical industry. *Management Science, 53*(9), 1452–1466.

Gustafsson, J., & Salo, A. (2005). Contingent portfolio programming for the management of risky projects. *Operations Research, 53*(6), 946–956.

Heidenberger, K., & Stummer, C. (1999). Research and development project selection and resource allocation: A review of quantitative modelling approaches. *International Journal of Management Review, 1*, 197–224.

Jackson, B. (1983). Decision methods for selecting a portfolio of R&D projects. *Research Management*, 21–26.

Kim, W. C., & Mauborgne, R. (1999). Strategy, value innovation and the knowledge economy. *Sloan Management Review, 40*(3), 41–54.

Liberatore, M. J., & Titus, G. J. (1983). The practice of management science on R&D project management. *Management Science, 29*, 62–974.

Loch, C. H., & Kavadias, S. (2002). Dynamic portfolio selection of NPD programs using marginal returns. *Management Science, 48*(10), 1227–1241.

MacMillan, I. C., & McGrath, R. G. (2002). Crafting R&D project portfolios. *Research Technology Management, 45*(5), 48–59.

Martino, J. P. (1995). *R&D project selection*. New York, NY: Wiley.

Mikkola, J. H. (2001). Portfolio management of R&D projects: Implications for innovation management. *Technovation, 21*, 423–435.

Noor, I., Martin, R., & Bowman, D. (2005). Implementation of successful risk-based portfolio management. *AACE International Transactions RISK, 02*, 1–6.

Remer, D. S., Stokdyk, S. B., & Van Driel, M. (1993). Survey of project evaluation techniques currently used in industry. *International Journal of Production Economics, 32*, 103–115.

Saaty, T. L., Rogers, P. C., & Pell, R. (1980). Portfolio selection through hierarchies. *Journal of Portfolio Management, 6*(3), 16–21.

Sanwal, A. (2007). *Optimizing corporate portfolio management*. Hoboken, NJ: John Wiley & Sons.

Souder, W. E. (1984). *Project selection and economic appraisal*. New York, NY: Van Nostrand Reinhold.

Stummer, C., & Heidenberger, K. (2003). Interactive R&D portfolio analysis with project interdependencies and time profiles of multiple objectives. *IEEE Transactions of Engineering Management, 50*(2), 175–183.

Sull, D. N., & Houlder, D. (2006). How companies can avoid a midlife crisis. *MIT Sloan Management Review, 48*(1), 26–34.

Triantis, A., & Borison, A. (2001). Real options: State of the practice. *Journal of Applied Corporate Finance, 14*(2), 8–24.

Vassolo, R. S., Anand, J., & Folta, T. B. (2004). Non-additivity in portfolios of exploration activities: A real options-based analysis of equity alliances in biotechnology. *Strategic Management Journal, 25*, 1045–1061.

Verma, D., & Sinha, K. K. (2002). Toward a theory of project interdependencies in high tech R&D environments. *Journal of Operations Management, 20*, 451–468.

Innovation Performance Measurement

Peter Schentler, Frank Lindner, and Ronald Gleich

Contents

1 Introduction 299
2 Performance Measurement 301
3 Performance Measurement for Innovation 304
4 Performance Levels in Innovation Performance Measurement 306
   4.1 Company Level (1) – Innovation Management Performance 306
   4.2 Multi-Project Level (2) – Innovation Portfolio Performance 309
   4.3 Single Project Level (3) – Innovation Project Performance 313
5 Conclusion 314
References 315

1 Introduction

Scholars and practitioners provide a great number of varying definitions for the term 'Innovation'. All definitions have in common that innovations can be regarded as something new[1] as they bring forward products or services which have not been available before, or which differ significantly from existing products and services.[2]

In order to be able to retain or increase the market share of a company despite high competitive pressure or saturated markets, innovations have become

---

P. Schentler (✉)
Strascheg Institute for Innovation and Entrepreneurship (SIIE), European Business School (EBS), Germany
e-mail: peter.schentler@ebs-siie.de

[1] See Littkemann and Holtrup (2008), p. 265.
[2] See Hauschildt (2004), p. 6.

increasingly important.[3] 89% of the respondents of a recent study answered that their company needs major innovations in order to meet financial goals.[4] Studies also show that companies with a high innovation rate are more successful than their competitors.[5] This cannot only be applied for the production of goods, but for the creation of services as well.[6]

Even though innovations are very important for a company's current and future success, a high percentage of innovation projects fail.[7] Hence, many companies invest huge amounts of resources in innovation projects which do not pay off in the end. In order to decrease risks of innovation projects and to minimize the squandering of resources, it is necessary to engage only in innovation projects which have the potential to lead to success and to ensure an efficient execution of the innovation projects. Considering the fact that innovations are characterized by high levels of risk and uncertainty[8] and the difficulty to plan their success, companies should establish an innovation controlling aimed at increasing the level of successful innovations.[9] By using innovation controlling, it should be possible to allocate resources to the most promising projects and stop projects, which do not fit corporate strategy, market needs or are unfeasible or unprofitable.

For planning and steering innovation activities, measurement systems are applied. These measurement systems often focus on financial targets or particular, either strategic or operational, levels of innovation or mix up different levels.[10] But it is necessary to cover all fields of innovation and of the innovation process – idea generation, research and development, market entry – as well as to find appropriate solutions for different levels of innovation. In addition, other facets of innovation management such as cooperation with suppliers, competitors or customers must be considered.

Due to these problems, existing measurement systems do not fit the requirements of companies. In order to establish a successful innovation controlling covering the discrepancies of existing systems, a concept covering all levels of innovation performance has to be developed. A performance measurement system can serve as a basis for such a controlling concept.

Therefore, the goal of this paper is to conceptualize a performance measurement system for innovation. To achieve this, the following procedure is used:

---

[3] See Storey and Kelly (2001), p. 71.
[4] Greiner, Römer, and Russo (2009), p. 18.
[5] Engel and Diedrichs (2006), p. 78; Greiner et al. (2009), p. 9; Hall and Mairesse (1995), pp. 286–288.
[6] Scholich, Gleich, and Grobusch (2006).
[7] Killen, Hunt, and Kleinschmidt (2008), p. 25; Granig (2007), pp. 133–135.
[8] Albala (1975), p. 154.
[9] Innovation controlling is an intensively discussed topic by scholars and practitioners; See e.g. Littkemann (2005); Schmeisser, Kantner, Geburtig, and Schindler (2006); Boutellier, Völker, and Voit (1999).
[10] Furthermore, the relations between different measurement levels are often not defined.

- In Chap. 2, the state of the art of performance measurement is put into focus.
- In Chap. 3, the application of performance measurement in the field of innovation management is introduced.
- In Chap. 4, the different levels of performance measurement are shown.

An overview about the results and the demand for further research are covered in Chap. 5.

2 Performance Measurement

Performance measurement systems were developed, because science and practice have come to the conclusion that traditional, financially oriented measurement systems provide limited use for the sustainable management and controlling of a company.[11] The rising criticism covers aspects like a disregard of nonfinancial parameters, the missing alignment to corporate strategy, the backward view, the short-term perspective, the insufficient customer orientation and misleading reference points for incentives.[12] Several new concepts have been developed, which are subsumed under the term 'performance measurement'. The best-known and most prominent performance measurement system, as studies show, is the Balanced Scorecard.[13]

Basically, performance measurement fosters two issues: target setting tailored to fit beneficiaries and performance levels; and the operationalization of strategy and its translation in quantified goals.[14] This involves the setup and use of several KPIs of various dimensions (e.g. costs, time, quality, ability for innovation, customer satisfaction and flexibility), which serve as a basis for the evaluation of the effectiveness and efficiency of the performance and the performance potentials of different objects, so-called performance levels (e.g. organizational units, processes and employees). Performance measurement should not only be used for evaluation and measurement of past actions, but mainly focus on future oriented management and controlling activities.[15]

Although a lot of papers have been written about this topic,[16] a definition of 'performance measurement' is still discussed in the literature. There is still no common

---

[11] Raake (2008), p. 30.

[12] Gleich (2001), pp. 5–10.

[13] Günther and Grüning (2002), p. 5. For a description of the Balanced Scorecard see Kaplan and Norton (1992) or Kaplan and Norton (2003). Other performance measurement systems are e.g. Data Envelopment Analysis, Tableau de Bord, Performance Pyramid, Quantum Performance Measurement, Productivity Measurement and Enhancement System and Business Management Window (see Gleich 2001, pp. 45–47).

[14] Gleich (1997), p. 114.

[15] Gleich (2001), p. 12; Raake (2008), p. 25.

[16] See Neely (2005), for an overview about the state of the art of performance measurement and further research perspectives.

meaning nor is there general agreement on its characteristics.[17] However, it can be derived that performance measurement systems should comply with the following requirements:[18]

- Provide past- and future-oriented management control information
- Reflect the demands of both internal and external stakeholders
- Provide basis, and on corporate and divisional levels
- Contain financial KPIs which can be extended by non-financial parameters which influence the long-term financial performance capabilities of a company
- Contain not only quantitative (hard facts) but also qualitative (soft facts) information
- Provide strategic and operational KPIs
- Support continuous improvement.

When applying this comprehensive understanding of performance measurement, the question if it is more than just *another* measurement concept, can be answered with a yes. But studies show that, in practice, performance measurement systems often do not meet the requirements formulated in theory.[19]

Ten determinants, which are also featured in a comprehensive empirical study,[20] clearly show the shortcomings that often exist in the practical implementation of performance measurement. In the past, applied controlling concepts were not structured along the lines of progressive performance measurement in most cases (see Fig. 1, left column). Currently, controlling systems in many companies are already equipped to meet these new requirements (central column).[21] In the future, the development and implementation of progressive performance measurement will be needed in order to tackle the increasing and multidimensional requirements of markets and customers on management and controlling structures of a company (right column).

Examples of a progressive performance measurement system can only be found in practice in isolated cases.[22] It can be assumed, and has been proven empirically for these cases, that progressive performance measurement leads to higher profitability in comparison to other companies in the same sector with less developed performance measurement systems.[23]

---

[17] Weiss, Zirkler, and Guttenberger (2008), p. 139; Raake (2008), p. 24.
[18] Gleich (2001), p. 194; Brown and Laverick (1994); Neely, Gregory, and Platts (1995), p. 80; Müller-Stewens (1998); Raake (2008), p. 25; Rummler and Brache (1990), p. 16; Gleich (1997), p. 115.
[19] Weiss et al. (2008), pp. 140–144.
[20] Gleich (2001).
[21] The results of an empirical study in Gleich (2001), pp. 362–363.
[22] Gleich (2001), p. 367.
[23] Gleich (2001), pp. 383–394.

	Yesterday	Today	Tomorrow
Universal differentiation between performance levels	None	Weak	Strong
Strategic and operative targets and indicators	Predominantly financial	Financial and partly non-financial	Balance between non-financial and financial
Linking of strategic and operative planning	Not established	Partially established	Established
Reconciliation of targets and strategies	Low	On some performance levels	On all performance levels
Stakeholder influence	Dominated by shareholders	Average influence	Strong influence
Choice of indicators and planned targets	Topdown	Partially autonomous on performance levels	Fulyautonomous on performance levels
Flexibility/adaptability for change in indicators	Low	Medium	High
Indicators for performance targets	Unbalanced	Hardly balanced	Balanced
Allocation of roles of participants in Performance Measurement	Unbalanced	Hardly balanced	Balanced
Use of new managerial tools in Performance Measurement	No use	Some tools used	Many tools used

Fig. 1 Development stages of performance measurement

At the end of this chapter it has to be noted that performance measurement and management control concepts in the English-speaking world as well as controlling concepts in German-speaking areas have many interdependencies. The underlying aim behind these concepts is to steer or influence the behavior of members of an organization, especially managers, in such a way as to increase the likelihood of achieving goals.[24] Performance measurement – which forms an interface between planning, control, and information systems in the same way as controlling systems do – does not only enhance controlling in terms of time and target groups, but also by means of the format information is delivered in (qualitative information instead of quantitative information) and by means of non-financial indicators.[25] Nevertheless, performance measurement should still be regarded as an element (subsystem) of the controlling system with a special focus on supporting strategy implementation.[26]

---

[24] Flamholtz (1996), p. 597.

[25] Müller-Stewens (1998), p. 37.

[26] Horváth, Arnaout, Gleich, and Seidenschwarz (1999), p. 290; Anthony and Govindarajan (1998), p. 8 and p. 461.

3 Performance Measurement for Innovation

A performance measurement system for a company is very complex. To reduce the complexity, a differentiation via subsystems such as performance measurement systems for different departments or main and cross-divisional corporate functions, is necessary.[27] Based on the insight that innovation management has to be both effective and efficient and that it demands particular attention besides other, more routine, activities, it can be assumed that innovation is one of these subsystems and an innovation performance measurement should be established.

However, the question arises why innovation is so important that a particular performance measurement system is necessary. To clarify this, innovations need to be defined. They can be characterized with the following attributes:[28]

- Strategic relevance
- Uncertainty of outcome
- Fundamental investments
- Complexity, cross-functional tasks
- Knowledge and collaboration intensive processes
- Involving internal and external stakeholders
- Difficult to plan because of the novelty.

It becomes evident that, on the one hand, innovations have a great importance to the medium- and long-term success of companies. Therefore, a company has to ensure that innovations are managed effectively. But, on the other hand, innovations are insecure, uncertain, and involve a lot of different internal and external stakeholders. Therefore, their success is difficult to predict. This leads to a dilemma: The more innovations a company pursues and the more fundamental innovations are, the more important planning and controlling of these innovations becomes. But the higher the number of parallel innovation projects and the more radical their scope, the more difficult planning and controlling are. Performance measurement for innovation should help to cope with this situation.

To develop a performance measurement system, Neely et al. suggest the following procedure:[29]

1. Decide what should be measured.
2. Decide how it is going to be measured.
3. Collect the appropriate data.
4. Eliminate conflicts in the measurement system.

---

[27] See e.g. the performance measurement system for supply management in Gleich, Henke, Quitt, and Sommer (2009).
[28] Vahs and Burmester (2002), pp. 51–57.
[29] Neely, Mills, Platts, Gregory, and Richards (1996), p. 425.

Points 1 and 2 are discussed in this paper. Points 3 and 4 are not included in the following explanations, because these steps are company specific.

To be able to conceptualize a performance measurement system for innovation and to decide what needs to be measured, a common understanding of innovation management is necessary. Innovation management is the conception and implementation of a company's innovation system.[30] It covers both R&D and non- R&D innovations, which can be products, services or processes or be concerned with the applied business model. Based on existing approaches to conceptualize innovation management ability, the following dimensions can be seen as parts of innovation management:[31]

- Innovation strategy (and portfolio)
- Innovation culture
- Innovation structure
- Innovation competences and learning.

Consequently, innovation management capability and performance (not equal to innovation performance) represents a multi-dimensional framework. Thus measuring the performance of innovation management needs to holistically picture different dimensions.

As mentioned above, the use of different dimensions and levels is a precondition for the success of performance measurement systems. Correlations within performance levels as well as level spanning correlations can be visualized and used for steering.[32] Figure 2 demonstrates the above dimensions of innovation management complemented by innovation projects and innovation fields. The innovation strategy plays a particular role, as fundamental strategic decisions have a major influence not only on the other dimensions, but also on the concrete innovation fields.

As seen in Fig. 2, Innovation performance measurement can be classified into three different levels:

- Company level (1), *innovation management performance*: This includes innovation culture, innovation competences/learning, innovation structure and innovation strategy.
- Multi-project level (2), *innovation portfolio performance*: Portfolio management is defined as a dynamic decision process in which a company's active innovation projects are constantly updated and revised.
- Single-project level (3), *innovation project performance*: A project represents a team-based approach to execute innovation processes. Practice shows that

---

[30] Hauschildt (2004), pp. 29–30.

[31] For example Adams, Bessant, and Phelps (2006), pp. 26–38; Lawson and Samson (2001), pp. 388–395; Sammerl (2006), pp. 197–204.

[32] Gleich (2001), p. 12.

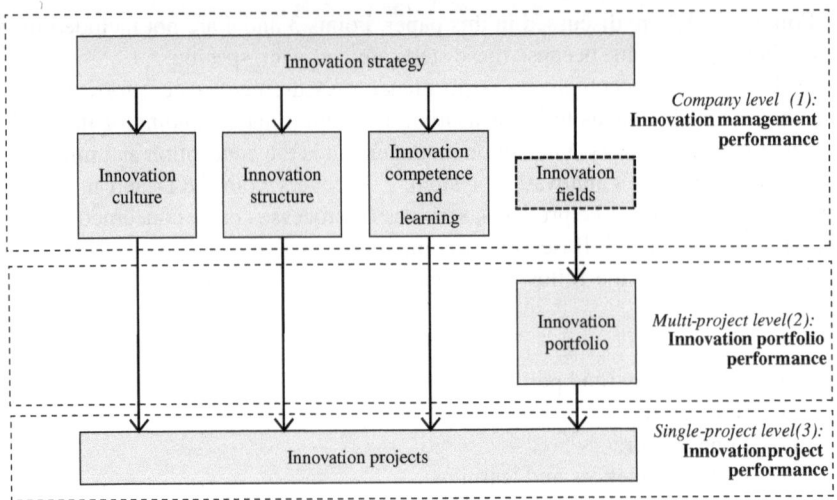

Fig. 2 Levels and dimensions of an innovation performance measurement system

- projects are the most common and important organisational form to put innovations into action. Each innovation project needs to be considered as a planning and controlling object.

The measurement of performance on all three levels allows a detailed understanding of innovation activities and results as well as of strategy implementation. It is of great significance to link the different levels and aspects to each other. Starting top down, the innovation strategy needs to be considered in the innovation culture, innovation competences/learning and innovation structure, as well as via the different innovation fields, in the innovation portfolio. The strategic decisions made on the first level need to be translated into specific goals and activities as input for the other dimensions and levels. The goals of the multi- project landscape need to be split up into different projects. Thinking bottom up, the status reports of single projects are aggregated as an input for the portfolio management on the second performance level, the portfolios themselves in the overall level.

4 Performance Levels in Innovation Performance Measurement

4.1 Company Level (1) – Innovation Management Performance

4.1.1 General Tasks and Contents

The definition of an *innovation strategy* represents a crucial element of a holistic innovation management system. The innovation strategy is derived from the company strategy and should thus be aligned with the latter. During the strategy process, innovation fields are determined. These are topics/fields in which the company wants to foster innovations. One innovation field can comprise one or a multitude of innovation projects.

According to Cooper et al. the innovation strategy has to embrace different kinds of decisions:[33]

- Selection of target markets, products and technologies to invest into
- Allocation of resources corresponding to each field of innovation
- Preselection of specific ideas and projects within the innovation fields
- Ensuring of a balanced innovation portfolio which fits the identified targets, available resources, and time horizons.

The *innovation culture* is the sum of the innovation-related attitudes, experiences, beliefs and values of the employees in an organization. Innovation culture has a coordinating function,[34] drives innovation activities and represents the environment in which they take place. A company can only be innovative, if the overall culture in the company allows and supports this.[35] "Companies that know how to innovate don't necessarily throw money into R&D. Instead they cultivate a new style of corporate behavior that's comfortable with new ideas, change, risk, and even failure."[36]

In order to measure the performance of the cultural dimension of innovation management, measures developed by Amabile et al. can be applied. They suggest criteria in the following dimensions:[37]

- Encouragement of creativity (organizational encouragement, supervisory encouragement, work group encouragement), e.g. readiness to take risks, fairness with idea evaluation, recognition and rewarding practice for creativity
- Freedom and autonomy as a prerequisite for innovative work
- Resource adequacy and its effect on motivation
- Pressure between fostering efficiency and inhibiting creativity
- Organizational aspects to impede creativity.

The *innovation structure* contains the innovation-related organizational aspects of a company. The organizational structure represents the backbone of innovation processes and of innovation projects. It links structured activities with

---

[33] Cooper, Edgett, and Kleinschmidt (1999), pp. 334–335.
[34] Sammerl (2006), p. 209.
[35] O'Reilly and Rao (1997), pp. 60–64; Amabile, Conti, Coon, Lazenby, and Herron (1996).
[36] O'Reilly and Rao (1997), p. 60.
[37] Amabile et al. (1996), pp. 1158–1162.

roles and responsibilities. Several authors suggest measures for innovation processes covering cost, time and quality dimensions as well as profit and customer satisfaction.[38] However not only process structure but also organizational structure should be considered in this dimension.[39] This includes the appropriateness of roles and responsibilities and of formal structures to execute innovation processes (e.g. decision boards, innovation teams and innovation project offices).

Innovation competences and learning represent the basis of innovation activities. Innovation derives from the combination of previously unconnected knowledge; thus both the ability and the performance in developing knowledge and building up competences form a crucial part of an innovation management system. Sammerl distinguishes between internal and external learning of an organization:[40]

- Internal learning refers to the creation of new knowledge within the company and is based on existing internal resources and people.[41]
- External learning refers to the integration of knowledge from outside the company, e.g. from partners, competitors, research institutes or customers. This demands certain learning processes and structures to achieve permeability of knowledge from the company's environment.

Measuring organizational learning ability and performance leads to a number of methodological problems due to the intangible nature of knowledge and learning.[42] Therefore, approaches to measure learning ability and performance should focus on the following factors:[43]

- Learning behavior of the members of the organization
- Management commitment to learning and knowledge management
- Openness and experimentation
- The exchange of knowledge
- Social networks used for knowledge transfer
- Systematic knowledge management.

---

[38] Cooper and Edgett (2008); Davila, Epstein, and Matusik (2004); Hauschildt (2004), pp. 537–542.

[39] Innovation processes are organized in the form of projects in many cases. Thus some authors do not draw a clear distinction between the innovation process and innovation projects.

[40] Sammerl (2006), pp. 197–198.

[41] Research and development is an example of a unit fostering internal learning of an organization.

[42] Probst, Raub, and Romhardt (2006), p. 221; Jerez-Gómez, Céspedes-Lorente, and Valle-Cabrera (2005), p. 715.

[43] For example Jerez-Gómez et al. (2005); Probst et al. (2006), pp. 221–222; Chiva, Alegre, and Lapiedra (2007), pp. 226–228; Kianto (2008), pp. 72–75.

4.1.2 Design of Performance Measurement System

Even if the 'soft' aspects that have been mentioned are difficult to plan and steer, they have to be considered in a holistic innovation performance measurement system. Therefore, a company needs to find measures which enable it to control culture, competences and structure. Suggestions for measures were given during the description of the different dimensions of innovation management above.

Among different performance measurement approaches (see Chap. 2), the balanced scorecard appears most suitable for the measurement of innovation performance on company level.[44] In order to consider the holistic approach of innovation management, the following dimensions of the innovation scorecard, regarding the different parts of the innovation management system, are recommended: portfolio, culture, structure (organization, processes), competences/learning and financials.

It has to be noted that companies should, despite all balanced scorecard euphoria, be careful with the isolated introduction of the balanced scorecard concept, as it is heavily focused on strategic performance measurement and many of the components mentioned in Chaps. 2 and 3 are not per se taken into consideration. A balanced scorecard certainly does, as a rule, form the basis for a comprehensive application of performance management. But there are more levels and aspects which have to be addressed. Only then can those ideas, concepts, and strategies be brought into (operative) action right down to the last performance level.

4.1.3 Link with Other Innovation Performance Levels

The aspects on this level of the innovation performance measurement system are the basis for the innovation projects as well as for the innovation portfolio. Therefore, they are a prerequisite for putting innovation management into action.

4.2 Multi-Project Level (2) – Innovation Portfolio Performance

4.2.1 General Task and Contents

The second innovation performance level is represented by the performance of the innovation project portfolios. The focus is placed on effectiveness – "doing the right projects".[45]

Portfolio management was first addressed in the area of financial investments,[46] but has been brought forward to other disciplines, and also has arrived in the field of innovation. In this paper, portfolio management is defined as a dynamic decision process in which a company's active innovation projects are constantly updated and

---

[44] See Gleich, Nestle, and Sommer (2009) or Möller and Janssen (2009), p. 95 for examples of innovation balanced scorecards.
[45] Cooper (1999), p. 115.
[46] Markowitz (1952).

revised. Within the scope of this activity, new projects are evaluated, prioritized, and selected. Furthermore, existing projects may be accelerated, stopped, or deprioritized. Therefore, portfolios help to bridge the gap between planned projects and required resources in the course of time. To sum up, the following goals are addressed with portfolio management:[47]

- The first goal of portfolio management is to maximize the financial value of the company's (innovation) portfolio.
- Secondly, it has to ensure a balance among the various projects of the company.
- Thirdly, it helps to fit the number of projects to the available capacity.
- Ensuring that the company's portfolio reflects the (innovation) strategy is the fourth major objective of portfolio management.

If companies do not establish an effective portfolio management system, it is likely that they will engage in innovation projects which are not promising. Thus, significant financial, human, and other resources are allocated to ineffective innovation projects.[48] In addition to that, companies run the risk of creating a project overload, resulting in decelerated innovation development, and missed targets.

Portfolio management builds up on strategic innovation fields which are derived from the innovation strategy. Within these fields, mentioned in performance level one, the portfolio of different innovation projects need to be planned and controlled. The following activities have to be carried out:

- Definition of the innovation project portfolio and initial selection of projects: evaluation (e.g. risk and profitability) and selection of ideas and innovation projects to pursue
- Planning and allocation of resources according to strategic objectives
- Continuous evaluation of target achievements (contribution to strategic innovation targets) of the projects in the portfolio
- Control of project portfolios
- Coordination of projects.

The controlling of project portfolios comprises both strategic and operational aspects. Strategic controlling of project portfolios implies the selection and evaluation of project alternatives, which feed the innovation pipeline, and the monitoring of goal achievements and thus the contribution to strategic innovation targets (effectiveness) of projects in the portfolio. Innovation projects have to be initially evaluated concerning risks, success probability, expected returns, competence enhancement and achievement of competitive advantages. The budgeting process represents the linking point of strategic and operational aspects of planning. According to strategic prioritizations of projects within the portfolio, the

---

[47] Killen, Hunt, and Kleinschmidt (2008), p. 27.
[48] Balachandra and Raelin (1980), p. 24.

resources for innovation activities are distributed. The operational controlling provides information about the progress and target achievement of the projects in the portfolio.

Since different companies have different strategies and objectives, some innovations may be appropriate for one company while they may be inappropriate for others. In addition to that, new innovation projects can differ in their level of novelty and consequently require different levels of resources to launch the innovation on the market. Consequently, an effective portfolio management has to consider this to support the company's management in making decisions about new innovation projects that are appropriate to assure the company's long-term survival and competitiveness.[49]

4.2.2 Design of Performance Measurement System

The innovation portfolio can be evaluated according to the (discounted) value of the projects or according to risk and strategic alignment with the corporate strategy. Cooper et al. developed six key requirements for project portfolio performance which should be measured:[50]

- Projects are aligned with business's objectives
- Portfolio contains very high value projects
- Spending reflects the business's strategy
- Projects are done on time
- Portfolio has good balance of projects
- Portfolio has right number of projects.

Besides the measurement, the main objectives of this level are the creation of an overview about the different projects and their prioritization, to allow the allocation of resources, the identification of gaps and the setting of actions. Criteria can be urgency, revenue impact and (strategic) importance. A possible solution is shown in the Fig. 3.

Another portfolio visualizes the project pipeline. A company needs to know which projects are positioned in which phases. This allows possible gaps in future to be predicted as soon as possible.

As shown in the Fig. 4, the phases of the innovation process are visualized on the horizontal axes, whereas the different innovation fields of the company are shown on the vertical axes. The size of the circle shows the planned turnover of different projects.

Another portfolio which can be used in this part of the performance measurement system is a mapping of the degree of cost consumption with the technical realization grade of a project (Fig. 5). If there is a mismatch, problems can occur.

---

[49]Calantone, di Benedetto, and Schmidt (1999), p. 66.
[50]Cooper, Edgett, and Kleinschmidt (2001), pp. 373–374.

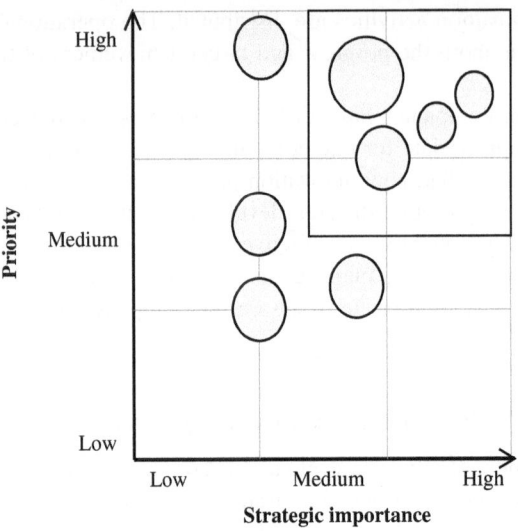

Fig. 3 Priority/strategic relevance portfolio

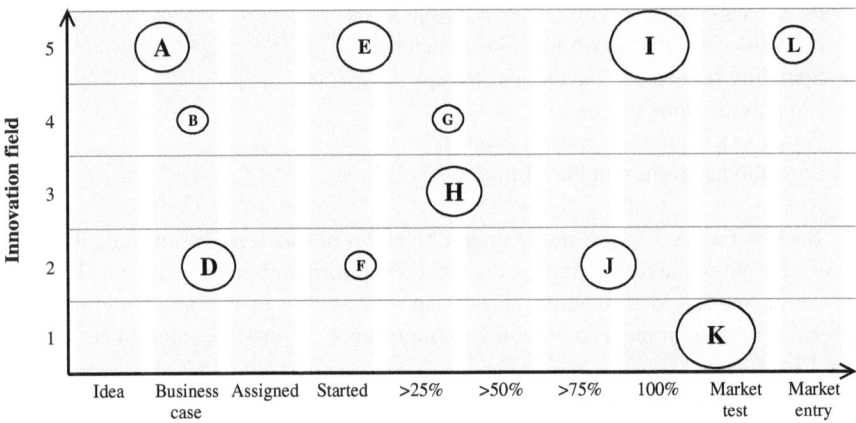

Fig. 4 Innovation pipeline

The portfolios shown on this level give an overview of the projects which allows the latter to be managed. Aggregated KPIs out of the portfolios should be used to integrate the results in the portfolio perspective of an overall Balanced Scorecard.

4.2.3 Link with Other Innovation Performance Levels

This level has a sandwich position between project and company level. It should enable different projects in different innovation fields of a company to be linked with the overall strategy.

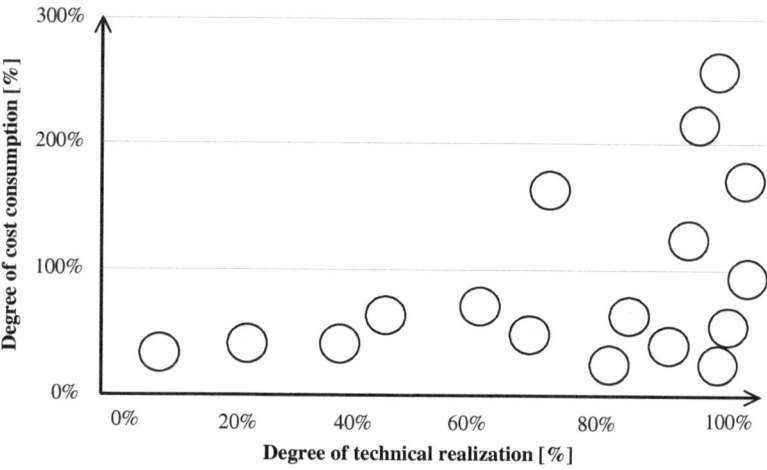

Fig. 5 Comparison between technical realization and cost consumption

4.3 Single Project Level (3) – Innovation Project Performance

4.3.1 General Tasks and Contents

Choosing the right projects is only half the way to ensure an organization's ability to survive in the long-term. Even if the right innovation projects are selected, it remains important to assess whether the execution of every single project is successful. More precisely, companies face the challenge of measuring the performance of innovation projects. It needs to be focused on

- Monitoring the efficiency of projects
- Monitoring the project progress, especially early-stage detection of deviation from plans
- Analysis of single project phases and overall project performance
- Comparisons of innovation projects with previous or parallel projects.

Performance measurement on a single-project level focuses on efficiency – "doing the projects right".[51] It refers to the different goals costs, time and quality but also to qualitative aspects such as stakeholder satisfaction, knowledge enhancement, learning and conformity to strategy.

4.3.2 Design of Performance Measurement System

In order to assure efficiency and effectiveness of projects, a number of tools can be applied. Examples are milestone trend analysis, project reporting, project status

---

[51] Cooper (1999), p. 115.

analysis or cost trend analysis. Another tool which can be applied is target costing. This strategic cost management allows the entire life cycle of a product and influencing the performance of innovation project in the early stages of product development to be considered.[52]

For the measurement of innovation project performance, approved approaches of project performance measurement can be resorted to. Some inherent characteristics of innovation projects such as goal uncertainty, difficulties in planning of results and demanded resources or a high degree of interdisciplinarity of project execution, represent possible obstacles for innovation project controlling, and need to be considered.

4.3.3 Link with Other Innovation Performance Levels

The aggregated project performance represents the input for the project portfolio level. Thus activities from early stages of the innovation process to the market launch of innovative products account for this level. Status reports of single projects are aggregated and used as an input for the second performance level, innovation portfolio performance.

5 Conclusion

The management of innovations demands deliberate controlling and performance measurement approaches. This is due to the nature of innovations, the organizational framework in which innovations are developed and the close connection between innovation and strategy. Furthermore, innovation activities do not only have to be effective and thus contributive to superordinate strategic objectives. Innovation activities need to be efficient and to be kept within the bounds of budgets and other resources. Innovation performance measurement addresses the three levels of innovation performance: innovation management performance on company level, innovation portfolio performance on a multi- project level and innovation project performance on a single-project level.

On the company level, an innovation scorecard to measure performance of the innovation system including culture, structure (organization and processes), competences/learning and financial aspects as well as the portfolio, is suggested. The innovation scorecard helps to evaluate a company's ability to manage innovation. On the multiproject level, the bundle of innovation projects is addressed. The innovation portfolio performance links the innovation strategy in different innovation fields with innovation activities carried out as projects. The innovation portfolio supports the definition of budgets for single innovation projects, tracks progress and goal achievement of these projects and allows the innovation portfolio to be steered. The third performance level focuses on the controlling of single innovation projects.

---

[52] Sakurai (1989).

For this purpose, common methods of project management and control can be used to improve goal achievement within budget and time. The performance measurement activities on these three levels must be linked. This link takes place in both an upward and downward direction.

A concept for innovation performance measurement has been introduced in this paper. However, there are still other aspects concerning this topic which should be analyzed in further research. This can be the identification of good practice approaches or the identification of success factors. A focus should also be placed on the development of KPIs for each performance level.

References

Adams, R., Bessant, J., & Phelps, R. (2006). Innovation management measurement: A review. *International Journal of Management Reviews, 8*, 21–47.
Albala, A. (1975). Stage approach for the evaluation and selection of R&D projects. *IEEE Transactions on Engineering Management, 23*, 153–164.
Amabile, T. M., Conti, R., Coon, H., Lazenby, J., & Herron, M. (1996). Assessing the work environment for creativity. *Academy of Management Journal, 39*, 1154–1184.
Anthony, R. N., & Govindarajan, V. (1998). *Management control systems*. Boston: McGraw-Hill.
Balachandra, R., & Raelin, J. A. (1980). How to decide when to abandon a project. *The International Journal of Research Management, 23*, 24–29.
Boutellier, R., Völker, R., & Voit, E. (1999). *Innovationscontrolling – Forschungs- und Entwicklungsprojekte gezielt planen und steuern*. München: Hanser.
Brown, D. M., & Laverick, S. (1994). Is your measurement system well balanced? *Journal for Quality and Participation, October/November*, 6–11.
Calantone, R. J., di Benedetto, C. A., & Schmidt, J. B. (1999). Using the analytic hierarchy process in new product screening. *Journal of Product Innovation Management, 16*, 65–76.
Chiva, R., Alegre, J., & Lapiedra, R. (2007). Measuring organizational learning capability among the work force. *International Journal of Manpower, 28*, 224–242.
Cooper, R. G. (1999). The invisible success factors in product innovation. *Journal of Product Innovation Management, 16*, 115–133.
Cooper, R. G., & Edgett, S. J. (2008). Maximizing productivity in product innovation. *Research Technology Management, 51*, 47–58.
Cooper, R. G., Edgett, S. J., & Kleinschmidt, E. J. (1999). New product portfolio management: Practices and performance. *Journal of Product Innovation Management, 16*, 333–350.
Cooper, R. G., Edgett, S. J., & Kleinschmidt, E. J. (2001). Portfolio management for new product development: Results of an industry practices study. *R&D Management, 31*, 361–380.
Davila, T., Epstein, M. J., & Matusik, S. F. (2004). Innovation strategy and the use of performance measures. *Advances in Management Accounting, 13*, 27–58.
Engel, K., & Diedrichs, E. (2006). Die Kunst, Innovationen erfolgreich zu Management. In: Garn, M. & Kalt, G. (Eds.), *Innovationstreiber am Standort Deutschland*. Unternehmen übernehmen Verantwortung, pp. 72–78.
F.A.Z.-Institut für Management-Markt- und Medieninformationen, Frankfurt am Main.
Flamholtz, E. G. (1996). Effective organizational control: A framework, applications and implications. *European Management Journal, 14*, 596–611.
Gleich, R. (1997). Stichwort performance measurement. *Die Betriebswirtschaft, 57*, 114–117.
Gleich, R. (2001). *Das System des Performance Measurement – Theoretisches Grundkonzept, Entwicklungs- und Anwendungsstand*. München: Vahlen.
Gleich, R., Nestle, V., & Sommer, L. (2009). Innovationsorientiertes performance measurement. Entwicklung und Implementierung einer Innovation Scorecard am Beispiel der Festo AG &

Co. KG. In: Fisch, J. H., Roß, J. M. (Hrsg.), *Fallstudien zum Innovationsmanagement – Methodengestützte Lösung von Problemen aus der Unternehmenspraxis*, Wiesbaden, pp. 187–204.

Gleich, R., Henke, M., Quitt, A., & Sommer, L. (2009). New approaches in performance measurement: Methods for specification and operationalisation within the context of supply management. *International Journal of Business Excellence, 2*, 105–123.

Granig, P. (2007). *Innovationsbewertung. Potentialprognose und -steuerung durch Ertrags- und Risikosimulation*. Wiesbaden: DUV.

Greiner, O., Römer, S., & Russo, P. (2009). *Innovationsstudie 2009: Das verschwendete Innovationspotenzial – geniale Ideen im Unternehmen finden und nutzen*. Stuttgart: Horváth & Partners Management Consultants.

Günther, T., & Grüning, M. (2002). Performance Measurement-Systeme im praktischen Einsatz. *Controlling, 14*, 5–13.

Hall, B. H., & Mairesse, J. (1995). Exploring the relationship between R&D and productivity in french manufacturing firms. *Journal of Econometrics, 65*, 263–293.

Hauschildt, J. (2004). *Innovationsmanagement*. München: Vahlen.

Horváth, P., Arnaout, A., Gleich, R., & Seidenschwarz, W. R. (1999). Neue Instrumente in der deutschen Unternehmenspraxis – Bericht über die Stuttgarter Studie. In: Egger, A., Grün, O., & Moser, R. (Eds.), *Managementinstrumente und -konzepte, Entstehung, Verbreitung und Bedeutung für die Betriebswirtschaftslehre* (pp. 289–328). Stuttgart: Schäffer-Poeschel.

Jerez-Gómez, P., Céspedes-Lorente. J., & Valle-Cabrera, R. (2005). Organizational learning capability: A proposal of measurement. *Journal of Business Research, 58*, 715–726.

Kaplan, R. S., & Norton, D. P. (1992). The balanced scorecard – measures that drive performance. *Harvard Business Review, 70*, 71–79.

Kaplan, R. S., & Norton, D. P. (2003). *The strategy-focused organization – how balanced scorecard companies thrive in the new business environment*. Boston: Harvard Business School Press.

Kianto, A. (2008). Development and validation of a survey instrument for measuring organisational renewal capability. *International Journal of Technology Management, 42*, 69–88.

Killen, C. P., Hunt, R. A., & Kleinschmidt, E. J. (2008). Project portfolio management for product innovation. *International Journal of Quality and Reliability Management, 25*, 24–38.

Lawson, B., & Samson, D. (2001). Developing innovation capability in organisations: A dynamic capabilities approach. *International Journal of Innovation Management, 5*, 377–400.

Littkemann, J. (Ed.). (2005). *Innovationscontrolling*. München: Vahlen.

Littkemann, J., & Holtrup, M. (2008). Evaluation von Dienstleistungsinnovationen – Möglichkeiten und Grenzen aus Sicht des Controllings, Der Controlling-Berater, pp. 261–284.

Markowitz, H. M. (1952). Portfolio selection. *Journal of Finance, 7*, 77–91.

Möller, K., & Janssen, S. (2009). Performance Measurement von Produktinnovationen. Konzepte, Instrumente und Kennzahlen des Innovationscontrollings. *Controlling, 21*, 89–96.

Müller-Stewens, G. (1998). Performance Measurement im Lichte eines Stakeholderansatzes. In: Reinecke, S., Tomczak, T., & Dittrich, S. (Eds.), *Marketingcontrolling*, Thexis, St. Gallen, pp. 34–43.

Neely, A. (2005). The evolution of performance measurement research. Developments in the last decade and a research agenda for the next. *International Journal of Operations & Production Management, 25*, 1264–1277.

Neely, A., Gregory, M., & Platts, K. (1995). Performance measurement system design. *International Journal of Operations & Production Management, 15*, 80–116.

Neely, A., Mills, J., Platts, K., Gregory, M., & Richards, H. (1996). Performance measurement system design: Should process based approaches be adopted. *International Journal of Production Economics*, pp. 46–47/423–431.

O'Reilly, B., & Rao, R. M. (1997). The secrets of America's most admired corporations: New ideas, new products. *Fortune, 135*, 60–64.

Probst, G., Raub, S., & Romhardt, K. (2006). *Wissen managen: Wie Unternehmen ihre wertvollste Ressource optimal nutzen*. Wiesbaden: Gabler.

Raake, A. (2008). *Strategisches Performance Measurement, Anwendungsstand und Gestaltungsmöglichkeiten am Beispiel des Öffentlichen Personennahverkehrs*. Lit, Berlin.

Rummler, G. A., & Brache, A. P. (1990). *Improving performance. How to manage the white space in the organization chart*. San Francisco: Jossey-Bass.

Sakurai, M. (1989). Target costing and how to use it. *Journal of Cost Management, 3*, Summer, 39–50.

Sammerl, N. (2006). *Innovationsfähigkeit und nachhaltiger Wettbewerbsvorteil: Messung, Determinanten, Wirkungen*. Wiesbaden: DUV.

Schmeisser, W., Kantner, A., Geburtig, & A., Schindler, F. (2006). *Forschungs- und Technologie-Controlling. Wie Unternehmen Innovationen operativ und strategisch steuern*. Stuttgart: Schäffer-Poeschel.

Scholich, M., Gleich, R., & Grobusch, H., (2006). Innovation Performance – Das Erfolgsgeheimnis innovativer Dienstleister. PricewaterhouseCoopers, Deutsches Zentrum für Luft- und Raumfahrt (DLR), European Business School (EBS).

Storey, C., & Kelly, D. (2001). Measuring the performance of new service development activities. *The Service Industries Journal, 21*, 71–90.

Vahs, D., & Burmester, R. (2002). *Innovationsmanagement: von der Produktidee zur erfolgreichen Vermarktung*. Stuttgart: Schäffer-Poeschel.

Weiss, M., Zirkler, B., & Guttenberger, B. (2008). Performance Measurement Systeme und ihre Anwendung in der Praxis. Ergebnisse empirischer Studien. *Controlling, 20*, 139–147.

Accounting for Innovation: Lessons Learnt from Mandatory and Voluntary Disclosure

Thomas Günther

Contents

1 Value Relevance of Intangibles and Innovation   319
   1.1 Intangibles and Financial Information   319
   1.2 Value Relevance of Innovation   320
2 Accounting for Innovation in Financial Statements   321
   2.1 Accounting for Innovation in German Gaap   322
   2.2 Accounting for Innovation in IFRS   323
3 Voluntary Disclosure   327
4 Lessons Learnt from Disclosure on Innovation   328
References   329

1 Value Relevance of Intangibles and Innovation

1.1 Intangibles and Financial Information

In recent decades, Western economies have experienced a shift from the industrial sector to the service sector. Consequently, the major production factors are no longer tangible assets such as materials, machines and equipment but knowledge and information. In this knowledge economy, intangible resources such as brands, customer and supplier relations, know-how, networks and patents play a major role.

Intangible resources are defined negatively as non-material and non-financial resources of an organization that can be exploited for longer than the current financial year (Guenther, Kirchner-Khairy, & Zurwehme, 2004, p. 162). Intellectual capital is diversely and widely defined (e.g. Stewart, 1997, Edvinsson & Malone, 1997; Ordóñes de Pablos, 2003). Rastogi describes *intellectual capital* "as the holistic or meta-level capability of an enterprise to coordinate, orchestrate and

---

T. Günther (✉)
Dresden University of Technology, Dresden, Germany
e-mail: THOMAS.GUENTHER@tu-dresden.de

deploy its knowledge resources towards creating value in pursuit of its future vision" (Rastogi, 2003, p. 230).

Intellectual property represents those intangible resources that are legally protected by a property right such as e.g. brand names, patents or licenses. However, intangible resources become *intangible assets* if they fulfill the comprehensive definitions and requirements of US Gaap, IFRS or local Gaap on abstract or concrete recognition as an asset (e.g. IASC F. 49, IAS 38.7 for IFRS or SFAC 6.25 and .26 for US Gaap and Section 248 (2) for German Gaap) (WGARIA, 2005). Following Lev, "*an intangible asset* is a claim to future benefits that does not have a physical or financial ... embodiment" (Lev, 2001, p. 5). However, this narrow definition of intangible assets determines what can be recognized on a balance sheet.

Looking at the sources or rather resources of a business, success results from the resource-based view in addition to a market-based view (Penrose, 1959; Wernerfeldt, 1984; Rumelt, 1984; Barney, 1986; Dierickx & Cool, 1989). The ability of a company to create innovations, here defined as innovation capital, can be regarded as a "resource", as it fulfills the requirements set out in the literature (e.g. Barney, 1991). It is strategically valuable, and it is seldom and rarely not mobile and not substitutable. On the other hand, innovation capital creates sustainable competitive advantages only if combined with other resources of the firm such as *physical capital resources*, *human capital resources* and *organizational capital resources* following the classification of Barney (1991).

Financial statements with their strong focus on financials are criticized because of their limited capability to deliver adequate information on intangibles (Lev & Zarowin, 1999; Breton & Taffler, 2001). Market-to-Book ratios are often seen as an indicator for the declining ability of financial statements to explain market value (Mouritsen, Bukh, Larsen, & Johansen, 2002; Kaplan & Norton, 2001). Nevertheless, this argument is partially misleading as market values are also driven by external factors such as prime rates or stock market hypes as could be seen during the new economy bubble. However, empirical studies give evidence of a declining value relevance of financial information (Francis & Schipper, 1999, Lev & Zarowin, 1999; Gu, 2007).

1.2 Value Relevance of Innovation

"An *innovation* is the implementation of a new or significantly improved product (good or service), or process, a new marketing method, or a new organisational method in business practices, workplace organisation or external relations (OECD, 2005, p. 46)". In contrast, innovation capital describes the renewal abilities of a company and the related results in the form of intellectual property rights and other intangible assets (Edvinsson & Malone, 1997). Comparing the two definitions, innovation capital comprises the entire stage level process from the invention through the R&D process to the successful introduction or implementation of the new idea. Following the OECD definition, innovation itself is always connected with any kind of business success of the invention. For this chapter, innovation is in

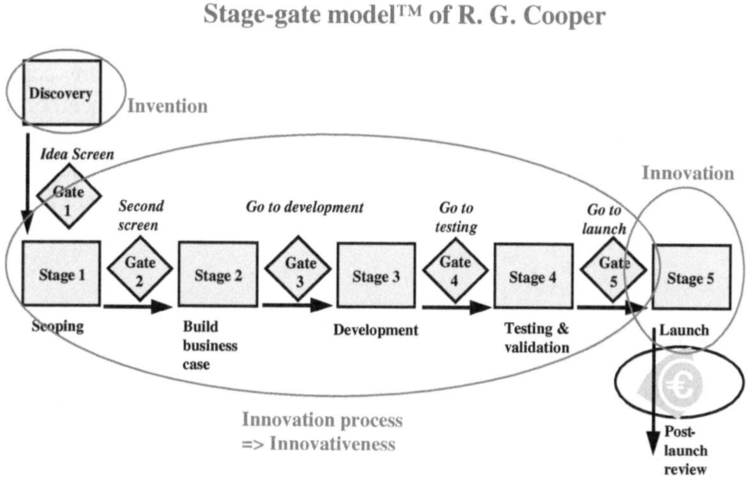

Stage-gate model™ of R. G. Cooper

Fig. 1 Stage-gate model for innovation

contrast to the OECD definition seen in a broader sense encompassing the entire stage-gate process (Cooper, 2008) (Fig. 1), thus allowing a look at the roots of innovations.

Since Schumpeter (1943), innovations have been recognized as strategic instruments to gain a competitive advantage leading at least temporarily to a monopolistic position and herewith to higher returns compared to non-innovative companies. Empirical research supports the positive effect of innovation on corporate performance (e.g. Ernst, 1995; Hall & Mairesse, 1995) and on market value (e.g. Griliches, 1981; Pakes, 1985; Deng, Lev, & Narin, 1999; Hall, Jaffe, & Trajtenberg, 2000). In addition, business-related studies also examine the drivers of innovation and deduct manifold and complex input and process drivers for the output "innovation" (e.g. meta-analyses of Montoya-Weiss & Calantone, 1994; Balachandra & Friar, 1997; Henard & Szymanski, 2001).

The impact of innovation on corporate performance and market value results in a demand of financial analysts and other stakeholders for information on innovation (Tasker, 1998; Eccles, Phillips, & Herz, 2001; Lev, 2001). Therefore, it is worthwhile looking at the ability of financial reporting to provide the requested information.

2 Accounting for Innovation in Financial Statements

In general, financial statements provide monetary mandatory information on innovation:

- Financial information on current expenses for innovation, like R&D expenses, if shown separately in the income statement,
- financial information on "investments" in innovation which are capitalized expenses or recognized acquisition costs, shown on the balance sheet, or
- additional financial information in the notes or in the MD&A, such as e.g. R&D expenses. Often non-financials in the general sections of annual reports, like the product pipeline in R&D or patents held or filed for application, are added on a voluntary basis and will be regarded later on.

2.1 Accounting for Innovation in German Gaap

Until 2009, according to Section 248 (2) HGB (*German Commercial Code*), non-acquired intangible fixed assets could not be recognized in the balance sheet. Therefore, any kind of self-created intangible assets, here especially innovation capital, could not be capitalized and consequently had to be expensed in the income statement. This reflected the primary orientation in German GAAP on debtors instead of shareholders and on the principle of prudence instead of decision usefulness and fair presentation in financial accounting. The same regulation still holds for tax purposes according to section 5 (2) EStG (German Income Tax Law). In 2001, the Working Group "Accounting and Reporting of Intangible Assets" recommended canceling section 248 (2) HGB (WGARIA, 2001). Due to the old legislation, German companies created separate legal entities for R&D activities, which sell know-how and intellectual property to the parent company where these intangibles can be recognized (and written-off) because they had been acquired from another legal entity. If customer-specific innovation is developed, these expenses can be recognized because these assets are allocated to current assets. Furthermore, rules for construction contracts (IAS 11) have to be applied to allocate earnings to periods.

In 2009, a new regulation for the German Commercial Code, *BilMoG*, was introduced which tends to harmonize the treatment of intangibles with IFRS and provide better investor orientation. According to BilMoG, section 248 (2) HGB now allows the option to recognize self-created intangible fixed assets. However, brands, print titles, copyrights, customer lists and similar intangible fixed assets (identically with IFRS) cannot be recognized. The option in section 248 (2) new HGB opens plenty of opportunities for a broader recognition of innovations on the balance sheet. According to section 255 (2a) new HGB, developments but not research costs for these intangible assets can be shown. This requires a reliable separation of research and development costs provided by adequate R&D accounting and cost allocations (WGARIA, 2008). To protect creditors, section 268 (8) new HGB forbids dividend payments due to recognition of these intangibles. To sum up, German HGB will exactly follow the current IFRS regulation. According to Regulation (EC) No. 1606/2002 of the European Parliament and the Council from July 19, 2002, all listed (stocks or bonds) EU companies have to disclose financial consolidated statements according to IFRS regulations. Consequently, the larger listed companies have to follow international law anyway. However, with the new regulations in local German Gaap, also SMEs and non-listed companies can follow but do not have to follow principles for innovation accounting set up by IFRS. The IFRS regulations will be analyzed next.

2.2 Accounting for Innovation in IFRS

In IAS 38.8, intangible assets are defined as "identifiable, non-monetary assets without physical substance". IAS 38.8 prescribes three critical attributes for an intangible asset: *identifiability* (separability from other assets or arising from contractual or other legal rights), *control* (the ability of a company to generate benefits from the asset) and *future economic benefits* (such as revenues or reduced future costs). This means for example that the human capital of excellent research or engineering staff is not an intangible asset because the company cannot have legal control of human beings. In addition, a good reputation (e.g. due to innovation) is not an intangible asset, as it cannot be separated from the company's other assets, even if it might be very valuable. Patents, trademarks and other intellectual property are legal rights and can therefore be identified and separated from other assets.

IAS 38 considers different ways of acquiring an intangible asset (purchase, part of a business combination, government grant, exchange of assets or self-creation) and therefore does not exclude self-created intangible assets per se. IAS 38.21 requires an enterprise to recognize an intangible asset only if it is probable that the future economic benefit will *flow to the company* and that the cost of the asset can be *measured reliably*. Both criteria are always satisfied in the case of an external acquisition. If both the definition of an intangible assets and the criteria for recognition are not met, IAS 38.68 requires the expenditure to be expensed when it is incurred and therefore to diminish net income of the current year. At a later date, a reinstatement, i.e., recognition as an intangible asset that had been expensed before is prohibited (IAS 38.71).

For self-created intangibles, such as know-how or patents created in R&D, IAS 38.40 differentiates between a research and a development phase. All research costs have to be immediately expensed as the standard takes the view that, in the research phase of a project, an enterprise cannot demonstrate that an intangible asset exists that will generate probable future economic benefits (IAS 38.43). In contrast to research which provides (general) new scientific or technical knowledge and understanding, development is the application of research findings or other knowledge to a plan or design for the production of new or substantially improved materials, devices, products, processes, systems or services prior to the commencement of commercial production or use (IAS 38.7). Disregarding this separation of research and development, which is difficult in practice, the split is crucial as "sheer knowledge" or "competences" are, in terms of IFRS, not regarded as an investment, which conflicts with the resource-based view of the innovation capital philosophy. In addition, development expenses can only be recognised if an enterprise can demonstrate that all of the following six criteria are met:

(a) The *technical feasibility* of completing the intangible asset so that it will be available for use or sale;
(b) its *intention to complete the intangible asset* and use or sell it;
(c) its *ability to use or sell* the intangible asset;
(d) the ability to show how the intangible asset will *generate probable future economic benefits*. Among other things, the enterprise should demonstrate the existence of a market for the output of the intangible asset or the intangible asset itself or, if it is to be used internally, the usefulness of the intangible asset;
(e) the *availability of adequate technical, financial and other resources* to complete the development and to use or sell the intangible asset; and
(f) its *ability to measure the expenditure attributable* to the intangible asset during its development reliably (R&D and project controlling).

If the research and development phases cannot be separated, expenses are generally treated as research costs and immediately expensed. This implies, on the one hand, a profound documentation of the development phase within the company and, on the other hand, an earnings management potential if a company intentionally designs its project and process organization to meet or not meet the requirements. Therefore it is no surprise that capitalization rates vary between industries and within industries (for DJ Stoxx 200 Hitz, 2007, for German IFRS prepares Hager & Hitz, 2007, Wulf, 2008, p. 139 f.) and might be influenced by management decisions as well as external factors such as industry culture or characteristics of the industry's stage-gate process.

An R&D project acquired in a business combination is recognised as an asset at cost (*In-Process R&D*), even if a component is research, which is not consistent with the treatment of self-created research output. Subsequent expenditure on that project is accounted for as any other research and development cost (expensed except to the extent that the expenditure satisfies the criteria in IAS 38 for recognising such expenditure as an intangible asset) (IAS 38.34).

However, some internally generated intangibles which are in general considered as "innovation" cannot be recognized as assets: brands, mastheads, publishing titles, customer lists and items similar in substance (IAS 38.63), internally generated goodwill (IAS 38.48), start-up, pre-opening, and pre-operating costs, training costs, advertising and promotional costs, including mail order catalogues and relocation costs (IAS 38.69). These items might fulfill the criteria of future economic benefits and control, but the standard rejects identifiability and separability (IAS 38.69). However, for brands there is at least a grey market where brand names without products are traded and, for Germany, the Trademarks Act (MarkenG) generates an intellectual property, which fulfills the definition of an intangible asset, but conflicts with the prohibition in IAS 38.63.

Initial measurement for intangible assets is at cost (acquisition costs or production costs) (IAS 38.24). Costs prior to the first-time fulfilment of the definition and recognition criteria cannot be recognized in the balance sheet and have to be expensed. So, the carrying amount on the balance sheet does not reflect the total costs of creation seen from an economic point of view. For internally generated intangibles like capitalized development costs, companies should have a solid cost accounting system for the development projects, clearly distinguishing between projects. All direct costs (like materials, services, salaries and wages etc.) and all overheads that can be allocated on a reasonable and consistent basis can be included, but selling, administration and other general overheads, inefficiencies and initial losses and training costs are definitely excluded (IAS 38.66). So, this full cost approach is in accordance with the measurement of inventories in IAS 2, but incorporates the same possibilities for earnings management (like e.g. the allocation of overheads and staff costs to projects).

Subsequent to an acquisition, measurement follows primarily a so-called benchmark treatment, the cost model, i.e., carried at cost less any amortization and impairment losses (IAS 38.74). Only if an active market exists, fair values less subsequent amortization and impairment losses can be applied (IAS 38.75–85). For amortization and impairment, intangible assets are classified as indefinite life (no foreseeable limit to the period generating cash for the enterprise, e.g. a broad, trend-setting technology like Li-ion batteries) and finite life (a limited period of benefit to the company, e.g. the maturity of a patent or the time span of a life cycle contract for a supplier in automotives). Whereas for finite lives an amortization on a systematic basis (e.g. straight-line method or amortization due to production volume in life cycle contracts) has to be calculated, intangible assets with indefinite lives are not amortized. However, both types of assets should be assessed for impairment in accordance with IAS 36. In addition, numerous financial information items on measurement, but no non-financials on these intangible assets, have to be disclosed (IAS 38.118 and 38.122).

To conclude, the legal opportunities to show "intangible capital" on the balance sheet are limited to development costs with clear foreseeable marketability, data is "only" financial and disclosure in notes does not require non-financial data. The recognition as well as the initial and subsequent measurement may be the subject of earnings management, which might restrict reliability and relevance in disclosure.

IAS 38 also reflects acquisition of intangible assets in the case of a business combination. However, IFRS 3 Business combinations determine how the purchase price of an M&A transaction is allocated to assets (purchase price allocation (PPA)) when using the so-called acquisition method (IFRS 3.14). After identifying acquired assets, which can be intangible assets in accordance with IAS 38, the assets are measured using their fair value. To determine the fair value, the benchmark approach is the market approach. If market values are not available, the income (e.g. relief from royalty or multi-period excess earnings approach) or cost approach can be used (IFRS 3.36). The difference between purchase price and identified assets results in goodwill. Guenther and Ott (2008) show in an explorative study of each 51 transactions of US Gaap and IFRS acquirers that more than 50% of total assets after the business combination are intangible assets (including goodwill) (Fig. 2).

The regulations in US Gaap, which are not focused in this article, are similar to IFRS 3 and IAS 38. Focusing on innovations, surprisingly enough, only 12.7% of identified intangible assets are technology-based assets (including in-process R&D) for US Gaap acquirers and 4.5% for IFRS acquirers (Fig. 3).

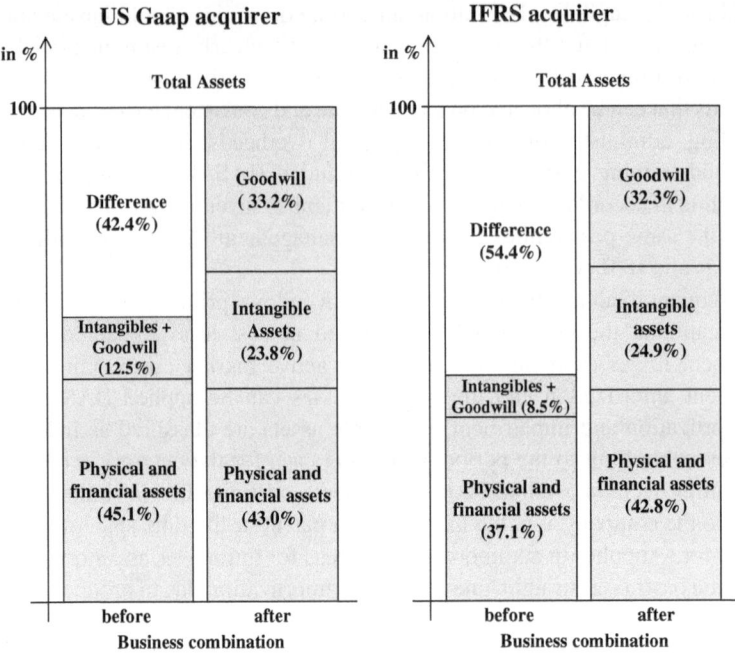

Fig. 2 Total asset pro and post mergers

The study (see also Glaum, Street, & Vogel, 2007) demonstrates the malfunctions of financial accounting, which is not able to disclose a true and fair presentation of intangibles before the business combination. However, the leeway for earnings management in recognition and measurement during PPA is considerable, raising doubts as to what the best of both ways of presentation might be.

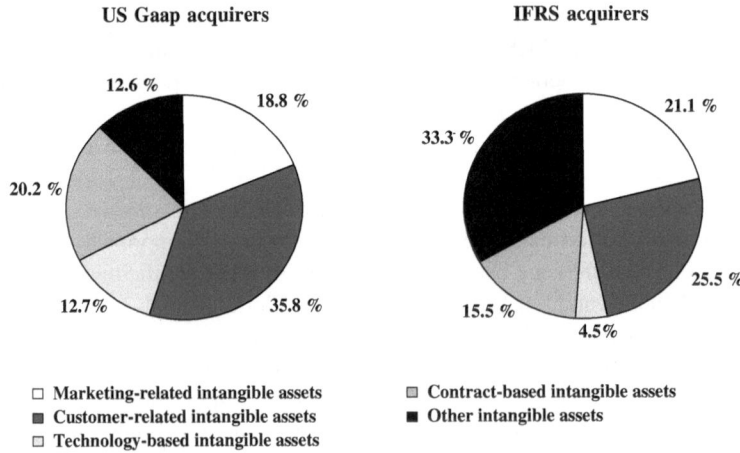

Fig. 3 Categories of intangibles

3 Voluntary Disclosure

In 1994, the Special Committee on Financial Reporting (Jenkins Committee) submitted a report demanding a stronger future orientation of business reporting and a focus on non-financials (AICPA, 1994). However, in 2004, the project "Disclosure about Intangible Assets" was removed from the agenda of FASB, and, at the IASB, the project "Intangible Assets" was not promoted for the active agenda of IASB. A broader mandatory reporting of financials and especially non-financials seems to be out of reach.

Nevertheless, numerous frameworks of so-called Intellectual Capital Statements have been suggested and discussed in the last decade (e.g. Skandia Navigator by Edvinsson & Malone, 1997; Intangible Assets Monitor by Sveiby, 1997; Intellectual Capital Navigator by Stewart, 1997; Analytical Accounting System for Intellectual Capital by Mouritsen, Larsen, & Bukh, 2001; Model for IC Statements for Austrian Universities by Leitner, Sammer, Graggober, Schartinger, & Zielowski, 2001; Value Chain Scoreboard by Lev, 2001; IC Statement of WGARIA, 2001, 2005; IC Statement Model "Wissensbilanz" of BMWA and AK Wissensbilanz, 2004). These models focus on intellectual capital as an entity; some are more focused on voluntary disclosure to investors and other stakeholders due to shortcomings of mandatory disclosure, and others focus more on performance measurement and management.

IC Statements follow the structure of performance measurement approaches. The IC Statement of WGARIA (2005), for example, consists of three sections. First, the general strategy of the management of intangibles in context with the long-term performance of the company has to be explained. Second, for each of the seven IC categories (innovation, human, customer, supplier, investor, process and location capital) the category-specific strategy is explained followed by a catalogue of indicators, an explanation and exact definition of the indicators and their interaction and finally comments on the IC category concerning further development. Third, a comment on all intangibles as an entity summarizes the IC statement (WGARIA, 2005, pp. 82–96). Figure 4 shows the indicators suggested by WGARIA for a voluntary disclosure on innovation capital. However, the percentage of companies disclosing IC statements is limited. Searching for IC Statements of often-cited frameworks (Austrian Research Centres Model, Danish Intellectual Capital Statement Guidelines, BMWA (2004), the Spanish Modelo Intellect and the framework Nordic Harmonised Knowledge Indicators), 260 IC Statements could be identified for European companies. Adjusting for ICS in English or German and for publicly available ICS, the sample is reduced to 126 ICS. Taking out public institutions and NPOs, only 52 ICS are available.

The disclosure of intangibles has to be seen in the light of a *cost-benefit consideration* of companies. Costs result from the preparation of additional information, its disclosure in public media and opportunity costs depending on the reaction of other stakeholders, especially external (damaging) effects from competitors. Benefits arise from lower costs of capital of investors (lower perceived risks or reduced agency costs) or other benefits from stakeholders (e.g. higher prices on markets because of perceived innovation, higher attractiveness for potential employees

Indicator	Explanation / Differentiation
R&D Expenses	Indication of R&D Expenses, R&D Expenses in relation to sales, information on deviation and concentration of R&D Ex-penses
Portfolio of patents and similar intellectual property rights	Number, structure and (residual) useful life of property rights and patents Presentation similar to fixed assets (initial value, increases, decreases, changes, final value), units only, no valuation intended
Patents and similar intellectual property rights filed for application	Number and structure of intellectual property rights and patents filed for application
Pending suits with patents and intellectual property rights	Number and importance of current pending suits with patents and intellectual property rights
Innovation rate	Net Sales of products introduced in the last three years in relation to Total Net Sales

Fig. 4 Indicators for innovation capital

etc.). For innovation capital, the major barriers are seen in the lack of measurability and objectivity (reliability) and the harm for the competitive position. Another reason is that often decision-useful indicators are missing in internal control systems (information gap). In the same survey of new economy industries, relevance of innovation capital was seen relatively high by top financial executives (Guenther, Beyer, & Menninger, 2005). Another reason for poor voluntary disclosure in ICS might be extended reporting on intangibles in annual reports (Abdolmohammadi, 2005; Vandemaele, Vergauwen, & Smits, 2005; Orens & Lybaert, 2007) and other channels of the company (e.g. Garcia-Meca, Parra, Larran, & Martinez, 2005 for 257 Spanish company presentations). Focusing on innovation capital within voluntary disclosure, empirical evidence shows, on the one hand, a relatively low level of reporting on innovation and, on the other hand, mostly qualitative or narrative reports. Reporting on human capital, and partially on customer and investor capital, is far more intensive than that on innovation (Haller & Dietrich, 2001; Speckbacher, Güldenberg, & Ruthner, 2004; Riegler & Kristandl, 2004; Hager & Hitz, 2007; Gerpott, Thomas, & Hoffmann, 2008). Looking at indicators, the most cited ones are patents, copyrights, "entrepreneurial spirit", active research projects, research cooperation ventures and intellectual property, i.e. output measured in property rights or input to the innovation stage-gate process. The reporting on process indicators as well as the general level of disclosure of innovation indicators is mediocre (e.g. the often replicated studies of Guthrie & Petty, 2000; Bozzolan, Favotto, & Ricceri, 2003; Bontis, 2003).

4 Lessons Learnt from Disclosure on Innovation

Summarizing the above analysis, a gap between the relevance of innovation for actual and potential investors and other stakeholders and the actual disclosure of information on innovation capital can be derived:

- Managing the innovation process and success of innovation in the marketplace is indisputably seen as a valuable resource of the company and a value driver for corporate performance and value.
- Current national (German Gaap) and international (IFRS) accounting limits the possibilities of disclosing innovation capital in a broader sense and is restricted to marketable, controllable development projects in accordance with IAS 38 Intangible Assets and IFRS 3 Business Combinations.
- Purchase price allocations generally increase total assets tremendously, thus revealing intangible assets that could not be shown before. However, innovation capital plays, on average, only a minor role. Empirical evidence shows that the disclosure of innovation varies between industries and from one firm to another.
- Various formats exist for voluntarily disclosing information on intangibles in IC statements, but usage is limited. The reasons are restricted measurability and objectivity of information and the fear of external effects by disclosing information to customers and competitors.
- If indicators on innovation are disclosed, they focus on legal property rights and input to the stage-gate process, process indicators being rarely reported.

References

Abdolmohammadi, M. J. (2005). Intellectual capital disclosure and market capitalization. *Journal of Intellectual Capital*, 6(3), 397–416.

AICPA. (1994). *Improving business reporting – A customer focus, meeting information needs of investors and creditors*. Jersey City: AICPA.

Balachandra, R., Friar, J. H. (1997). Factors of success in R&D projects and new product innovation: A contextual framework. *IEEE-Transactions on Engineering Management*, 44(3), 276–287.

Barney, J. B. (1986). Strategic factor markets: Expectations, luck and business strategy. *Management Science*, 32, 1231–1241.

Barney, J. B. (1991). Firm resources and sustained competitive advantage. *Journal of Management*, 17(1), 99–120.

BMWA/AK Wissensbilanz (2004). Wissensbilanz – Made in Germany Leitfaden 1.0 (IC Statement – Made in Germany, Guideline 1.0), Bundesministerium für Wirtschaft und Technologie BMWi, Berlin 2004, http://www.bmwi.de/. BMWi/Redaktion/PDF/W/ wissensbilanz-made-in-germany-leitfaden,property=pdf, bereich=bmwi, sprache=de, rwb= true.pdf. Accessed 30 June 2008.

Bontis, N. (2003). Intellectual capital disclosure in Canadian corporations. *Journal of Human Resource Costing and Accounting*, 7(1/2), 9–20.

Bozzolan, S., Favotto, F., & Ricceri, F. (2003). Italian annual intellectual capital disclosure – an empirical analysis. *Journal of Intellectual Capital*, 4(4), 543–558.

Breton, G., & Taffler, R. J. (2001). Accounting information and analysts stock recommendation decisions: A content analysis approach. *Accounting and Business Research*, 31(2), 91–101.

Cooper, R. G. (2008). Perspective: The Stage-Gate-Idea-to-Launch Process-Update, What's New and NexGen Systems. *Journal of Product Innovation Management*, 25(3), 213–232.

Deng, Z., Lev, B., & Narin, F. (1999). Science and technology as predictors of stock performance. *Financial Analysts Journal*, 55(3), 20–32.

Dierickx, I., & Cool, K. (1989). Asset stock accumulation and sustainability of competitive advantage. *Management Science*, 35, 1504–1510.

Eccles, R. G., Phillips, D., & Herz, R. H. (2001). *The value reporting revolution – moving beyond the earnings game*. New York: Wiley.

Edvinsson, L., & Malone, M. S. (1997). *Intellectual capital – realizing your company's true value by finding its hidden brainpower*. New York: HarperCollins.

Ernst, H. (1995). Patenting strategies in the German mechanical engineering industry and their relationship to company performance. *Technovation*, 15(4), 225–240.

Francis, J., & Schipper, K. (1999). Have financial statements lost their relevance? *Journal of Accounting Research*, 37(2), 319–352.

Garcia-Meca, E., Parra, I, Larran, M., & Martinez, I. (2005). The explanatory factors of intellectual capital disclosure to financial analysts. *European Accounting Review*, 14(1), 63–94.

Gerpott, T. J., Thomas, S. E., & Hoffmann, A. P. (2008). Intangible asset disclosure in the telecommunications industry. *Journal of Intellectual Capital*, 9(1), 37–61.

Glaum, M., Street, D. L., & Vogel, S. (2007). Making acquisitions transparent – an evaluation of M&A-related IFRS disclosures by European companies in 2005, Fachverlag Moderne Wirtschaft, Frankfurt am Main.

Griliches, Z. (1981). Market value, R&D, and patents. *Economics Letters*, 7(2), 183–187.

Gu, Z. (2007). Across-sample incomparability of R^2s and additional evidence on value relevance changes over time. *Journal of Business Finance & Accounting*, 34(7), 1073–1098.

Guenther, T., & Ott, C. (2008). Behandlung immaterieller Ressourcen bei Purchase Price Allocations – Ergebnisse einer explorativen empirischen Studie (Treatment of Intangible Assets in Purchase Price Allocations – Results of an Explorative Empirical Study), *Die Wirtschaftsprüfung*, 61(19), 917–926.

Guenther, T, Kirchner-Khairy, S, & Zurwehme, A (2004). Measuring intangible resources for managerial accounting purposes. In: Horvath, P., & Möller, K. (Eds.), *Intangibles in der Unternehmenssteuerung* (pp. 157–185). Munich: Vahlen.

Guenther, T., Beyer, D., & Menninger, J. (2005). Does Relevance Influence Reporting About Environmental and Intangible Success Factors? – Empirical Results from a Survey of "New Economy" Executives, *Schmalenbach Business Review*, Special Issue 2/2005, pp. 101–138.

Guthrie, J., & Petty, R. (2000). Intellectual capital – Australian annual reporting practices. *Journal of Intellectual Capital*, 1(3), 241–251.

Hager, S., & Hitz, J.-M. (2007). Bilanzierung und Berichterstattung über immaterielle Werte – eine empirische Bestandsaufnahme für Geschäftsberichte deutscher IFRS-Bilanzierer 2005, (Accounting and Reporting on Intangible Assets – Empirical Evidence for Annual Reports of German IFRS Preparers 2005), Zeitschrift für Kapitalmarktorientierte Rechnungslegung (KoR), 2007 (4), pp. 205-218.

Hall, B. H., & Mairesse, J. (1995). Exploring the relationship between R&D and productivity in French manufacturing firms. *Journal of Econometrics*, 65(1), 263–293.

Hall, B. H., Jaffe, A., & Trajtenberg, M. (2000). *Market value and patent citations: A first look, national bureau of economic research*. Working paper series, No. 7741, Cambridge.

Haller, A., & Dietrich, R. (2001). Intellectual Capital. Bericht als Teil des Lageberichts (IC Report as Part of the MD&A). *Der Betrieb*, 54(20), 1045–1052.

Henard, D. H., & Szymanski, D. M. (2001). Why some new products are more successful than others. *Journal of Marketing Research*, 38(3), 362–375.

Hitz, J.-M. (2007). Capitalize or expense? Recent evidence on the accounting for intangible assets under IAS 38 by STOXX 200 firms. *Zeitschrift für internationale Rechnungslegung IRZ*, 5, 319–324.

Kaplan, R. S., & Norton, D. P. (2001). Transforming the balanced scorecard from performance measurement to strategic management – Part I. *Accounting Horizons*, 15(1), 87–104.

Leitner, K.-H., Sammer, M., Graggober, M., Schartinger, D., & Zielowski, Ch. (2001). Wissensbilanzierung für Universitäten (IC Statements for Universities), Seibersdorf Research Report ARC-S-0145, www.systemforschung.arcs.ac.at/ Publikationen/21.pdf, Access 10 April 2004.

Lev, B. (2001). *Intangibles – management, measurement, and reporting*. Washington, DC: Brookings.

Lev, B., & Zarowin, P. (1999). The boundaries of financial reporting and how to extend them. *Journal of Accounting Research, 37*(3), 353–378.
Montoya-Weiss, M. M., & Calantone, R. (1994). Determinants of new product performance: A review and meta-analysis. *Journal of Product Innovation Management, 11*(5), 397–417.
Mouritsen, J., Bukh, P. N., Larsen, H. T., & Johansen, M. R. (2002). Developing and managing knowledge through intellectual capital statements. *Journal of Intellectual Capital, 3*(1), 10–29.
Mouritsen, J., Larsen, H. T., & Bukh, P. N. (2001). Reading an intellectual capital statement: Describing and prescribing knowledge management strategies. *Journal of Intellectual Capital,* 2, 359–383.
Ordóñez De Pablos, P. (2003). Intellectual capital reporting in Spain – A comparative view. *Journal of Intellectual Capital, 4*(1), 61–81.
Orens, R., & Lybaert, N. (2007). Does the financial analysts' usage of non-financial information influence the analysts' forecast accuracy – some evidence from the belgian sell-side financial analyst. *The International Journal of Accounting, 42*(3), 237–271.
Organisation for Economic Co-Operation and Development (OECD) (2005). Oslo Manual – Guidelines for collecting and interpreting innovation data. http://www.oecd.org/document/23/0,3343,en_2649_37417_35595607_1_1_1_37417,00.html. Accessed 13 March 2009.
Pakes, A. (1985). On patents, R&D, and the stock market rate of return. *Journal of Political Economy, Vol. 93*(2), 390–409.
Penrose, E. T. (1959). *The theory of the growth of the firm.* Oxford: Blackwell Publishers.
Rastogi, P. N. (2003). The nature and role of IC – rethinking the process of value creation and sustained enterprise growth. *Journal of Intellectual Capital, 4*(2), 227–248.
Riegler, C., & Kristandl, G. (2004). Value Reporting in österreichischen Unternehmen – Beobachtung des Berichtverhaltens von ATX und ATX-Prime Unternehmen (Value Reporting in Austrian Companies – Observations of the Disclosure of ATX and ATX Prime Companies). In: Seicht G. (Ed.), Jahrbuch für Controlling und Rechnungswesen 2004, LexisNexis, Vienna, pp. 245–267.
Rumelt, R. P. (1984). Towards a strategic theory of the firm. In: Lamb, R. (Ed.), *Competitive strategic management* (pp. 556–570), Prentice Hall: Englewood Cliffs.
Schumpeter, J. A. (1943). *Capitalism, socialism, and democracy.* London: Allen & Unwin.
Speckbacher, G., Güldenberg, S., & Ruthner, R. (2004). Externes Reporting über immaterielle Vermögenswerte (External Reporting on Intangible Assets). In: Horváth, P. & Möller, K. (Eds.), *Intangibles in der Unternehmenssteuerung* (pp. 435–455). Munich: Vahlen.
Stewart, T. (1997). *Intellectual capital – the new wealth of organizations.* London: Broadway.
Sveiby, K.-E. (1997). *The new organizational wealth – managing and measuring knowledge-based assets.* San Francisco: Berrett-Koehler.
Tasker, S. (1998). Technology company conference calls: A small sample study. *Journal of Financial Statement Analysis, 4*(1), 6–14.
Vandemaele, S., Vergauwen, P., & Smits, A. (2005). Intellectual capital disclosure in The Netherlands, Sweden and the UK. *Journal of Intellectual Capital, 6*(3), 417–426.
Wernerfeldt, B. (1984). A resource-based view of the firm. *Strategic Management Journal, 5*(2), 171–180.
Workgroup "Accounting and Reporting of Intangible Assets" (WGARIA) (2001). Kategorisierung und bilanzielle Erfassung immaterieller Werte (Categorization and Recognition of Intangible Assets), *Der Betrieb, 54*(19), 989–995.
Workgroup "Accounting and Reporting of Intangible Assets" (WGARIA) (2005). Corporate Reporting on Intangibles – A Proposal from a German Bachground. *Schmalenbach Business Review*, Special Issue 2/2005, pp. 65–100.
Workgroup "Accounting and Reporting of Intangible Assets" (WGARIA) (2008). Leitlinien zur Bilanzierung selbstgeschaffener immaterieller Vermögensgegenstände des Anlagevermögens nach dem Regierungsentwurf des BilMoG (Guidelines for the Recognition of Self Created Fixed Intangible Assets Due to Government's Draft for BilMoG), *Der Betrieb, 61*(34), 1813–1821.

Wulf, I. (2008). Immaterielle Vermögenswerte nach IFRS – Ansatz, Bewertung, Goodwill-Bilanzierung (Intangilbe Assets According to IFRS – Recognition, Measurement, Goodwill Accounting), Erich Schmidt, Berlin.

Integrated Financing Strategies for Innovation-Based Growth

Ervin Schellenberg

Contents

1 Introduction	334
2 A Financier's Perspective on Innovation	334
2.1 Types of Innovation	334
2.2 Innovation and Risk	335
2.3 Stages of Development and Financing	335
3 Structural Alternatives	337
3.1 Corporate (Recourse) Financing	337
3.2 Non-Recourse Financing	337
4 Sources of Expansion Capital	339
4.1 Capital Structure Illustrated	339
4.2 Equity Capital	340
4.3 Debt Capital	341
4.4 Mezzanine Capital	342
4.5 Funding Alternatives	342
5 Case Study on the Solar Industry	343
5.1 Sulfurcell	343
5.2 Sulfurcell's Innovation	344
5.3 Market Penetration	345
5.4 Competitive Landscape	346
5.5 Stage of Development	346
5.6 Integrated Financing Transaction	346
6 Summary and Outlook	348
References	348

E. Schellenberg (✉)
EquityGate Advisors GmbH, Wiesbaden, Germany
e-mail: schellenberg@equitygate.de

1 Introduction

Innovation is a major driver of sustainable corporate growth and will largely determine the success and the future profitability as well as the value of a company. Success of innovations depends on the quality of the organization and the people driving a systematic process. The process itself should be intentional, measurable, sustainable, well-documented and securely financed.

Today, obtaining financing on competitive terms and commensurate to the risk profile of an investment can be a daunting task. Private capital markets are fragmented, complex, and constantly evolving. The past few recessions have been prompting a massive reorganization of the global capital markets. From across the financial centers of our globalized world, a broad range of venture capitalists, growth capital, and mezzanine investors, private equity groups, and large family estates as well as traditional international investment banks, finance companies, subordinated debt funds, and hedge funds are providing a variety of sources of capital on different terms.

In order to achieve the best overall transaction conditions in the light of the company's objectives, it is necessary to understand the investment/lending criteria of capital providers in order to identify and address issues that could impact the financing structure. It is also vital to outline realistic expectations in the early process.

In this chapter, I will explore how to develop integrated financing strategies for innovation-based growth. Further I discuss the financier's perspective on innovation; outline the different stages of development of a company as well as the associated cash generation profile. I will derive a perspective on the possible types of capital accessible for financing.

2 A Financier's Perspective on Innovation

Principally a financier's perspective on innovations will depend on the type of innovation, which in turn drives the risks associated with an investment. In this chapter the principal types of innovation and the risk associated with the innovation process are defined.

2.1 Types of Innovation

Innovation can be defined as a process of putting ideas into valuable action in order to improve products, processes or technologies. Christensen and others[1] define three "types" of innovation

- Incremental innovation is focused on making small but significant improvements to existing products or services.

---

[1] Christensen, C. M. (1997). *The Innovator's Dilemma: When new technologies cause great firms to fail*. Harvard Business School Press, p. xv et sqq.

- Breakthrough innovations introduce an existing technology into a new market or a new technology into an existing market
- Game changers disrupt an existing market or create an entirely new market.

While all innovations are uncertain by nature and involve new ways to expand a business, e.g. the application of established proven technologies, the visibility of marketability determines the risk associated with an investment decision in a company.

2.2 Innovation and Risk

A company's risk is generally composed of financial risk, which is linked to the level of debt financing, to the associated risk of defaulting on a payment, and to business risk, which is linked to the nature and maturity of a company's business operations. If a company were entirely financed by equity, it would face almost no financial risk, but it would still be susceptible to business risk.

Innovations, i.e. new untested products, processes, technologies etc, often offer great opportunities, but always involve the risk of landing in a dead end:

- *Timing* – an innovation does not create a competitive advantage, e.g. due to bad timing
- *Acceptance* – not enough "early adopters" pave the way toward mass acceptance of a new technology
- *Misinterpretation of the market* – the new product is excellent in terms of technical specifications; however, no customer will pay for it since its costs are higher than its expected benefits
- *Pace of penetration* – nobody can forecast how fast a new product or process will penetrate the market
- *Copies/Time to market* – e.g. a follower copies the innovation, learns more, is better at manufacturing or distribution,

In addition, the ability to get financing will be spurred by the quality of the organization and by the people driving a systematic progress which should be intentional, measurable, sustainable, well documented, and financed.

2.3 Stages of Development and Financing

Financing the establishing of a market for an entrepreneurial idea and its maturation can be divided into two phases: early-stage capital and expansion-stage capital.

Early-stage capital includes seed capital, i.e. financing provided to study, assess and develop an initial concept and start-up capital, such as financing provided to firms for product development and initial marketing.[2] Firms may be in the process of

---

[2]Ross, S. A., Westerfield, R. W., & Jaffe, J. (2005). *Corporate Finance 7* (Ed.,, p. 563). New York: McGraw Hill.

being set up or may exist without having used their product or service commercially. Equity at this stage of the firm is typically provided by business angels, early stage venture/technology funds, and corporate ventures. Early-stage financing is connected with high entrepreneurial risk and therefore requires equity funding; integrated financing strategies are not applicable at this stage of a company's development. I will therefore focus on the expansion stage of a company.

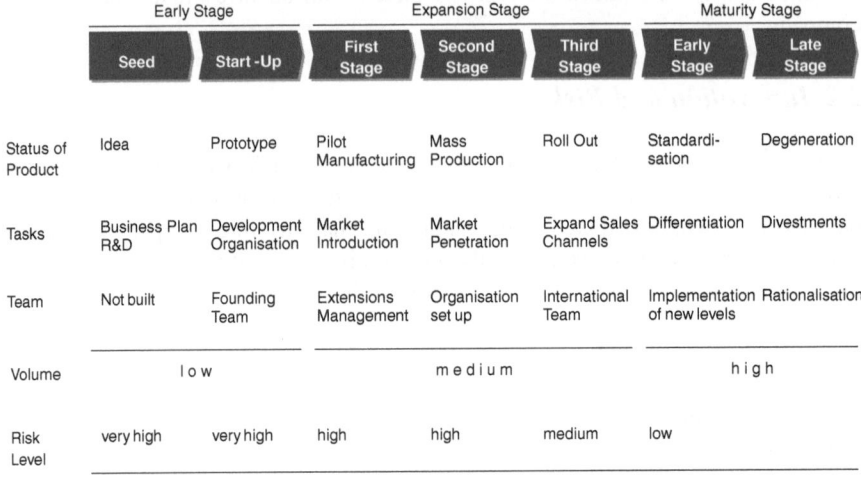

Fig. 1 Typical development through corporate life cycle. Source: EquityGate Advisors

The following paragraph illustrates a typical development of a company:

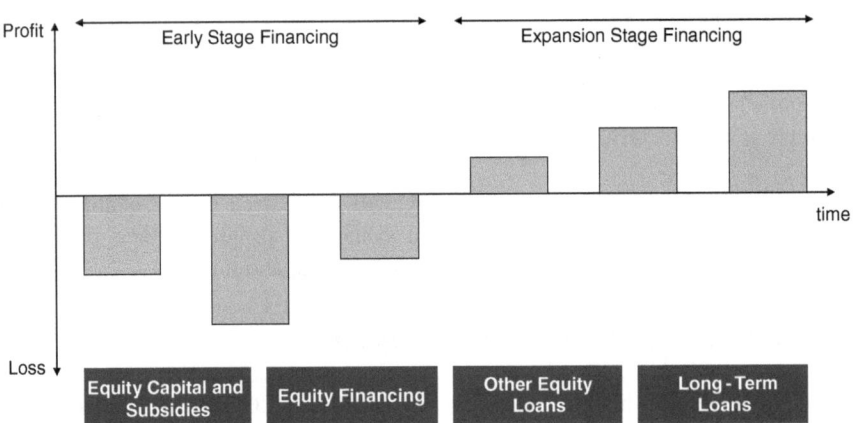

Fig. 2 Typical cashflow profile through lifecycle

Expansion capital: financing provided for the growth of a firm, which may or may not break even or be profitable. The capital may be used to finance increased production capacity, market or product development, or to provide working capital.

Depending on the type of innovation, the stage of development, which defines the risk reward profile and visibility of cash flows, a variety of financial instruments and capital sources can be tapped to finance a company's development.

Innovation-based growth can either be financed on the balance sheet of a sponsor or in a special-purpose company. The latter offers the advantage that there is no or only limited recourse to the sponsor, the SPV has to live with what was budgeted. Furthermore, prior to each financing round the concept has to be validated before fresh money is put into the company. At a certain stage of development (e.g. past seed capital), it is common for capital-intensive innovations to be spun out and financed without recourse.

3 Structural Alternatives

3.1 Corporate (Recourse) Financing

In order to assess risks and solvency development corporate financing generally refers to an established company's historical balance sheet. Therefore, corporate financing is generally not – or to a very limited extent – available (e.g. as an asset-based loan secured against highly marketable assets) to new companies. However, a large "parent company" can finance innovations by using the debt capacity of its existing business and assets to fund innovation. In this case, the credit relationship is established directly between the parent company and a bank.

3.2 Non-Recourse Financing

Non- or limited recourse financing is an alternative very common in financing innovations at a certain point in the development of new product and process. In this case, agreements are completed between the capital providers, such as growth financiers,

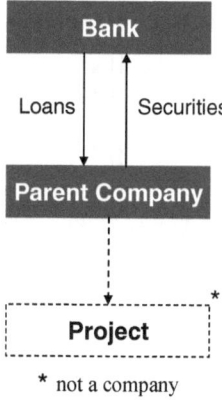

Fig. 3 Corporate financing

and/or banks and the project company which are therefore "off balance" of the sponsor's balance sheet.[3] This results in a non-recourse or limited recourse structure. The equity providers and lenders rely on the projected cash flows of this company and its assets as the source of repayment and return.[4]

The following simplified chart shows the participating parties and their functions within the project.

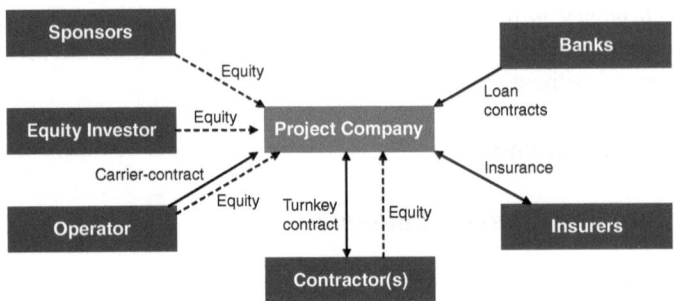

Fig. 4 Typical contractual relationship

The outlined structured non- or limited recourse financing can be used as a competitively priced source of capital at the expansion financing stage where, contrary to start-up financing, a certain calculability and probability of occurrence of the planned costs and revenues is materialized.

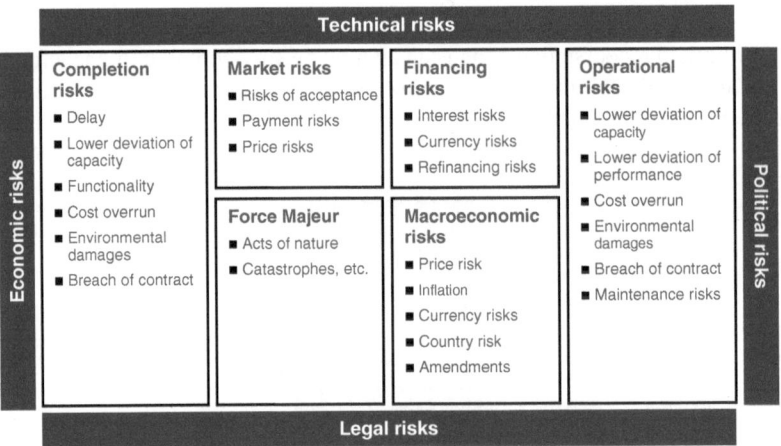

Fig. 5 Framework for integrated risk analysis. Source: EquityGate Financing Guide

---

[3]Tinsley, R. (2000). *Advanced project financing*. London: Euromoney Books, p. 2 et sqq.
[4]Arnold, G. (2002). *Corporate financial management* (2nd Ed., p. 491), Harlow: Pearson Education.

If the risks discussed in the table above can be addressed appropriately, financing with a lower weighted average cost of capital (WACC)[5] of the whole funding can be achieved through the introduction of "cheaper" debt capital.

However, the distribution of cash flows will then take place according to a "waterfall-principle". This means to cover firstly the expenses needed to operate and maintain the project, secondly the debt service and it lastly results in a payout to investors. Financial covenants govern the distribution of cash flows, e.g.:

- Debt Service Cover Ratio
- Interest Cover Ratio
- Gearing and Leverage Ratios
- Debt Service Reserve Account
- Capital Expenditures Reserve Account

4 Sources of Expansion Capital

Across the full spectrum of capital sources, due to the considerable risk associated with the business, equity capital is the primary source for expansion stage companies. It ranges from venture capital to equity provided by the capital market (through an exchange). If public interest is involved, subsidies provided by a regional or national government might be available. The availability of debt capital depends on the actual progress of a company and on a sufficient level of equity. Mezzanine capital with its flexible characteristics is another form of capital for expansion stage companies. It can be structured as an equity type or debt type of capital and might be an attractive alternative when bank financing is limited. For further improving the financing structure, innovative funding alternatives like contracting, leases or investor model are also taken into consideration.

4.1 Capital Structure Illustrated

A solid expansion stage company has three principal funding alternatives. Equity capital is the most expensive type of capital. In general, capital increases, particularly in capital-intensive businesses, result in massive dilutions to founding shareholders at the expansion stage.

Therefore, all operational and contractual possibilities should be explored from the outset of the innovation process in order to enable the company using as much cheap capital as possible in a later stage of the innovation process. This approach will preserve the upside to early shareholders in the project.

---

[5]WACC: Often the weighted average of the cost of equity and the cost of debt. The weights are determined by the relative proportions of equity and debt in a firm's capital structure. It is used as a hurdle rate for capital investment.

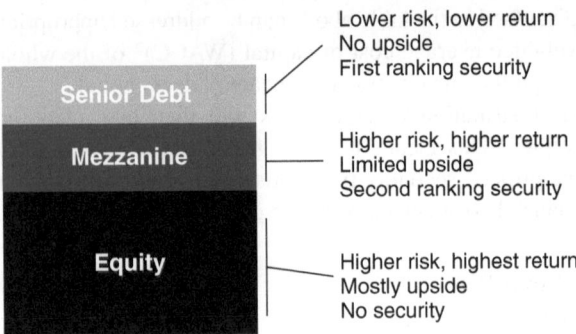

Fig. 6 Capital structure

4.2 Equity Capital

However, due to the considerable business risk described above, the primary sources of capital for innovation-based growth are equity capital, i.e. a cash contribution or a commitment to invest in the ordinary share capital of a company. There are two principal sources of equity: private equity and the equity capital market.

Venture capital funds usually invest large amounts in firms with potential for rapid growth, focusing on later stages of company development – at the expansion stage a company has survived long enough to arouse the interest of many venture capital funds.[6]

Private equity: Private equity provides equity capital to enterprises not quoted in a stock market. Private equity (also called venture capital) can be used to develop new products and technologies, to expand working capital, to make acquisitions, or to strengthen a company's balance sheet. It can also resolve ownership and management issues. A succession in family-owned companies or the buyout and buying of a business by experienced managers may be achieved by using private equity funding.[7]

Corporate venturing: For strategic or financial reasons or because of social responsibility a larger company takes a direct minority stake in a smaller unquoted company. Corporate venturing is predominantly used by large corporations in order to support external technology development.[8]

Equity capital market or stock exchanges are part of the capital market and provide capital to listed companies. It is a market in which long-term capital is raised

---

[6] Arnold, G. (2002). *Corporate financial management* (2nd Ed., p. 425). Harlow: Pearson Education.
[7] Jesch, T. A. (2004). *Private equity beteiligungen* (p. 22). Wiesbaden: Gabler-Verlag.
[8] Arnold, G. (2002). *Corporate financial management* (2nd Ed., p. 433). Harlow: Pearson Education.

by industry and commerce, government, and local authorities. They are no appropriate markets for early expansion companies. However, an Initial Public Offering (flotation, going public), i.e. the process of launching a public company for the first time by inviting the public to subscribe in its shares,[9] is one of the primary exit channels for private equity companies and venture capitalist after their company has achieved a positive earnings track record.

Subsidies/subventions are often available to finance innovation, particularly if public interest is involved, e.g. in green-tech, or if significant employment is created in economically underdeveloped regions. They will reduce the investment size and the weighted cost of capital for an investment. Therefore, a subsidy is a form of financial assistance paid to a company to encourage activities that otherwise might not take place. Subsidies typically stem from a government, e.g. the European Union, a national government or a regional government. Depending on the location of a new business, the form of subsidy may vary according to local legislation. From a bank's perspective subsidies can replace equity.

4.3 Debt Capital

Depending on the risk reward profile and the actual progress of a company, debt capital can be introduced for expansion financing stage companies. As a precondition, sufficient equity from one or several of the above sources has to be made available by the sponsor and by third party equity investors.

Principally debt financing can be broken down into three generic groups of capital: asset-based loans, cash flow-based loans, and mezzanine capital.

Asset-based loans: Asset-based lending or asset-based financing refers to loans secured by a broad variety of assets. Businesses can obtain asset-based lending by using the company's liquid, current assets (such as accounts receivable and/or inventory) or the fixed assets of a business (such as plant, property, and equipment) as collateral. In order to minimize the loan's credit risk, asset-based financing relies on the value of the underlying collateral. The so-called commercial lending market is the most accessible market for expansion stage companies.

Cash flow-based loans: Generally cash flow-based loans will only be granted to profitable businesses in the expansion stage. Typically, only companies with predictable and historically sustainable cash flows (operating performance and enterprise value based on brand, franchise value, technology, or customer base) may succeed in obtaining cash flow lending. However, if the company was able to secure off-take agreements from quality counterparties, project type loans might be available.

---

[9]Berk, J., & DeMarzo, P. (2007). *Corporate finance* (p. 757). Harlow: Pearson Education.

4.4 Mezzanine Capital

Mezzanine capital is another form of capital available to expansion stage companies, particularly pre-IPO. It describes a whole basket of financial instruments. Just as the mezzanine lies between two conventional floors in a building, mezzanine finance fills the gap between 'senior' bank loan (which has first right of repayment in the event of financial difficulty) and equity capital.[10]

Also in terms of risk and reward, mezzanine capital holds an intermediate position between senior debt and equity and is often used in conjunction with both in order to increase the total loan without significantly diluting ownership. A great advantage of mezzanine is its flexibility. It can be structured as a secured subordinated loan with equity warrants but it can also take the form of preference shares or a convertible loan. Mezzanine is a particularly attractive option when bank financing is limited; perhaps because the deal exceeds the credit the banks will extend, or because of a lack of tangible assets or of an unduly downbeat view of the prospects. Maturities can range from seven to ten years. Repayments are made in one payment at final maturity and therefore – unlike bank debt – do not absorb the borrower's cash flow. This can set free cash for growth and development of the business. Moreover, mezzanine is cheaper than unquoted equity and allows the existing shareholders to maintain a higher percentage of the ownership.

4.5 Funding Alternatives

Financing can be further improved by replacing expensive venture or growth equity by innovative alternatives including leasing, sale and rent back of real estate and contracting models for e.g. energy distribution or generation equipment.

4.5.1 Contracting

A good example for the frequent application of contracting model is the supply and distribution of energy. This often constitutes a considerable CAPEX amount for a new company. Lenders are reluctant to lend against these assets as they have limited "third party" marketability. Recently, a number of infrastructure service providers and subsidiaries of energy suppliers have started with comprehensive services, i.e. energy supply is provided by a third party (the contractor) who engineers, procures, constructs, and operates/maintains the required energy facilities. This results in a lower CAPEX for the new company which in turn pays the contractor a rental/services (for heating, steam, power and water) type fee over the lifetime of the assets.

---

[10] Krimphove & Tytko (2002). *Praktiker-Handbuch Unternehmensfinanzierung*, p. 887.

4.5.2 Leases

Leases are contracts under which a lessee has committed himself to pay stipulated cash amounts for the use of an asset (either equipment or real estate) for a specific period of time. Leases can be broadly divided into capital and operating leases and are typically considered as long-term financial obligations of the lessee and therefore are using the lessee's debt capacity. However, tapping the leasing market may prove advantageous as it is a distinct market segment with own liquidity. Also some capital providers prefer holding real estate in a separate legal entity in case of insolvency.[11]

4.5.3 Investor Model

This model offers an alternative to a leasing model whereby property investors, such as real estate companies and dedicated property funds, set up a special-purpose company and either acquire a used real estate property (office building, logistic center or light industrial manufacturing plant) or appoint a construction company to build a new property for the company. Typically, full service including the arrangement of bank financing is offered. The company then rents the property under a standard rental agreement for a fixed period. In some cases a purchase option can be obtained at the end of the rental period. The advantages for the company include a reduction of the total investment as well as a positive impact on all balance sheet ratios. The disadvantage is that if no purchase option can be agreed upon the location, it can only be secured for the duration of the rental agreement.

5 Case Study on the Solar Industry

Sulfurcell is a good example of significant integrated financing ($130m equity and debt) of a breakthrough innovation at the expansion stage in a fully non-recourse structure to the sponsors Vattenfall, Gaz de France, Ventegis, and Masdar. This deal was voted "Venture Investment of the Year 2008" by Private Equity International Awards online – an agency of good standing – and was advised by EquityGate as financial advisors to Sulfurcell and their shareholders. In the following, company, technology, innovation, and market are described.

5.1 Sulfurcell

Sulfurcell was established in 2003 as a spin-off from Europe's largest research institute for thin-film photovoltaics, the Hahn-Meitner Institut ("HMI") in Berlin

---

[11] Arnold, G. (2002). *Corporate financial management* (2nd Ed.). Harlow: Pearson Education, p. 532 et sqq.

and develops, manufactures, and distributes thin-film solar modules using a proprietary technology based on Copper-Indium-Sulfide (CIS). The unobtrusive and sleek appearance of each module meets the highest aesthetic standards especially in the high-priced and high-margin market for integrated PV Systems (BIPV).

Fig. 7 Sulfurcell pilot plant, Berlin

Fig. 8 Application of Sulfurcell BIPV modules

5.2 Sulfurcell's Innovation

Conventional photovoltaic-modules use a specific type of semiconductor in order to convert sunlight into electrical power. Thin-film modules are one hundred times thinner than conventional solar modules and absorb as much sunlight as crystalline silicon modules. These characteristics, combined with the fact that thin-film solar modules allow simple manufacturing processes and lower energy consumption, offer the clear potential to achieve far higher cost reductions than be conventional, crystalline silicon-based technologies for certain markets and applications.

5.3 Market Penetration

Sulfurcell had established a simple CIS-based technology featuring high throughput, robust processes, and methods well established in the glass industry. The use of environmentally friendly materials and the aesthetic excellence of the modules clearly differentiate Sulfurcell from its competitors.

Within the thin-film segment, CIS is considered to be the most promising technology compared to alternative thin-film technologies, including Cadmium Telluride (CdTe) and amorphous and microcrystalline silicon (a-Si/μ-Si). This modules provide significantly higher efficiencies in both laboratory and industrial environment. However, CIS-based technologies' superior efficiencies have not yet been converted to larger scale commercial success. To date, the alternative cadmium telluride ("CdTe") and amorphous silicon technologies have the largest installed production capacities. In CdTe, big players include First Solar, a U.S.-based public company with very substantial manufacturing capacities in Germany.

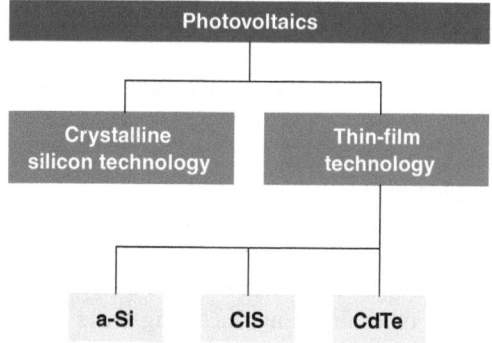

Fig. 9 Analysis of photovoltaic technologies

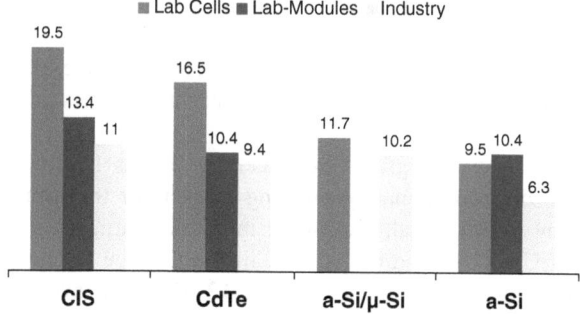

Fig. 10 Confirmed maximum efficiencies compared to best industrial manufacturer

5.4 Competitive Landscape

Sulfurcell is the fourth largest CIS manufacturer, with modules achieving efficiency at the average level of all thin-film modules available on the market. Sulfurcell's key competitors in CIS technology include Würth Solar and Showa Shell, both with higher efficiencies but, on the other hand, with more complex technologies.

Company		2006	Efficiency
CIS			
Global Solar	USA	4.2	4.8%
Würth Solar	D	3	11.0%
Shell Solar	USA	0.7	9.3% 4)
Sulfurcell	D	0.2	7.4%
Daystar	USA	0.1	
Miasole	USA	< 1	
Avancis	D	0	
Honda	J	0	
Johanna	D	0	
Nanosolar	USA	0	
Showa Shell	J	0	
Scheuten	D	0	
Terra Solar	USA	0	
Total		8.2	

1) Produkt K 120 V
2) Produkt MA 100 T2
3) First Solar, Annual Report 2006: 64 W durchschnittlich
4) Produkt PowerMax 80

Company		2006	Efficiency
Amorphes Silizium			
Unisolar	USA	32	6.3%
Kaneca	J	29	6.3% 1)
Mitsubishi	J	13	6.4% 2)
Shenzhen	China	7	
Bangkok	Thai	6	
Kanto San	J	5	
Schott Solar	D	3	
Sinonar	Taiw	3	
Tianjin	China	2	
Sharp	J	1	
CSG	D	1	
API	D	0	
Ersol	D	0	
Total		102	
Cadmium Telluride			
First Solar	USA	50	8.8% 3)
Antec	D	7.5	7.4%
Calyxo	D	0	
Total		57.5	

Fig. 11 Manufacturing volumes for thin-film solar modules 2006. Source: Publicly available manufacturers' information

5.5 Stage of Development

Since inception, Vattenfall, Gaz de France, and the Credit Suisse-managed Masdar Clean Tech Fund, and other investors had invested a significant amount of capital, comprising venture and other financing, including loans and subsidies. This was invested in three financing rounds from seed to early stage expansion.

5.6 Integrated Financing Transaction

The transaction was to raise debt and equity financing for scale-up investment in a 75 MW manufacturing facility substantially increasing volumes produced at Berlin Adlershof.

This successful financing grants Sulfurcell to use its first-mover technology advantage in the disproportionately growing market for thin-film modules. The round was led by Intel Capital, the global investment arm of Intel Corporation, which invested EUR 24 M and co-led by Climate Change Capital Private Equity (CPE, London) with an investment of EUR 12 M. Both investors were joined by a group of leading European clean tech investors, AIG Investments (Zurich), DEMETER (Paris), Zouk Ventures (London), BankInvest Group (Copenhagen),

Fig. 12 Groundbreaking

Fig. 13 The new production site

and Conetwork (Hamburg). In addition, existing investors BEU Berliner Energie Umweltfonds GbR (a joint venture of Vattenfall Europe and Gaz de France), Vattenfall Europe Venture GmbH, Ventegis Capital AG, Masdar CleanTech Investments Ltd. (New York), IBB Beteiligungsgesellschaft mbH, and other individual investors participated in the financing round.

In addition, a comprehensive commercial finance debt package was provided by DAL / Landesbank Berlin, Deutsche Leasing, and Volksbank Berlin. Lenders were able to lend against marketable securities. Risk was mitigated by the fact that the new facility will substantially replicate key machinery and equipment used in the pilot plant and "smart copy" proven manufacturing processes, whilst using standard equipment and processes established in the semiconductor and glass industries. Sales were on the horizon, as off-take contracts to secure a significant portion of total planned manufacturing volumes had been agreed with customers. Equity returns will be driven by further capacity expansion beyond the 75 MW facility, further capacity expansion will be available by near exact replication of this plant, offering an attractive future growth path with further substantial scale benefits.

Sulfurcell is now on the road to become a leading manufacturer of CIS-type thin-film solar modules. The company plans an initial public offering within a three to four year timeframe.

6 Summary and Outlook

Availability of financing instruments and capital sources for innovation-based growth hinges on the type of innovation and on the stage of development which defines a company's risk/reward and its cash flow profile.

Integrated financing is available for companies having passed the product development and initial marketing phase and is then provided for a company's growth. Growth can be financed on or off the balance sheet; the ultimate selection of an appropriate structure, individual instruments, and investor/lender depending on the individual company. Competitively priced non- or limited recourse financing, which is very common, is based on a company's profile of cash flows/assets as the origin of repayment and return.

Accessing capital in the equity and debt market can prove a daunting task. Capital markets are fragmented, complex and constantly evolving. That holds especially true in a credit crunch phase where lower leverages with a higher level of equity are required and higher margins are observed in the market. Key success factors for an optimal funding structure comprise a sound knowledge of the financier's perspective on investment/lending decisions at the actual stage of a company as well as of the variety/interdependences of capital instruments combined with an adequate capital market preparation of the business. The funding structure should be flexible and forward-looking, e.g. (if possible) assets should not be bound to a financing agreement; non- or limited recourse structures should be negotiated, and massive dilutions to founding shareholders should be avoided.

In order to achieve the best overall funding structure in terms of "cheap" capital, flexibility, and secure financing, innovation-based companies must consider the full spectrum of capital sources that ranges from the variety of equity capital to mezzanine, debt, contracting, leases, and investor models.

References

Arnold, G. (2002). *Corporate financial management* (pp. 385–509). Harlow: Pearson Education.
Berk, J., & DeMarzo, P. (2007). *Corporate finance* (pp. 757–770). Harlow: Pearson Education.
Christensen, C. M. (1997). *The Innovator's Dilemma: When new technologies cause great firms to fail* (ix–xxvii), Harvard Business School Press.
Jesch, T. A. (2004). *Private equity beteiligungen* (pp. 21–24). Wiesbaden: Gabler.
Krimphove, D., & Tytko, K. (2002). *Praktiker-Handbuch Unternehmensfinanzierung*, pp. 879–890.
Ross, S. A., Westerfield, R. W., & Jaffe, J. (2005). *Corporate finance* (pp. 540–568). New York: McGraw Hill.
Tinsley, R. (2000). *Advanced project financing* (pp. 1–14). London: Euromoney Books.

Corporate Venture Capital

Malte Brettel

Contents

1 Introduction 349
2 Designing Corporate Venture Capital Funds: The Perspective of Large Corporations . 350
3 Corporate Venture Capital from the Perspective of Young Firms Seeking
   Investment Capital 354
4 Evaluation of Corporate Venture Capital from a Scientific and a Practical Perspective 356
References 357

1 Introduction

Corporate venture capital (CVC) is venture capital (VC) that is administered by regular enterprises and corporations. Both CVC and VC target firms in early stages of the company life cycle. In their investments, corporate venture capital funds (CVCF) pursue financial and strategic goals. It is important for young firms to explore these goals and to examine the extent to which the goals of CVCF can help them in accomplishing their own future vision. Yet, as opposed to regular VC, CVC has to date only been applied in negligible volumes. It is noticeable that while a number of corporations already have sophisticated CVC vehicles available, the capital in such programs is usually relatively low and significantly less than in other investments or shareholdings.

M. Brettel (✉)
RWTH Aachen University, Aachen, Germany
e-mail: brettel@win.rwth-aachen.de

2 Designing Corporate Venture Capital Funds: The Perspective of Large Corporations

Including a number of significant ups and downs, CVC has been used in the United States for almost five decades. The first CVC initiatives had been launched in the early 1960s. While most of these initiatives disappeared in the wake of the oil crisis, CVC experienced a revitalization in the 1990s (Rind, 1981). It was at that time that the first German company, Deutsche Telekom, stepped forward with concrete efforts to build a CVCF. Other German corporations, like DaimlerChrysler AG (today: Daimler AG), Siemens AG, BASF AG and Bertelsmann AG quickly followed suit.

CVCFs are usually established as independent units and in most cases are equipped with a capital of less than € (EURO) 100 million. One of the advantages of setting up an independent structure for a CVCF is that it allows investors outside the corporation founding the CVCF to also participate in the fund. In addition, such a setup enables the CVCF to design incentive systems that are independent of those of the parent company, for instance, allowing the managers of a CVCF to also become shareholders in the CVCF. However, apart from these advantages, formal independence also entails challenges with respect to the maintenance of the link to the parent company and the top-to-bottom enforcement of strategic guidelines by the parent company.

Generally, there are two prime motivations for corporations to start making use of CVC.

It has been ascertained that radical innovations in particular require an entrepreneurial spirit in order to grow and prosper. The fact that such a spirit is difficult to maintain in relatively large and formal organizational structures makes the case for the use of CVC more compelling.

Exciting innovations are oftentimes developed outside the realm of big and established entities. CVC allows corporations to participate in such innovations despite these constraints and gives them extra time to deal with the technology risk and market risk that is sometimes associated with new and innovative products.

Therefore, in a nutshell, following in particular Schween (1996), it appears best to analyze CVC in the context of innovation management in large corporations.

In Fig. 1, CVC is illustrated as the basic framework with which the venture capital activities of industrial corporations should be supported. Both this framework and the introductory remarks to this article clearly emphasize the relevance of defining strategic goals for CVC.

According to a study by Brunner, Fahlbusch and Hundertmark (2000), three types of objectives of CVCF can be distinguished: strategic, financial and social and PR goals. The interplay between these different goals is depicted in Fig. 2.

With respect to strategy, four different goals of CVC can be distinguished:

1. *Technology observation* means that employees of CVCF are used to track new technological developments. However, technology observation must not necessarily result in an equity investment. It is also possible that efforts are

Entrepreneurship	Corporate Venturing
• **Core concept**: Motivation • **Objectives**: Finding, developing and organizing new business areas in existing companies • **Area of application**: Internal projects • **Resources**: Project budget, internal equipment and management capacity	• **Core concept**: Organisation • **Objectives**: Increasing turnover, productivity or quality • **Area of application**: Internal, new activities with higher loss risk than in the core business at least partly managed separately • **Resources**: Project budget, internal equipment and management capacity

Venture activities of corporations

Venture Management	Entrepreneurship
• **Core concept**: Management • **Objectives**: Securing business development by founding a new company or division • **Area of application**: Internal projects, but also spin-offs or existing or newly-founded companies • **Area of application** : Internal projects, but also spin-offs or existing or newly-founded companies • **Resources**: Internal budget, VC funds, involving third parties, management capacity, material and immaterial resources	• **Core concept**: Financing • **Objectives**: Primary goal is to observe Markets & technologies a high return is a secondary goal • **Area of application**: External, existing or new, small, innovative companies or spin-offs • **Resources**: Internal budget, VC funds, Involving third parties, management capacity, material and immaterial resources

Fig. 1 Classification of CVC into the venture activities of corporations (adapted from Schween, 1996)

exerted to entice those who had a share in developing specific technological knowledge, such as researchers in academic institutions, to work in the parent company of the CVCF. While employees in the parent company could also perform mere technology observation, it is unlikely that young firms would establish relationships with big corporations if they did not have specialized investment units like CVCFs. Accordingly, CVCFs constitute an important enabling element for corporations aiming at comprehensive technology observation.

2. *Growth enhancement* means that the parent company of the CVCF can achieve growth by taking equity stakes in young technology-based firms with promising

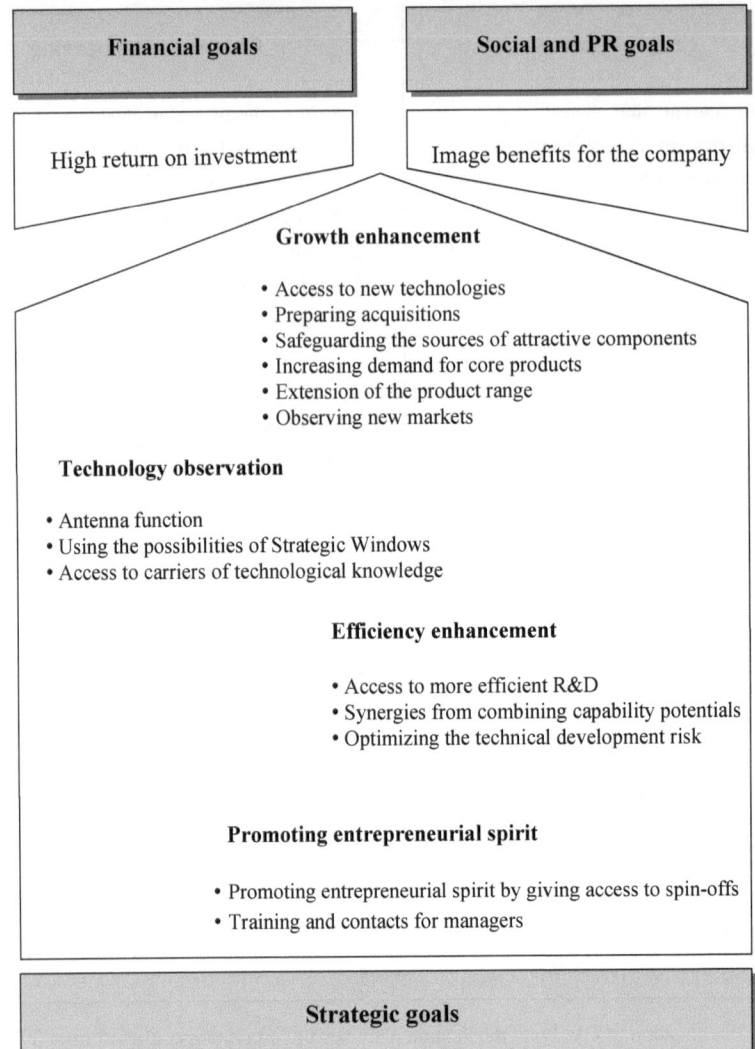

Fig. 2 Overview of goals associated with CVC (Brunner et al., 2000)

growth prospects. Equity stakes are expected to give the parent company access to new technologies that can then be used to develop new products or services.

3. *Efficiency enhancement* means that it can be more efficient for a corporation to procure certain technologies on the market than to develop the technologies themselves. This is particularly relevant when internal coordination and execution costs can be expected to be higher than the transaction and execution costs that are associated with market procurement. For instance, this could be the case when competition and animosity between departments leads to motivational differences and higher coordination costs.

4. The fourth *goal* encompasses the motivational impact of CVC activities and its impact on the entrepreneurial spirit. On the one hand, taking stakes in young firms in a way implies the integration of entrepreneurial people into the parent company. On the other hand, CVC policies that also allow employees of the parent company to pursue their own entrepreneurial ambitions could increase employee motivation even in projects that appear only of secondary importance.

In addition to strategic goals, CVCFs also pursue financial and social and PR goals. For instance, by having a CVC arm, a big corporation can develop an entrepreneurial image both vis-à-vis the public and its own employees.

Financial goals are of course relevant for all units of a corporation. A CVCF can achieve these goals in a way that is similar to that of other investment corporations: It takes a stake in the target company and works towards increasing the value of its investment with the goal of arriving at a profitable exit – by selling its stake to other investors, by means of an IPO, or by arranging some form of buy-out. Compared with regular venture capital funds, the advantage of CVCFs is that in many cases they do not have a predetermined maturity or duration. Thus, they are freer to choose the best time for an exit, which can be advantageous when the overall market climate is not favorable. On the other hand, a clear disadvantage of CVCFs is that the parent company constitutes an ideal buyer for the investment of the CVCF; particularly so, when the strategic focus of the CVCF is closely aligned with that of the parent company. This could entice the parent company to influence the return of the CVCF.

If parent companies of CVCFs are primarily interested in pursuing the strategic goals of technology monitoring and growth enhancement with their CVCF, then one alternative to employing CVC could be the complete purchase of a young firm in an early phase. Cisco is a prime example of this modus operandi. However, proceeding in such a way would leave little room to pursue other strategic or social and PR goals. Moreover, taking only a partial stake in a young firm is a relatively cheap way to gain access to new technologies which might be associated with technology and market risk, as the full price for these technologies does not have to be paid immediately.

The motivation to establish CVC facilities has demonstrated that such programs have their origins primarily in technology and innovation management. Accordingly, the strategic goals of such programs should be accorded equally great importance.

Ironically, these programs have exhibited developments that are similar to those of pure venture capital funds. Gompers and Lerner (1998) show that most CVC programs in the United States were terminated in the course of the capital market crisis in the 1980s at the latest, often before. Similar observations were made in Germany during and following the boom of the new economy (Witt & Brachtendorf, 2002). Essentially, this shows that while CVCFs try to act strategically, they in fact display a behavior that is predominantly oriented towards the achievement of financial goals.

However, experience shows that it is not advisable for parent companies of CVCFs to let financial goals dominate. If the sole goal of equity investment is financial, then it would be preferable to invest in pure venture capital funds, since investment managers in such institutions generally possess more experience than the heads of CVCFs. A CVCF can try to evade this problem by recruiting former venture capitalists. Still, if there is the slightest doubt regarding the financial objectives of the CVCF, then a CVCF will always be an inferior working alternative for a capable VC manager. Moreover, as Witt and Brachtendorf (2002) show, the remuneration of managers in CVCFs is often move aligned towards that of colleagues in the parent company, and not towards that of comparable managers in the VC industry.

3 Corporate Venture Capital from the Perspective of Young Firms Seeking Investment Capital

For young firms seeking investment capital, a corporate venture capitalist will initially always act in a way that is largely comparable to that of a regular venture capitalist. In both cases, the investment process (ideally) follows four major phases:

1. Establishing contact: The young firm establishes contact with a CVCF by sending an executive summary and a request for equity investment.
2. Rough analysis: Based on a business plan and first inteviews/management presentations, the CVCF performs a rough analysis regarding a potential investment.
3. Due diligence: The target company is analyzed in detail by the CVCF.
4. Investment negotiation: based on a term sheet, which concludes the due diligence, the final details of the investment are negotiated.

It is important that young entrepreneurs are informed about the goals of CVCFs before they establish contact. This is particularly relevant because differing strategic goals can also pose a danger for the CVCF that invests in the start-up.

- It is possible that technology observation by a CVCF is not necessarily associated with an interest in a target company as a combination of technology and entrepreneurs. A CVCF could be interested only in the technology or only in the entrepreneur who has the technological knowledge. However, this danger is not very significant, because technology and technological knowledge can often not be disentangled easily.
- Ideally, the investment of a CVCF in a target company should make the CVCF develop its investment in the best possible way. But it is also feasible that the CVCF simply envisages adopting the technology or taking it off the market in a cheap way. Entrepreneurs are unlikely to favor either of the two options. Fortunately, in most cases to date such a behavior has not been ascertained. Such cases occur more frequently with smaller corporations, who often enter into collaborations and joint ventures with the sole purpose of adopting a particular technology. Paramount for the success of such a strategy is the investment contract and the size of the investment.

As was emphasized earlier, these dangers are only theoretical in the context of CVCFs and can hardly be observed in reality. The reason is that if such occurrences were ever to become public, the deal flow of any CVCF would quickly dry up and the attainment of any strategic goal would become sustainable more difficult.

Nonetheless, it is important to discuss the question of whether young firms really want to give strategic investors access to their own company and core technology by means of a partial equity investment, or whether it would be preferable for them to try to sell the whole firm immediately. While investors benefit from taking stakes that are below 100% in the sense that they gain access to a technology without having to bear full market and technology risk, this advantage is a disadvantage for the target company, which in that case would have to bear the full risk itself. This can be compensated for only if the participation of the CVCF enables the target company to achieve above average growth rates that benefit it immediately. It is here that limited exit opportunities come into play again. Given the interests of the parent company of the CVCF, exit opportunities become highly regulated. Therefore, it is advisable for young entrepreneurs to already include in the initial investment contract stipulations with respect to certain exit opportunities or structure the contract in such a way that pullout or drag-alone clauses are included.

In addition, the investment of a CVCF, with which the parent company is often clearly identifiable, can close the door towards collaborations with other partners. For instance, it is not evident for competitors of the parent company whether or not they should enter into collaboration with the target company of the CVCF. Doing so could mean the indirect strengthening of a potential future competitor. Young firms therefore have to carefully reflect upon their readiness to sacrifice their freedom to choose business partners for the benefits that are brought along by the involvement of a CVCF.

Despite the disadvantages that are described above and are associated with the involvement of a CVCF and its parent company, there are also significant advantages, which can propel the development of the target company. Four areas where CVCFs and their parent companies can add significant value can be distinguished (Brettel, Rudolf, & Witt, 2005):

- Provision of industry, market and technology knowledge. This is particularly applicable where technological proximity can supply the target company with unique technological knowledge from the parent company of the CVCF.
- Opening up of new distribution channels and potentially provision of an existing customer base.
- Image transfer from the parent company of the CVCF to the target company. This gives the target company new opportunities to establish relationships with a multitude of institutions.
- Opening up of networks that would have been closed to the target company, were it not for its relationship with the parent company of the CVCF.
- Provision of administrative resources.

To sum up, an equity investment by a CVCF can provide young firms with significant advantages. Yet, any participation by a CVCF should be carefully studied in order avoid potential pitfalls. The investment contract is theoretically more difficult to negotiate, for it also needs to take account of the strategic goals. In reality, however, it is not noticeable that an involvement of CVCF is associated with complex contractual arrangements.

4 Evaluation of Corporate Venture Capital from a Scientific and a Practical Perspective

Although, from a theoretical perspective, it makes sense both for big corporations and young firms to employ CVC, it has not been widely applied to date.

Theoretically, CVCFs have the opportunity to act without the time constraints of regular venture capital funds, which should enable them to avoid sub optimal exits that are associated with specific fund maturity dates. Interestingly, this advantage has not largely been exploited by CVCFs. It even appears that setting up a predefined maturity date for CVCFs can be advantageous, for these programs are mostly set up on a cyclical basis and are often closed relatively quickly. By setting specific maturity dates, the issuing corporations give much consideration to the development of the capital markets, despite the underlying strategic motivation that initiated and should guide CVC activities.

Also, in practice, significant doubts have been raised as to whether the goal of strengthening entrepreneurship can really be achieved by means of CVC. Oftentimes the distance between a CVCF and its parent company is so big that no behavioral impact on the parent company can occur. Therefore, academics have commenced to discuss other concepts for fostering entrepreneurship more intensely.

Another unresolved problem in the application of CVC is the simultaneous pursuit of financial and strategic goals. It is not sensible for a parent company to first exert efforts to integrate the technologies of target companies and then to pay a high price for these technologies when the exit is prepared. However, achieving a high price is the very goal of the management of CVCFs and the founders of the target company. As Sykes puts it, "the impact of possible capital gains on total corporate results is viewed as minor compared with the potential for development of new business." This raises serious doubts with respect to the financial objectives of CVCFs and makes the link between CVCFs and the conditions of the capital markets all the more astounding.

The disadvantages associated with CVCFs for young firms render it more difficult for CVCFs to generate enough deal flow, even if the disadvantages described above are only hypothetical.

One of the big theoretical advantages of CVC for young firms is the exploitation of relevant know-how from the parent company. Unfortunately, in reality it has shown to be very difficult for young firms to benefit from know-how of the parent company. The reason is that employees of the parent companies in many instances are not interested in a positive development of the target company, because the technology it develops could turn out to become a substitute or threat to them in one way or the other. In such cases it has to be taken into account that employees of the parent company could attempt to even negatively impact the target company.

The exit of the CVCF and/or the founders of the target company is an additional problem that has been addressed in the context of the relationship between the CVCF and the parent company.

All in all, the studies that tackle the issue of CVC in Germany have clearly demonstrated the problems associated with CVC and their absence from the early financing scene. Even if Schween (1996) draws an interesting picture of the strategic goals of CVCFs in Germany, as Witt and Brachtendorf (2002) show, the goals of CVCF remain ambiguous (see also Witt, 2005). The two authors clearly illustrate the challenge underlying the quest of corporations to establish CVCFs with a clear direction and unambiguous goals. For instance, during their analysis of 29 CVCFs, 8 of these funds disappeared from the market. These disappearances, however, did not affect the business of the parent companies, which continued with their business as usual. It therefore seems that CVC is unlikely to establish itself either as an important element in early-phase financing or as an important building block in innovation management, even if the positive exceptions to this negative outlook are taken into consideration.

Thus, it appears that the analysis run by Gompers and Lerner in the United States in 1998 also holds true in Germany. They found that CVC has not established itself in the United States for three reasons. First, the goals of CVC initiatives were not clear, but oftentimes rather confusing. Second, many CVCFs were closed so early that they did not have the opportunity to develop a strategic contribution to the parent company. Third, one of the reasons CVCFs did not perform well was the inappropriate incentive system of CVCF managers and the lacking remuneration for the assumption of risks. Unfortunately, it is for the very same reasons that the American experience has repeated itself in Germany in a similar fashion (Witt, 2005)

References

Brettel, M., Rudolf, M., & Witt, P. (2005). *Finanzierung von Wachstumsunternehmen*. Wiesbaden.
Brunner, M. F., Fahlbusch, N., & Hundertmark, N. (2000). *Corporate venture capital-programme: Die Beziehung zwischen Corporate und Venture.*
Gompers, P. A., & Lerner, J. (1998). *The determinants of corporate venture capital success: Organizational structure, incentives and complementarities.* Working Paper, No. 6725, National Bureau of Economic Research, NBER Working Paper Series, Cambridge.
Rind, K. W. (1981). The role of venture capital in corporate development. *Strategic Management Journal, 2,* 169–180.
Schween, K. (1996). *Corporate venture capital: Risikokapitalfinanzierung deutscher Industrieunternehmen.* Wiesbaden.
Witt, P. (2005). Corporate venture capital. In: Börner, C., & Grichnik, D. (Eds.), *Entrepreneurial finance, Kompendium der Gründungs- und Wachstumsfinanzierung* (pp. 259–276). Heidelberg.
Witt, P., & Brachtendorf, G. (2002). Gründungsfinanzierung durch Großunternehmen, Die Betriebswirtschaft, pp. 681–692.

Knowledge-Based Financing Strategies for Innovation Output

Thomas Rüschen and Frank Rohwedder

Contents

1 Introduction 359
2 We Have Arrived in the Knowledge Society! 361
3 Innovation Output 363
4 Financing Strategies for Companies 367
   4.1 Increase Revenue 367
   4.2 Reduce Cost 368
   4.3 Transfer Risk 369
5 Conclusion 369
References 370

1 Introduction

Today, developed economies are characterised by the transition from the industrial to the knowledge society. According to a study by Deutsche Bank Research, this is verified by two macro-trends: (1) *Knowledge* becomes a pivotal production factor and (2) companies increase cooperation to develop innovative products (*open innovation*). At the same time, the importance of classic production factors such as land and machinery decreases. These trends require a profound review of current business approaches within innovative companies.[1]

This paper will show innovative companies how to finance the increasingly important production factor of knowledge by applying a simple systematic approach

---

T. Rüschen (✉)
Deutsche Bank AG, Frankfurt, Germany
e-mail: thomas.rueschen@db.com

[1] Hofmann (2005).

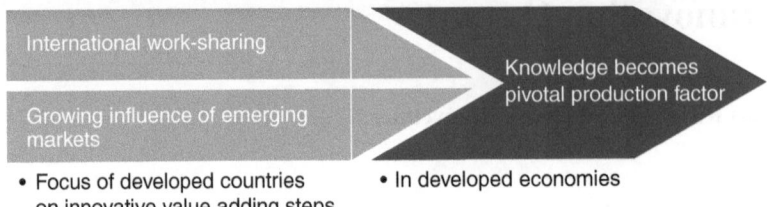

Fig. 1 Knowledge becomes a pivotal production factor
Source: Deutsche Bank Research

towards their existing R&D value-adding process. The authors will also demonstrate how particular financing instruments affect company's profits. Thus, companies are offered, means to improve their competitiveness and to accelerate corporate growth.

The production factor *knowledge* manifests itself in various ways. When a company succeeds in creating property rights from knowledge, these rights (*innovation output* or *intangible assets*) can add value by forming the basis for the introduction of new products.

In this paper, therefore, *innovation output* is understood as codified knowledge that was secured legally thereby enabling its tradability. As a rule, codified knowledge can be divided into technical rights (patent rights), artistic rights in the form of copyrights (music rights, literature rights, etc.) and brands. As patent rights in particular play an ever more important role in company's value-generation they will be discussed in greater depth. Wherever appropriate, examples will be provided.

Section 2 sets out by showing how in developed economies intangible assets are of increasing importance in transitioning industrial societies into knowledge societies. In the industrial society financing focused on the classic production factors such as land, property, and machinery. In the knowledge society, however, financing concentrates on the factor knowledge. The requirements of knowledge financing will be discussed as a further crucial issue.

Section 3 will depict how innovative companies adapt themselves to this new development. Then, the concept of *Open Innovation* will be presented. The concept raises two questions: How can the R&D budget be optimized? And: How can profits be optimized in this context?

In Sect. 4, the two optimization approaches will be combined into a *decision matrix* for companies. Starting from this matrix, banks may derive financing instruments that innovative companies can use to finance growth and to generate competitive advantages.

Finally, the authors summarize the results and assess to what extent intangible assets in their entirety can already be financed in the same way as tangible assets (e.g. property, machinery, and land) are.

2 We Have Arrived in the Knowledge Society!

Today the most important production factors of developed economies are intangible assets. Intangible assets such as patents, trademarks and copyrights provide a rapidly increasing contribution to a company's growth and competitiveness. Companies invest huge amounts of capital into these assets: According to an estimate by Leonard Nakamura from the Federal Reserve Bank of Philadelphia, the investments in the United States amount to approximately one trillion U.S. dollars, which is about 9 percent of the U.S. GDP. In 2003, U.S. companies for the first time invested more in intangible than in tangible assets. Whereas investments in tangible assets over the last 50 years have remained relatively constant at between 9 and 12% of the GDP, the proportion of intangible values has increased continuously from approximately 4 to 9% in the course of the last 25 years. The long term trend is rising (see Fig. 2).[2]

This trend of an ever growing importance of intangible assets is also reflected in the international trade between developed economies. Cross-border payments linked to patents and licenses have increased significantly since the beginning of the nineties. This results from the growing importance of research-intensive industries and knowledge-intensive service providers in developed economies. Between 1991 and 2002, the contribution of these service providers to GDP rose from 5 to 26% of the GDP within the G6 Group (United States, Japan, Germany, France, Great Britain and Italy). Including the research-intensive industries raises the proportion to one third of the GDP. At the same time, the importance of the classic processing industry with tangible assets such as property, machinery and land is declining.[3]

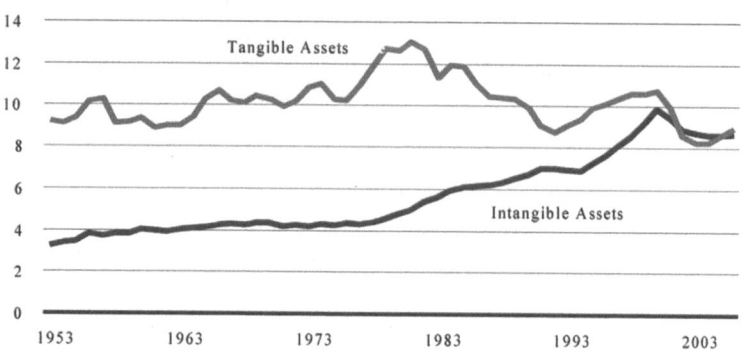

Fig. 2 USA: Intangible assets compared to tangible assets
Source: L. Nakamura (2007), Federal Reserve Bank of Philadelphia

---

[2] Nakamura (2003) updated 2007.
[3] Hofmann (2005).

Breaking down these figures on the level of individual companies, the pre-eminence of intangible assets will be revealed. Companies such as Siemens, IBM or Philips have several thousand patent families in their patent portfolios.[4]

These figures illustrate the rapid progress of the knowledge society and underline the importance these assets have for individual companies.

However, this importance does not – in all cases – become evident from corporate balance sheets, credit decisions by banks and investment decisions by investors. The overwhelming majority of decision processes are currently based on the evaluation of a company's tangible assets. There are four possible reasons: (1) companies with internally generated intangible assets are not allowed – under either national or international accounting principles – to show these values on the balance sheet to the full extent. However, when those intangible assets are not shown on the balance sheet they can hardly serve as collateral for banks. (2) It is difficult to price intangible assets. Currently, there is no standardized, recognized valuation method comparable to methods used for example in the real estate markets. As innovations are new by definition it is difficult or even impossible to determine a fair market value. Without market prices, however, investors cannot make decisions on investments and banks cannot grant credit based on the loan-to-value concept. (3) In many industries, innovations are seen as core value-drivers and are therefore treated as industrial secrets. Accordingly, no patents are filed, patent rights for innovations are not granted. In the absence of codified patent rights, there is no trading activity with patents, which logically hampers price-building for the intangible asset. (4) Furthermore, innovations often cannot be used by third parties as they represent special industrial solutions in a company's production process or because they have product-specific features ("special use assets"). However, banks need a third party buyer in order to monetize a pledged patent in the event of a loan default.

Only when intangible assets can be assessed by generally accepted valuation methods and when trade takes place in the form of organized markets, will intangible assets be suitable for financing in the same way as tangible assets.

Anyhow, during the last 15 years globalization and the emergence of intermediaries have helped lay the groundwork for the knowledge society. Globalization and the increasing pressure of competition have urged companies to fundamentally rethink not only the organization and execution of R&D, but also its trade. Furthermore, intermediaries that have specialized in valuation and utilization of intangible assets have emerged in the market. These asset intermediaries provide markets in the form of auction platforms, the bilateral matching of supply and demand and the analysis and evaluation of rights. This can be illustrated by; on the one hand, the Swiss company Interbrand in the sector of brands and, on the other hand, the IP Bewertungs AG in Hamburg for patent rights. Furthermore, finance intermediaries such as Deutsche Bank AG are starting to offer capital and structured products.

---

[4]Hofmann (2005).

Politicians have also recognized the increasing importance of intangible assets in companies. In Germany, for example, the accounting principles will be revised and amended to the effect that internally generated intangible assets can, at least partially, be shown on the balance sheet.

3 Innovation Output

In many industries engaged in research, R&D departments operated as *Closed Shops* up until the 1990s. Often, a mere 10% of their R&D results were used in the form of granted patents to legally secure the company's own products on the market. For the remaining 90% of the patent portfolio, the maintenance fees were paid to the patent offices but the patents themselves were never used.

The R&D landscape has changed. As a result of globalization, the product cycles were shortened enormously by innovation and the cost pressure on R&D is increasing. Thus, the old innovation model of the *Closed Shop* has had its day and open innovation processes (*Open Innovation*) are gaining ground.

The term *Open Innovation* is used to describe innovation processes in which knowledge from several companies, universities or other organizations is combined and used for developing new products. *Open Innovation* manifests itself in three different forms: Outside-in Innovations, Inside-out Innovations and Coupled Innovations.[5]

Outside-in Innovation means that knowledge from partner companies, as for example customers, suppliers or even competitors is integrated into the company. Otherwise inaccessible knowledge can be acquired by licensing in or by purchasing patents. The advantage of Outside-in Innovations lies in eliminating the risk of the company's own research and development and thus reducing the risk of misallocated investments. For instance the U.S. company Procter & Gamble, is increasingly buying in or licensing in patents.

When using *Inside-out Innovations*, a company sells its own patents to external companies. This can be done by selling or licensing out. This option is particularly interesting for companies conducting a lot of research, but not using all of their research results themselves. Thus, for example, one quarter of all German patents is unused, although more than 50% of these could be sold. Selling or licensing out these patents can unveil new sources of income. For example, since the beginning of the 1990s the U.S. technology company IBM has been conducting active patent management by selling or licensing out patents that it is not using itself. In 2006, IBM achieved revenues of more than one billion dollars from selling and licensing out unused patents.

Coupled Innovations refer to joint research and development of two or more partners. Coupled Innovation can be both: vertical cooperation between suppliers

---

[5]Gassmann and Enkel (2006).

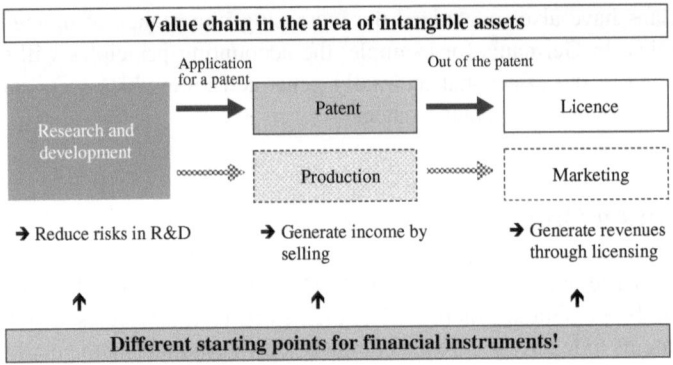

Fig. 3 Optimization of the production factor knowledge
Source: Rüschen/Rohwedder

and manufacturers and horizontal cooperation between competitors. Such cooperation models prove profitable when financial resources and the knowledge required for the development of new products are not at an individual company's disposal. It also makes sense when market participants fail to recognize a new trend early enough and can catch up only by combining resources. This is currently apparent in the automobile industry, where the competitors Daimler and BMW are jointly developing components for a hybrid engine.

Evidently, these examples underline to what extent the production factor knowledge has become a tradable good. Companies are able to (i) optimize this production factor by means of active management and (ii) at the same time to positively influence corporate growth by means of profit optimization. These two optimization approaches will be illustrated below.

The various practiced forms of open innovation processes show a trend towards the effective utilization of codified knowledge in the form of *Intellectual Property Rights*. The initial challenge for companies is to determine what partial processes are to be conducted in the actual R&D process and what R&D results are to be licensed in or bought in. Conceptually, the processes in the *production-based industry* can be split up into three phases: Research and development – manufacturing of the product – marketing of the product.

By analogy, these three phases can also be applied to the *knowledge-based industry*. Research and development – grant of a patent (corresponds to the tradable product) – licensing out or sale of the patent (corresponds to the marketing of the product).

For companies conducting R&D, such process is very time- and cost-intensive and the outcome is uncertain. R&D bears a high risk in terms of capital and time. Therefore, companies have to decide in a *first phase* whether they want to take these risks or whether this knowledge should be acquired from third parties.

The *second phase* assumes that the company has taken the risk of its own R&D, the results have proved positive and a patent has been granted by the patent office.

In this case, the company has created added value by transforming a R&D outcome into a tradable good. The next decision has to address the question, whether the patent is needed strategically for legally securing the company's own products and thus for maintaining its competitive position. If the patent is not required it can be capitalized elsewhere. Apart from maintaining the current competitive position, decision-makers should also take into account future competition. With a well-outlined patent strategy potential competitors can be prevented from entering the market. In other cases patents representing the basis for further technological developments, may be preserved.

If the company decides to use the patent externally (*third phase*), the necessary decision will focus on licensing out versus selling. If the patent is not needed, but is to remain in the company because of strategic considerations, the company may extend licenses under the patent. The patent right remains with the company, but is used by external parties. In the event of a patent being sold, the patent right is transferred to the purchaser.

Ultimately, the R&D management has to decide which patent strategy is pursued:

1. Knowledge can be generated by the company itself through its own R&D (Inside-out),
2. Knowledge can be bought in (Outside-in) or
3. Research can be done jointly with other partners (Coupled). Furthermore, it has to be decided in which of the three phases of the value-adding process action should be taken. As mentioned at the beginning of the section, *Open Innovation*, compared to *Closed Shop Innovation*, results in additional balance sheet effects. Below, we will describe how purchase, sale or granting a license could affect the balance sheet and improve profits.

In our systematic approach to the optimization of innovation processes, a company has to determine not only the phase in which the patent portfolio is to be optimized; it must also identify the economic goals to be achieved. A company that maximizes profit is assumed below. Profit is defined as revenues minus costs. Both for increasing revenues and for reducing costs (and also risks), there are various financial instruments a company could use.

When *revenues* are to be increased, there are several options available within the Inside-out strategy. By selling the patents, it is possible to generate revenues in the form of a one-off payment. If, however, a continuous cash flow is desired over a period of several years, there is the possibility of licensing out. The latter option has the advantage that the company remains the owner of the patent and can therefore monetize the patent in the future. Both options increase equity.

A company that generates the intangible assets (e.g. patents) may not be able to fully record such assets on balance sheet in most jurisdictions. Selling intangibles to unlock such hidden reserves is a third option to increase revenues. This happens for example when a company sells a patent to a subsidiary at arm's length and receives a one-off payment that is booked as profit. In a second step, the subsidiary licenses the patent back to the company, which continuously pays royalties for it.

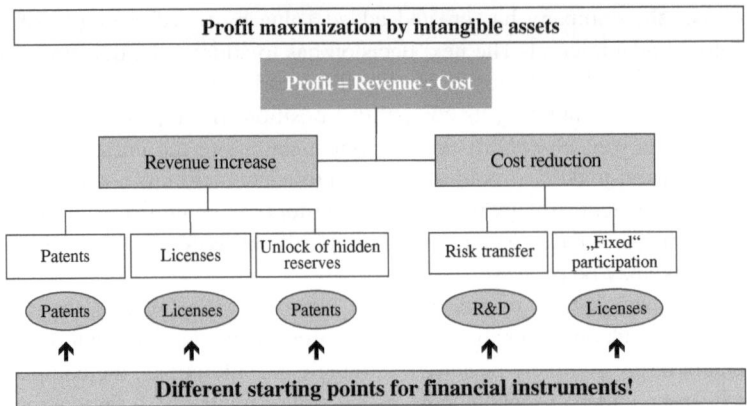

Fig. 4 Profit optimization
Source: Rüschen/Rohwedder

The patent right is recorded on the subsidiary's balance sheet at a value equivalent to the purchase price. This kind of transaction is known as sale and license-back.

By lowering the risks of the R&D process, the *costs* of the entire innovation process could be reduced. The risks arise due to potential delays in the R&D projects, implementation difficulties and complexity in estimation of financial costs of such projects.

A company that applies the Outside-in strategy, acquires licenses or parts of research from external partners and thereby reduces the risk inherent to R&D. The development risks discussed above can be avoided by directly acquiring research results. If such results are not available in the market, R&D risks can still be reduced by inviting additional strategic investors (Coupled innovation) or financial investors (venture capital investors) to share the costs and the risks (but also the rewards) of self-generated knowledge. This will result in a favorable effect on company's performance.

Finally, in addition to the classic methods of financing, the patent owner can generate additional financing based on cash flows received from licenses. These cash flows are either sold on directly or used as underlying for securitizations, which are then placed in the capital markets. The company does not need to post any collateral, as it may be required for a loan, and therefore does not restrict its equity capital basis.

Ultimately, it is the responsibility of the Chief Financial Officer to decide in what form revenue or cost items are to be optimized. Generally, it is possible to enhance profitability by selling the patents or avoiding risks via reduction of refinancing costs.

To sum up: Companies increasingly meet the challenges of the emerging knowledge society with an open innovation processes. In comparison to the Closed Shop Innovation, additional balance sheet effects result from an acquisition and sale respectively licensing in and licensing out of R&D results. Innovative companies can use these effects in order to improve their competitive position and to finance their growth initiatives. To properly manage these processes, companies should focus on an up-front assessment from a technical and economic perspective. This assessment may reveal numerous opportunities for financing instruments, which will be described in the following section.

4 Financing Strategies for Companies

The systematic approach to optimize the innovation process is continued in this chapter. The two optimization approaches presented in the preceding chapter are merged into a *decision matrix* (see Fig. 5) for innovative companies. Optimization of the R&D value-adding process is shown in the first column of the figure and profit optimization in the first line.

Financial instruments offered by banks need to match the profit and R&D optimization methods discussed in Sect. 3. The financial instruments listed below are well known from the world of tangible assets. In general, these instruments can also be applied to the financing of intangible assets. However, they have not been frequently used. Examples of financial instruments that have been employed are listed below. Detailed explanations are given why other instruments may not have been offered so far.

4.1 Increase Revenue

As shown in Sect. 3, in order to *increase revenues*, a company can sell its own patents, grant licenses, or unlock hidden reserves via sale and license-back transactions.

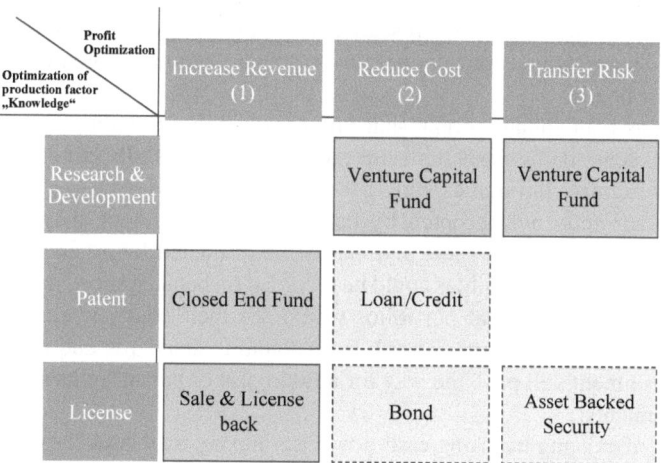

Fig. 5 Decision matrix for businesses
Source: Rüschen/Rohwedder

If a company decides to sell, a closed-end fund can be a viable instrument. These funds buy patent rights and then monetize them in the market place. Investors in these funds can be both institutional and private financial investors. The advantage for the selling company is the option to capitalize patent rights. Furthermore, there is no need to pay maintenance fees anymore. Additionally, the company can sell patented R&D results it does not or no longer need itself.

Deutsche Bank AG, for example, has structured patent rights funds for its private clients. The objective of these funds was to acquire, refine and market patent right portfolios. Each individual patent right went through a legal, technical and economic due diligence process before being accepted into the fund and marketing concepts were developed. The duration of the funds is 7 to 8 years.

If the company intends to use the patent itself, a sale and license-back transaction may be feasible. This financial instrument allows the company to use the patent continuously by means of licensing back. Such transactions have been carried out with trademarks in the past.

4.2 Reduce Cost

If the company is looking at possibilities to *lower costs*, banks can offer venture capital funds, patent-based loans and patent-based bonds.

If the company decides to reduce R&D costs, it can invite further strategic investors (Coupled Innovation) or finance investors (venture capital) through venture capital funds. Venture capital funds provide shareholder equity thereby sharing the high research and development costs. However, these investors then also participate in the successful marketing of the results in the form of profit sharing.

As a rule, venture capital funds are structured in such a way that they do not invest in a company's individual research areas, but inject risk capital into the entire company. This is why venture capital funds typically require (a) board seat(s) on the supervisory board.

Patents as loan collateral represent further possibility to reduce financing costs. This makes sense particularly if patents are already successfully utilized by the company – for example through existing license agreements. However, as a rule, patent rights are currently not accepted by banks as collateral since they are typically not shown on the balance sheet and cannot be evaluated by a generally accepted appraisal method. If intangibles could be recorded on balance sheet, innovative companies with valuable patent portfolios would see their equity increase and hence their ability to borrow. Further work on accounting standards and changes in the legal environment will pave the way for a future use of patent rights as the basis for loan instruments.

In case of existing licensing cash flows, a company may issue a cash flow-based (non-recourse) bond. The company sells its cash flows at a discount to financial investors who then receive the future cash flows (royalties). The company exchanges a regular (uncertain) cash flow for an immediate one-off payment. The advantage for a company is the reduced use of borrowing capacity from its banks.

Doubtlessly, the most well-known transaction was the placing of the David Bowie bond in 1997. The transaction was based on the David Bowie song rights. The song rights were sold for USD 55 million to financial investors who in turn receive the royalties over the term of the transaction.

4.3 Transfer Risk

It can be pivotal for companies to decide on the ways of *transferring* the high legal, technical and financial research and development *risks* to other market participants.

Taking into consideration the high litigation risks in the United States for example, the question arises how legal conflicts in the case of expensive patent disputes can be mitigated with insurance or bank products. The "Blackberry" case is a striking example in this context.

In order to reduce technical and financial risks in the field of R&D, a company may use venture capital funds as an instrument. Venture capital funds provide risk capital from investors. Companies may transfer risks to other market participants, but, in case of success, they have to share the upside that is in line with the risk involved.

If cash flows are pooled, there is the possibility to bundle risk tranches and offer them as asset backed securities (ABS). Such instruments can also be structured using patent-based loans (collateralized loan obligations = CLOs). However, these instruments require transparent and functioning patent markets. These types of transactions will take some time to emerge, as patent-based loans will have to be established on a big scale first.

5 Conclusion

Entrepreneurial success in developed economies is increasingly based on intangible assets, such as technical know-how (patents), artistic knowledge (copyrights), trademarks and the know-how of employees (human capital). Companies must therefore invest in intangible assets. In brief: Knowledge as a production factor has to be financed.

We have tried to show how companies can capitalize their knowledge and to identify opportunities of growth for innovative companies. Banks as partners for innovative companies have some instruments at their disposal to finance such opportunities. All the instruments that have been presented are known from the tangible world of the industrial society. However, they have to be adapted to the knowledge society. The stage and the challenge are set: Now it is up to politicians, banks and companies alike to establish *knowledge markets*.

References

Gassmann, O., & Enkel, E. (2006). Open Innovation: Externe Hebeleffekte in der Innovation erzielen. *Zeitschrift Führung und Organisation, 3*, 132–138.

Hofmann, J. (2005). Value Intangibles: Deutsche Bank Research, International Topics, Frankfurt.

Nakamura, L. (2003). A trillion dollars a year in intangible investments and new economy. In: Hand, J., & Baruch, L. (Eds.), *Intangible assets – values, measures and risks* (pp. 19–47). Oxford University Press.

The Value Impact of R&D Alliances in the Biotech Industry

Hady Farag and Ulrich Hommel

Contents

1 Introduction 371
2 Background and Prior Research 372
3 Research Approach 377
4 Results 378
5 Conclusion 387
6 Appendix 388
References 389

1 Introduction

The extraordinary importance of collaboration for biotechnology manifests itself in an over-proportionally high frequency of alliance formation vis-à-vis other research-intensive industries. As Fig. 1 indicates, biotechnology accounted for over half of all technology alliances in the U.S., compared to having for instance only 7% of all R&D employees. That is, whereas other R&D-intensive industries rely on in-house research, alliances are an integral part of business models in biotechnology. Due to their relatively small size and early stage of development, dedicated biotechnology firms (DBFs) often depend on collaboration partners to finance their ongoing research activities and to commercialize the results of their work. Concurrently, established pharmaceutical firms (Big Pharma) rely on DBFs to complement their internal R&D efforts and to fuel their drug development pipeline (cf. Zucker & Darby, 1995; Pyka & Saviotti, 2001). As a consequence, collaboration between Big Pharma and DBFs has become an essential element of modern drug discovery operations (cf. Arora & Gambarella, 1990; Whittaker & Bower, 1994).

H. Farag (✉)
The Boston Consulting Group (BCG), Frankfurt, Germany
e-mail: farag.hady@bcg.com

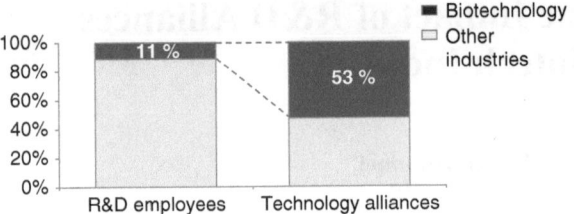

Fig. 1 Biotechnology industry alliance activity (2003). (Source: National Science Board (2006))

The present study aims to provide a detailed account of collaborative value creation in the European biotechnological industry.[1] More specifically, the objective of this work extends to assessing the overall value created by biotechnology alliances as well as understanding the evolution of value over the lifetime of an alliance.

The present study represents the first inquiry into the value of European biotechnology alliances in general as well as the first contribution extensively analyzing the value of alliances across numerous lifecycle stages. This endeavor requires using a homogeneous (single-industry) yet diverse (multi-country) data set of European biotechnology firms and alliances.

2 Background and Prior Research

BIO (2005) defines "new biotechnology [as] the use of cellular and biomolecular processes to solve problems or make useful products" (p. 1).[2] In turn, "red biotechnology" is largely synonymous with biopharmaceutical research and development (R&D), i.e., the development of diagnostics and medications using biotechnologies. Most researchers with an interest in the economics of biotechnology either explicitly (e.g., Grossmann, 2003) or implicitly (e.g., Häussler, 2005) limit their work to this segment. So do we.

Governed by technological necessities and regulatory obligations, biopharmaceutical R&D follows a stringent process model. Figure 2 provides an overview

---

[1] As biotechnology is arguably the industry most reliant on collaborative activities, it provides a suitable setting for studying the value of strategic alliances. Indeed, much alliance-related research has used it as a background (e.g., Lerner, Shane, & Tsai, 2003; Baum & Silverman, 2004). Some studies have even addressed the effect of collaborative agreements on pharmaceutical and biotechnology firm value (e.g., Campart & Pfister, 2003; Karamanos, 2002). While biotechnology collaboration thus has been extensively addressed in general, industry-specific aspects have been largely neglected.

[2] More broadly defined, biotechnology would encompass all applications of biological systems and processes (cf. Christensen, Davis, Muent, Ochoa, & Schmidt, 2002), including fermentation (e.g., in beer brewing) and the cultivation of crops or breeding of animals.

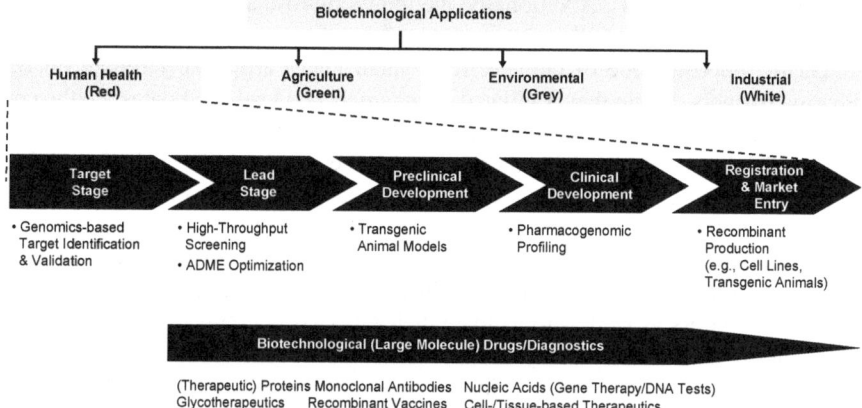

Fig. 2 Overview of biotechnology R&D process. (Source: Own Illustration)

of various biotechnological applications and the main stages of biopharmaceutical R&D.[3]

Along the biotechnological product development process, the core capabilities of biotechnology and Big Pharma firms are highly complementary. While biotechnological skills are best applied during (early) drug discovery stages, the commercial experience and resources of Big Pharma firms give them an edge in (later) clinical drug development, registration, and commercialization.

As the rate of attrition among drug candidates is extremely high,[4] external sourcing of pre-validated targets and leads also provides an essential tool for Big Pharma firms to fill their drug pipelines. In addition, the major pharmaceutical firms were slow and ineffective in picking up new biotechnological concepts, primarily due to their almost exclusive focus on chemistry-based drug discovery. From their perspective, collaboration with biotechnology firms was used as a substitute for internal R&D in these new research areas (cf. Zucker & Darby, 1995; Prevezer & Toker, 1996).[5] Hence, biotechnology collaboration has become an essential tool for Big Pharma to overcome the shortage of new product introductions limiting its continuing growth and endangering its relatively high stock valuation (e.g., Mallik,

---

[3] For more in-depth reviews of biopharmaceutical applications in drug discovery, see (Tollman, Guy, Altshuler, Flanagan, & Steiner, 2001), (Grossmann, 2003), and (Ng. 2004).

[4] About 10,000 substances evaluated in the drug discovery stage on average correspond to one new drug eventually introduced to the market (cf. PhRMA, 2003).

[5] As Gambanos and Sturchio (1998) find, several Big Pharma firms did not even build general biotechnological know-how internally, thus effectively relying on collaboration with biotech firms. Those, who chose to internalize such knowledge, often acquired biotech firms as a basis of their biotechnology activities.

Zbar, & Zemmel, 2004). Additionally, the long time-to-market renders in-sourcing an attractive solution for filling internal gaps.[6]

On the opposite side of the spectrum, small DBFs are equally reliant on Big Pharma partners. While their validated drug targets and lead candidates are the cornerstones of successful new drug development, they often lack the resources to independently progress these projects through the development cycle. In particular, the large-sample trials required in later clinical stages as well as the scale-up of manufacturing processes and the fixed costs of building a proprietary sales force are often outside their capabilities. As a consequence, Big Pharma alliances have long been an important mechanism for DBFs to refinance and commercialize their scientific progress as well as to validate their otherwise unobservable quality characteristics (e.g., Pisano & Mang 1993; Nicholson, Danzon, & McCullough, 2002). Young and small biotechnology firms may also rely on collaborative service agreements and technology out-licensing to gain the financial means necessary for further firm development. As such, drug discovery, service provision, and platform technologies are the prevalent business models in the biotechnology industry (e.g., Höger, Fuchs, & Bähr 2004).[7]

In addition to explaining the occurrence of pharma-biotech alliances, these motives also affect their structure and the relative bargaining power of collaborating firms. On the one hand, a relatively small number of Big Pharma (or large biotech) firms possess the skills and resources to successfully lead drug candidates to the market (cf. Roberts, 1999; DiMasi, 2000; Malerba & Orsenigo, 2002). On the other hand, extraordinary drug candidates may allow their originators to negotiate favorable collaboration terms. Along these lines, (Coombs & Deeds, 2000) observe substantially higher compensation for advanced stage projects, which have already proven themselves in clinical trials. Conversely, platform technologies and research services may be offered by a broader range of firms and may be more easily replicable, which makes them less valuable and decreases the provider firms' bargaining power (cf. Höger et al., 2004; Fisken & Rutherford 2002).[8]

---

[6]Definitions of corporate collaboration are quite diverse. The present study considers as collaborative ventures (or alliances) all *voluntary agreements between independent firms to jointly pursue complementary objectives*. For a similar definition, see (Häussler, 2005). To distinguish between the different organizational modes, the terms 'joint venture' (or JV) or equity-based alliance as well as contractual alliance (or collaboration) will be used.

[7]Alternative business model classifications may be cruder (e.g., Fisken & Rutherford, 2002), who only distinguish products (i.e., drug), platform technologies and hybrids) or even more fine-grained (e.g., Grossmann, 2003).

[8]Technology or service provisions were earlier deemed equally viable business models vis-à-vis proprietary drug discovery. Recent trends have seen most firms abandon pure-play technology or service strategies in favor of own drug discovery activities or hybrid forms (e.g., Anonymous, 2003). Some firms, however, compete successfully by providing state-of-the-art process technologies (e.g., Qiagen) or biopharmaceutical services (e.g., Evotech OAI) due to superior skills and technologies.

The impact of corporate collaboration on firm value has been extensively measured in the existing literature.[9] While some authors (e.g., Koh & Venkatraman, 1991; Anand & Khanna, 2000) consider both JVs and contractual alliances, others focus on JVs, since they are more formally institutionalized.

The expected value impact of entering into strategic alliances and joint ventures is clearly positive. Event studies regularly observe significantly positive valuation effects upon their announcement.[10] Mean abnormal returns on the announcement day range from 0.5% (Das, Sen, & Sengupta, 1998) to over 2.6% (Park & Mezias, 2005) and are different from zero at standard significance levels. Moreover, (Kale, Dyer, & Singh, 2001, 2002) for alliances and (Koh & Venkatraman, 1991) for joint ventures document a high level of correlation between event-study ARs and ex-post alliance success measured by managerial assessment years after the transaction.

Studies in other high-growth settings (e.g., Neill, Pfeiffer, & Young-Ybarra, 2001; Park, Mezias, & Song, 2004) also observe particularly high collaborative gains. Combined with the evidence that technology-related collaboration tends to earn above-average ARs (Chan, Kensinger, Keown, & Martin, 1997), this nourishes the expectation that biotechnology alliances should have significantly positive value impact.

> Hypothesis 1: The abnormal returns to strategic alliance (and JV) formation are significantly positive.

Prior event-study evidence on the termination of collaborative ventures is mixed, in particular showing a potential for wealth gains, when collaborative ventures are bought-out or sold-off (Reuer, 2000). However, the mere discontinuation of alliances may reflect non-performance or inter-alliance rivalry and appears to consistently hurt firm value (e.g., Häussler, 2006). Excluding internalization or external sale, alliance termination thus should lead to significantly negative abnormal returns in biotechnology alliances.

> Hypothesis 2: The ARs to announcements of alliance (or JV) termination are significantly negative.

---

[9] The approaches available in this context can be roughly classified into studies of short-term (or announcement) and long-term effects. In the domain of alliance-related research, the former approach has been prevalent. While long-run effects have been considered in M&A research, such an approach may not be feasible in a collaborative context due to (a) the greater number of similar transactions per firm and (b) the substantially smaller operational magnitude of these events. Complementarily, indirect value effects, i.e., interactions of collaborative portfolios with other events such as IPOs, will be considered.

[10] See Farag (2009) for a more detailed review of prior research into the value created by strategic alliances and joint ventures.

Aside from alliance formation (Hypothesis 1) and termination (Hypothesis 2), no other events along the alliance lifecycle have been studied in an event-study context. Yet, such intermediate steps may reflect the evolution of collaboration and the successive disclose of value-related information to the capital market.

As suggested by (Harrigan, 1985), the adaptation as well as progression of an existing alliance may reflect successful (re)alignment of the alliance with company objectives and environmental requirements. Given that firms tend to adapt their alliances activities to changing technological and market development (Madhavan, Koka, & Prescott, 1998), such realignment may further increase the alliance's value. Moreover, firms may adapt or expand well-performing alliances (e.g. by securing commercialization rights), such that adaptation announcements may also signal collaborative success.[11]

> Hypothesis 3a: Announcements of alliance adaptation exhibit significantly positive ARs.

In addition to formal restructuring of alliance terms, collaborative value may continuously evolve over the duration of an alliance. Announcements relating to alliance operations thus may update the market's assessment of collaborative benefits and thus affect firm value. Similarly, prior research has documented significant valuation effects for announcements of technological progress, including R&D activities (e.g., Chan, Martin, & Kensinger, 1990), patenting (e.g., Austin, 1993), and new product introductions (e.g., Chaney, Devinney, & Winer, 1991; Sharma & Lacey, 2004).[12] As these indicators may reflect improved competitiveness (e.g., successful market entry ~ Glaister and Buckley 1996), they are generally associated with positive effects on firm value [see (Brockhoff, 1999) for a review and summary]. Consequently, the progression of an alliance, i.e., the achievement of milestones and the commencement of new activities, should increase firm value.

> Hypothesis 3b: Announcements of alliance progression exhibit significantly positive ARs.

The present research sets out to empirically test these hypotheses.

---

[11] In this context, the internalization of successful collaborative projects (Kogut, 1991) may be considered as an expansive form of alliance adaptation. As such events are rare in the biotechnology setting, internalization will not be considered separately.

[12] Liu (2000) distinguishes six types of innovation-related news announcements and documents significantly positive announcement ARs, but negative long-run returns (BAHR).

3 Research Approach

So far, no universally accepted listing of European biotechnology firms exists, as evidenced by substantial differences in the firms included in standard industry publications, such as Biocentury's "The Bernstein Report®", BioScan, BioVenture View, the annual Ernst&Young biotechnology industry reports, or Recombinant Capital. Since no individual classification scheme can claim universal acceptance, sample selection was based on the consensus of valid classifications. Specifically, firms had to be included in at least four out of five listings to be included in the final sample. A comprehensive LexisNexis search on the 46 sample firms from 01/01/1997 to 12/31/2003 resulted in 2572 news items, of which 690 (or 26.8%) were related to collaborative activity (379 alliance formation announcements).[13]

Studying the evolution of value across the alliance lifecycle (Hypotheses 1–3) requires a segmentation of alliance-related announcements into subgroups of lifecycle events. While alliance formation and termination have commonly been distinguished (e.g., by Reuer, 2000; Häussler, 2005), the present study extends this line of research by considering a greater variety of alliance-related news, namely extensions of existing agreements, expansions in the scope of collaboration, and other modifications (which also comprise structural modifications). On an operating level, alliance activity (i.e., collaborative projects advancing into subsequent phases of development) and outcome-related news (i.e., completion of specific collaboration stages) can be distinguished. Based on this classification scheme,[14] the overall sample was divided into event categories. Figure 3 provides an overview of the announcements assigned to each event category.[15]

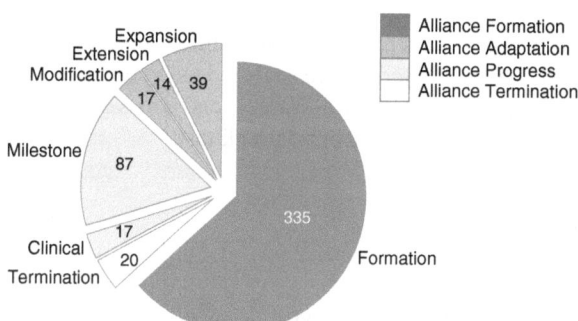

Fig. 3 Sample split by event category. (Source: Own illustration)

---

[13] Note that not all sample firms are publicly traded over the entire period of study. However, only events related to stock-listed companies have been considered in the empirical analysis.

[14] Table 1 of the appendix provides the explicit definitions for the types of alliance-related news considered as part of this study. It also takes account of a number of complementary news items to validate the findings regarding the primary event types. These include reruns of previously announced alliances, publicized alliance-related rumors, and news including information on multiple alliances or alliances outside the human healthcare sector.

[15] For confounding events and missing return data has been corrected ex ante.

It emphasizes the importance of new alliance formation, which represents over 60% of all relevant news items.

The present study analyzes the announcement effects of alliance-related news based on the event-study methodology. A wealth of prior literature has generated, discussed and analyzed various versions of this approach to determine the value impact of unanticipated news announcements. Generally speaking, the analysis requires correcting the actually observed stock returns [$R_{i,t}$ in Eq. (1)] for the expected return on the security assuming the absence of the event.

We apply the most commonly used approach to estimating expected returns, going back to (Fama, Fisher, Jensen, & Roll, 1969).[16]

Their market model accounts for the overall market return on the event day [$R_{m,t}$ in Eq. (1)] as well as the average sensitivity of focal firm returns to market movements [β estimated in Eq. (2)].

$$AR_{i,t} = R_{i,t} - (\alpha_i + \beta_t R_{m,t}) \quad (1)$$

with

α, β – coefficients of the following one-factor OLS regression model estimated over a relevant estimation period:

$$R_{i,t} = \alpha_t + \beta_t R_{m,t} + \varepsilon_{i,t} \quad (2)$$

We use both general and biotechnology-specific stock indices as benchmarks to approximate market movements. Similar to other alliance-related event-studies (e.g., Park et al., 2004; Janney & Folta, 2003), we also employ a two-factor approach combining a general market and an industry index. Finally, an equal-weighted intra-sample index was used to reflect the most similar peer group available.

The calculated abnormal returns were analyzed using four different test statistics: The parametric Dodd-Warner, Brown-Warner, and cross-sectionally-standardized tests as well as the non-parametric Corrado rank test.[17]

4 Results

The present study first calculated the ARs for each day of the observation period ranging from 10 days prior to the announcement to 10 days after the announcement. Figure 4 exhibits the aggregate ARs (AARs) for all three return models.

---

[16] Several authors have provided overviews of the methodologies used in event-study research. Bowman (1983), Peterson (1989), Strong (1992), Armitage (1995), MacKinlay (1997), McWilliams and Siegel (1997), and Bhagat and Romano (2001), among others. Another line of literature provides evidence on the performance of alternative event-study methodologies. This includes (Brown & Warner, 1980, 1985) as well as (Cable & Holland, 1999).

[17] See Farag (2009) for details.

The Value Impact of R&D Alliances in the Biotech Industry 379

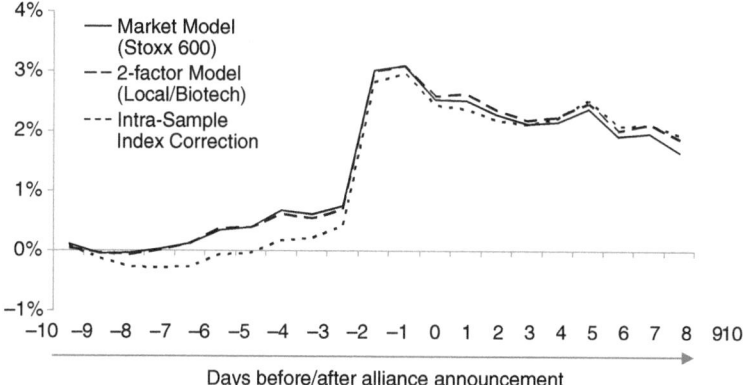

Fig. 4 Summed daily AAR over event period. (Source: Own illustration)

On an aggregate level, these findings support the notion of alliance-related news having a substantially positive impact on firm value. Overall, the statistical evidence is highly congruent across return models and testing procedures, which reflects the great robustness of these results. The main effect occurs on the event day itself (day 0), indicating that the market recognizes the announcements as new information. In addition, there appears to be some activity before and after the actual announcement date:

- The ARs on the day of the announcement itself are significantly positive. Their magnitude ranges from 2.26% (STOXX600 market model) to 2.42% (intra-sample model) reflecting a substantive effect of alliance-related news items. Firms on average also experience positive ARs during the two days surrounding the announcement (days −1 and +1), but these returns remain far from statistically significant. Similarly, ARs for day −3 are positive, but only significant for some tests.
- Some information may already be incorporated during the days leading up to the announcement, but such anticipatory effects appear limited.
- While day +1 returns may still be positively affected by the announcement, firms loose between 1 and 1.5% during the ensuing period.[18] In particular, the ARs 2 and 8 days after the announcement are significantly negative. However, their magnitude remains small compared to the initial announcement. For all three models, they are in the range of 0.5% each. Also, the net effect of the announcements over the entire 21-day period remains positive, in the range of 1.5–2%.

---

[18]This negative AR trend following the announcement may be attributed to market overreaction or investors cashing-in on the announcement gains and is not uncommon in event-study research. Similar patterns have also been observed in the ARs to M&A announcements (e.g., Bae, Kang, & Kim, 2002)].

Next, the individual securities' abnormal returns were aggregated and averaged into cumulated abnormal returns (CARs) across 11 different multi-day windows.[19] These event windows were chosen to allow comparison to prior research in the area of collaborative value creation. Table 1a and b presents the main findings.

The results are consistent across return models and test statistics (the three parametric significance tests and the Corrado rank test):

- The aggregate event-study analysis provides a clear picture regarding the valuation effects of alliance-related news. CAARs are positive and, with the exception of the 21-day (−10 to +10) and one of the two 12-day (−1 to +10) event windows, are statistically different from zero.
- Similarly, longer event periods are associated with greater ARs than the announcement day itself, if limiting the weight of the post-announcement period. In particular, all event windows up to 11-days in length (centered around the announcement date) consistently exhibit significant ARs.

The insignificance of ARs in the two specific cases thus may be due to the profit-taking behavior towards the end of the observation period (days +8 and +10) in combination with the difficulties of detecting statistical significance in longer event windows (sometimes also called "buy the rumor, sell the news").

Overall, as the 5-day event window ranging from day −3 to +1 reveals the numerically highest CAARs, some value-related information may be processed before or immediately after the event. However, the magnitude of test statistics steadily declines with expanding event windows, indicating that the main value effect is sufficiently reflected in shorter time spans.

In order to assess the evolution of collaborative value over the lifetime of an alliance, seven major types of alliance-related news were distinguished. In addition to alliance formation, these reflect the adaptation (i.e., expansion, extension, or modification), progression (i.e., milestone and clinical advancement) and termination of alliances. Furthermore, announcements relating to multiple alliances, alliances outside the human health sector, rumors on alliance formation and announcements previously publicized alliances are used as control cases.[20]

The evidence in Table 2a and b provides a comprehensive assessment of collaborative value creation. The main pieces of evidence are quite consistent across different event windows, return models, and statistical tests.[21] In particular:

---

[19] Note that the summed ARs exhibited in Fig. 4 differ from the CARs referred to in Table 1 and further analyzed in this paper. Specifically, the daily AR forming the basis of Fig. 4 only were corrected for confounding events on the announcement date itself. Contrarily, CARs are only calculated based on events without any confounding items during the entire observation period.

[20] Specifically, *multiple* and *other* alliance announcements (as well as some *rumors*) should entail a significant market reaction, whereas restatement of existing information (*follow*) should not.

[21] While the tables included in the text are only based on the standard market model, the results obtained for the 2-Factor-Model and the Intra-Sample-Model are highly similar.

Table 1a Results of event study by event windows (all events)

Window	N	Market Model(Stoxx600)						2-Factor-Model(Local/StoxxBiotech)			
		CAAR	Z(DW)	Z(BW)	Z(CS)	Z(Corr)	CARR	Z(DW)	Z(BW)	Z(CS)	Z(Corr)
−10 to +10	277	0.0067	0.37	0.76	−0.58	0.51	0.0099	0.82	1.18	−0.35	0.66
−10 to +3	398	0.0261	2.78***	3.62***	1.68*	2.11**	0.0283	3.15***	4.12***	1.89*	2.17**
−10 to +1	427	0.0316	4.26***	4.74***	2.61***	2.71***	0.0322	4.39***	5.07***	2.64***	2.66***
−1 to +10	417	0.0066	0.89	0.99	0.33	0.36	0.0098	1.31	1.55	0.68	0.74
−5 to +5	428	0.0237	3.12***	3.72***	1.56	2.29**	0.0250	3.36***	4.12***	1.69*	2.26**
−3 to +3	500	0.0254	4.73***	4.99***	2.39**	3.24***	0.0273	5.19***	5.63***	2.71***	3.30***
−2 to +2	524	0.0238	5.16***	5.52***	2.31**	2.71***	0.0252	5.64***	6.15***	2.58**	2.95***
−3 to +1	532	0.0320	7.70***	7.43***	4.37***	4.51***	0.0321	7.78***	7.82***	4.37***	4.39***
−2 to +1	547	0.0280	7.40***	6.50***	3.85***	3.97***	0.0286	7.64***	6.96***	3.95***	4.08***
−1 to +1	564	0.0285	9.30***	8.54***	4.71***	4.80***	0.0289	9.54***	9.10***	4.79***	5.04***
−1 to 0	597	0.0252	10.37***	9.28***	5.18***	5.88***	0.0254	10.60***	9.79***	5.20***	6.18***

Z(DW)/(BW)/(CS)/(Corr) – Z-Statistic for Dodd-Warner/Brown-Warner/Cross-sectionally standardized/Corrado tests. ***/**/* indicates significance at 1/5/10% levels

Table 1b Results of event study by event windows (all events)

Window	N	Market Model (Intra-Sample Index)				
		CAAR	Z (DW)	Z (BW)	Z (CS)	Z (Corr)
−10 to +10	277	0.0128	1.15	1.54	−0.26	0.77
−10 to +3	398	0.0268	2.92***	3.86***	1.61	1.66*
−10 to +1	427	0.0312	4.11***	4.95***	2.43**	2.16**
−1 to +10	417	0.0157	1.76*	2.38**	1.52	1.48
−5 to +5	428	0.0295	3.99***	4.79***	2.19**	2.70***
−3 to +3	500	0.0284	5.46***	5.77***	2.72***	3.44***
−2 to +2	524	0.0282	6.43***	6.80***	3.12***	3.50***
−3 to +1	532	0.0345	8.33***	8.31***	4.74***	4.64***
−2 to +1	547	0.0312	8.32***	7.52***	4.50***	4.51***
−1 to +1	564	0.0303	9.95***	9.43***	5.10***	5.25***
−1 to 0	597	0.0265	10.99***	10.10***	5.48***	6.43***

Z(DW)/(BW)/(CS)/(Corr) – Z-Statistic for Dodd-Warner/Brown-Warner/Cross-sectionally standardized/Corrado tests. ***/**/* indicates significance at 1/5/10% levels

- Alliance formation induces significantly positive wealth gains across all event windows, with the exception of the post-announcement period (here: -1/+10 window). This effect is similar in significance and potentially larger in magnitude for announcements relating to *multiple* alliances or taking *other* forms (e.g., asset sales).
- Alliance expansion is associated with substantial and significant announcement gains around the announcement date (up to 7-day event window), but aggregate returns over longer time horizons are insignificant. The announcement ARs on alliance extensions and modifications are minimal and insignificant across all event windows.
- The achievement of collaborative *milestones* yields significant ARs on and around the announcement days. The magnitude of these wealth gains even appears to exceed that of alliance announcements. However, the level of significance is not consistent across event windows and test statistics. In particular, CAARs are insignificant for the 7-, 11-, and one of the 12-day (−1/+10) windows, indicating that the value quickly degenerates in the post-formation period. *Clinical* advancement (i.e., the initiation of further alliance activities) does not yield significant ARs.
- Alliance *termination* results in significantly negative ARs on the announcement day. This effect, however, is limited to the Dodd-Warner and Corrado test statistics. Additionally, the negative significance does not persist across all event windows. In particular, medium length (3–7 days) windows report insignificant results, whereas short- and longer-term perspectives show significantly negative results on standardized CAARs.[22]

---

[22] Note that while non-standardized CAARs are positive for the 11- and 12- event windows, the Dodd-Warner-type test results are significantly negative. As this approach standardizes daily

Table 2a Results of event study by event category for different event windows (based on Stoxx600 market model)

Event	12-Day Window (-10/+1)						12-Day Window (-1/+10)						11-Day Window (-5/+5)					
	N	CAAR	DW	BW	CS	Corr	N	CAAR	DW	BW	CS	Corr	N	CAAR	DW	BW	CS	Corr
Formation	244	0.0373	***	***	**	***	228	0.0118					243	0.0239	***	***		***
Expansion	25	0.0173					26	-0.0094					25	0.0267				
Extension	9	-0.0106					10	0.0221					8	-0.0364				
Modification	15	-0.0007					12	-0.0478					16	-0.0262				
Milestone	62	0.0335	*	*			56	-0.0089					60	0.0396	**	**		
Clinical	9	0.0136				*	13	-0.0119					11	0.0043				
Termination	15	0.0089	**				15	0.0532					14	0.0381	*			
Multiple	8	0.0772	**	**			6	0.0397				*	6	0.0786	**	**		
Other	16	0.0990	***	***	**	*	14	0.0139					15	0.1055	***	***	***	
Rumor	6	-0.0559	**				7	0.0831		**			6	0.0051				*
Follow	22	-0.0013					30	-0.0175					27	-0.0220				

DW/BW/CS/Corr – Significance of Dodd-Warner/Brown-Warner/Cross-sectionally standardized/Corrado tests. ***/**/* indicates significance at 1/5/10% levels

Table 2b Results of event study by event category for different event windows (based on Stoxx600 market model)

Event	7-Day Window (-3/3)						4-Day Window (-2/1)						2-Day Window (-1/0)					
	N	CAAR	DW	BW	CS	Corr	N	CAAR	DW	BW	CS	Corr	N	CAAR	DW	BW	CS	Corr
Formation	279	0.0258	***	***		***	301	0.0282	***	***	***	***	311	0.0296	***	***	***	***
Expansion	32	0.0490	**	**			35	0.0414	***	**			37	0.0391	***	***	**	***
Extension	11	0.0068					13	-0.0067					14	0.0025				
Modification	17	0.0065					17	0.0149					17	0.0171				
Milestone	69	0.0210					76	0.0284	***	**			81	0.0324	***	***	**	***
Clinical	12	0.0207					13	0.0109				**	14	-0.0027	***			
Termination	17	0.0462					19	0.0620			***		19	-0.0213	***	***		*
Multiple	9	0.0904	***	***		*	9	0.0945	***	***	**	**	9	0.0591	*			
Other	16	0.0604	**	***	**		17	0.0450	***	**	*		18	0.0171				**
Rumor	6	0.0906		***			8	0.0543		**			9	0.0080				
Follow	35	-0.0281	*	*			39	-0.0086					39	0.0068				

DW/BW/CS/Corr – Significance of Dodd-Warner/Brown-Warner/Cross-sectionally standardized/Corrado tests. ***/**/* indicates significance at 1/5/10% levels

- Finally, among the news items included for control purposes, the reiteration of previously announced alliances at a later point in time does not affect corporate value, which supports the basic premise that efficient capital markets only react to actual news. Conversely, public *rumors* on alliances (generally concerning alliance formation) are associated with some wealth implications. The pattern of these ARs is peculiar, since short-to-medium windows around the announcement date (i.e., 4 and 7 days) exhibit significant value gains, although the announcement-day returns themselves are insignificant. A comparison of the two 12-day windows indicates that stock returns are negative prior the "rumored news", whereas they are highly positive for the post-formation period.[23]

These results need to be interpreted with some caution, since some of them are based on reasonably small subsamples. Additionally, the significance of some findings varies across testing procedures. In particular, the cross-sectionally standardized and Corrado rank statistics appear more powerful in medium- to long-term event windows. This suggests that findings may be sensitive to cross-sectional heterogeneity in the ARs, which also may be the root of non-normality. That is, high-AR events are the primary source of positive ARs as opposed to a general but moderate increase in firm value. For shorter event-windows (e.g., 1–5 days for alliance formation), however, findings are consistent across all test statistics.

Overall, the event-study thus supports Hypotheses 1–3, which argue that alliance formation, termination, and post-formation events (i.e., adaptation and progression) substantially impact corporate value. In particular, these findings create a colorful picture of post-formation value dynamics.

First, with regard to the flexibility to adapt alliances, the present study documents significantly positive returns to alliance expansion but neither to extension nor modification. This may reflect that alliance expansion is a more pronounced indicator of collaborative success than mere alliance extension or modification. The expansion of an alliance may build on technological progress, signal mutual trust, and open additional sources of value generation. In contrast, alliance extension may indicate work-in-progress and alliance modification may be indicative of mutual understanding, but also suboptimal performance or changes in the intra-alliance power structure. Most generally, the insignificance of extension and modification announcements suggests that the flexibility inherent in strategic alliances may not be valuable per se. This would be contrary to the general hypothesis that a substantial share of alliance value arises from them being flexible organizational schemes

---

returns, the wealth gains reported by some firms around alliance termination are smaller relative to the firms' historic volatility than the wealth losses experienced by other firms.

[23] Similar to the case of alliance termination, the effects of alliance rumors are not homogeneously significant across test statistics. Specifically, the Brown-Warner-type test consistently provides a more positive assessment than the Dodd-Warner-type statistic (long-window negative significance). Consequently, the positive value of rumors is mostly driven by securities also underlying a higher volatility in general (i.e., during the estimation period). Additionally, significantly positive abnormal returns following the announcement (esp. day +1) render CAARs positive on average.

(cf. e.g. Jones & Hill, 1988). Alternatively, these flexibilities may have been correctly "priced" at the time of alliance formation.[24] The (expected) flexibility value of extending or modifying an alliance will have been part of the formation ARs.

Second, with regard to the operational progress of collaborative ventures, the results indicate positive valuation effects for the completion of tasks (milestones) as opposed to insignificant returns on the continuation of activities. This may again indicate that the market only values actual news. The achievement of milestones reveals new technological information concerning the stage and state of the collaboration. Given such prior information, the advancement of collaborative activities (e.g., into subsequent stages of development) may be anticipated. Additionally, the achievement of milestones is often associated with financial payments between alliance partners. The observed value gains thus reflect increases in the value of collaborative projects, in which the focal firm either continues to hold an interest or participates through milestone payments.

Third, alliance termination results in short-term wealth losses. This is congruent with the limited prior evidence on alliance termination (e.g., Häussler, 2006). Beyond the actual announcement-day effect, however, termination may not substantially destroy value. In the given context, this contradictory evidence may be attributed to a variety of factors. On the one hand, alliance termination may not always be actual news to informed market participants. Since DBFs regularly report on collaborative progress, the market may anticipate alliance termination following less than satisfactory progress reports.[25] This explanation would also be in line with the evidence that short-term wealth losses are primarily significant for firms subject to low volatility during the estimation period (i.e., Dodd-Warner test). On the other hand, alliance termination may also create new opportunities, which may partially compensate for the disadvantages of alliance discontinuation. In particular, the termination of outbound alliances, i.e., collaboration providing the partner with rights to proprietary drug candidates or technologies, usually results in these rights being returned to the focal firm. Given that termination decisions may result from factors other than outright technological or market failure (e.g., misfits within partner portfolio), the focal firms may continue working on the project alone or in collaboration with new partners. This represents an intermediate case between discontinuation of failed projects and the alliance internalization, which may be valuable. A further analysis of these issues would require cross-sectional analysis taking into account

---

[24]While a strategic perspective highlights the adaptive advantages of collaboration, transaction cost economics argue that the need to adapt represents a source of coordination costs. The findings presented here, however, do not distinguish between these two elements, representing a joint hypothesis test. Consequently, the overall insignificant ARs may also result from adaptive gains being mitigated by intra-alliance rivalry. Note that high levels of uncertainty may increase the benefits of hierarchical (rather treuhan hybrid) coordination.

[25]Consequently, milestone and termination announcements may not be fully independent from each other, which would violate the basic assumptions underlying event-study analysis. In the present study, however, the share of non-positive milestone announcements is negligibly small. Moreover, such information is often first reported as part of the termination announcement itself. Nonetheless, the market may have anticipated alliance termination without explicit news on collaborative failure, e.g., based on rumors or the lack of positive milestone announcements.

the direction of resource flows and the existence of (negative) milestone announcements, among other items. Given the small size of the termination subsample, however, this was not feasible in our case.

5 Conclusion

The present article has reported and discussed the results of the event-study analysis on different types of alliance-related news announcements. Across all types of alliance-related news, the present study has documented a positive and significant value impact using a variety of event windows, estimation and testing procedures. In particular, the observed wealth gains only become insignificant when fully considering a 10-day post-announcement period, during which investors may have capitalized (i.e., cashed-in) on the announcement returns.

The findings are highly similar for the subsample of alliance formation announcements, which represents about half of all events. With regard to other types of collaborative news, the event study has observed positive ARs to announcements on the expansion of collaborative activities and on the achievement of collaborative milestones. Similarly, alliance termination results in significantly negative (short-term) announcement returns. Conversely, alliance extension, modification, and the advancement of collaborative activities do not result in significant valuation effects. All-in-all, these findings support the basic notion that alliance-based value is realized sequentially, although this effect is limited to announcements reflecting clearly positive (or negative) developments and having an air of novelty, whereas the realization of standard flexibilities may be well anticipated. Figure 5 summarizes these findings.

The first three hypotheses (No. 1–3) addressed the evolution of collaborative over the lifecycle of an alliance relationship and were tested using the event-study method itself. More precisely, the study applied significance tests to the ARs associated with alliance formation (Hypothesis 1), termination (Hypothesis 2), adaptation and progression (Hypothesis 3a/b) announcements to assess the value impact of these lifecycle events. The results supported the hypothesized relevance of alliance formation, expansion, and milestone announcements, whereas alliance extension, modification, termination, and activities were not associated with significant announcement effects. In summary, these findings suggest that the capital market adequately anticipates many lifecycle-events at the time of alliance formation. Only substantial new developments then provoke a significant reaction during the post-formation period.

While the event study provided a clear and consistent picture of value evolution along the alliance lifecycle, the extent of information trickling down into the market aside from such formal announcements is difficult to assess. In particular, the insignificance of alliance termination (hypothesized to be value-reducing) and extension (hypothesized to be value-enhancing) could be due to such leakage effects. A comprehensive model of collaborative value creation thus should account for both one-time AR (such as alliance formation, expansion, and milestones) as

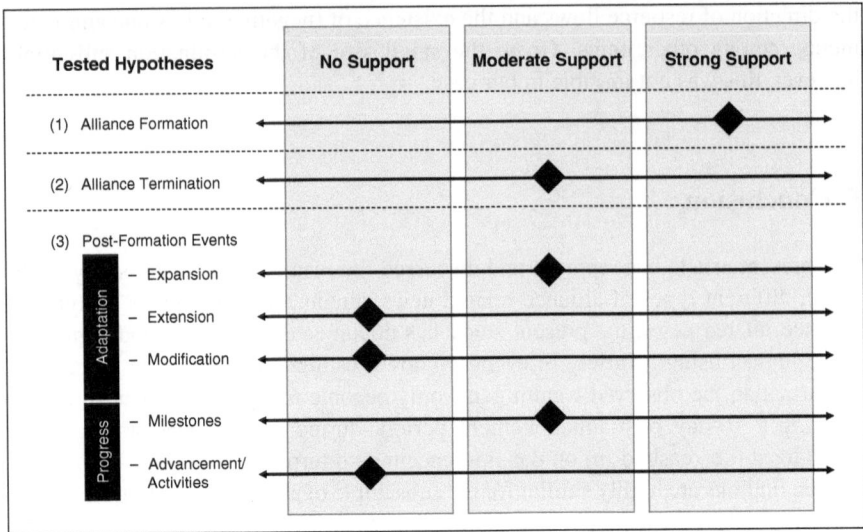

Fig. 5 Summary of findings (Hypotheses 1–3). (Source: Own illustration)

well as more continuous value development. Clearly, this is easier said than done given the general delicacy of long-term event-study research and the multitude of alliances (as well as other value-affecting events) occurring in the biotechnology industry. Clinical studies may present a starting point for identify the value dynamics in some selected alliances.

6 Appendix

Table 3 Coding of alliance-related news

Transaction type	Definition	Including
Hypothesized Effects (Substantive)		
Formation	Announcement of new collaboration	Alliances between previously related partners
Expansion	Announcement of increase in the scope of alliance activities	Joint expansion and extension
Extension	Announcement of Prolongation (with largely unaltered scope)	
Modification	Announcement of changes to the structure/terms of the alliance without extension or expansion of activities	

Table 3 (continued)

Transaction type	Definition	Including
Milestone	Announcement of collaborative achievements (e.g. research results)	Natural end point of alliance
Clinical / Advancement	Announcement of collaborative activities (e.g. entry into clinical trials)	
Termination	Announcement of premature end of collaborative activities (i.e., not satisfaction of collaborative objective)	Non-exercise of existing options (especially to expand or extend collaboration)
Complementary Events and Non-Substantive News		
Other	Announcement of activities akin to collaboration or not related to human health (red biotechnology)	Asset Sales Environmental Protection/Animal Health/Agriculture
Multiple	Announcements on multiple unrelated alliances	
Rumor	News on potential alliance formation (or other alliance-related activities)	
Follow	Reiteration of already published information on alliance formation (or other alliance-related activities)	

References

Anand, B. N., & Khanna, T. (2000). Do firms learn to create value? The case of alliances. *Strategic Management Journal, 21*, 295–315.

Anonymous. (2003). Climbing the Helix staircase – a survey of biotechnology. *The Economist* (March 29th, 2003), 3–18.

Armitage, S. (1995). Event study methods and evidence on their performance. *Journal of Economic Surveys, 8*(4), 25–52.

Arora, A., & Gambarella, A. (1990). Complementarity and external linkages – the strategies of the large firm in biotechnology. *Journal of Industrial Economics, 38*, 361–379.

Austin, D. H. (1993). An event-study approach to measuring innovative output – the case of biotechnology. *American Economic Review, 83*, 253–258.

Bae, K.-H., Kang, J.-K., & Kim, J.-M. (2002). Tunneling or value added? Evidence from mergers by Korean business groups. *Journal of Finance, 62*(2), 2695–2740.

Baum, J. A. C., & Silverman, B. S. (2004). Picking winners or building them? Alliance, intellectual, and human capital as selection criteria in venture financing and performance of biotechnology startups. *Journal of Business Venturing, 19*, 411–436.

Bhagat, S., & Romano, R. (2001). Event Studies and the Law – Part 1 – Technique and Corporate Litigation, Yale International Center for Finance Working Paper No. 00–31 / Yale Law School Program for Studies in Law, Economics, and Public Policy Research Paper No. 259.

BIO BIO. (2005). *2005–2006 Guide to Biotechnology*. Washington, DC.

Bowman, R. G. (1983). Understanding and conducting event studies. *Journal of Business Finance and Accounting, 10*(4), 561–584.

Brockhoff, K. K. (1999). Technological progress and the market value of firms. *International Journal of Management Reviews, 1*(4), 485–501.

Brown, S. J., & Warner, J. B. (1980). Measuring security price performance. *Journal of Financial Economics, 8*, 205–258.

Brown, S. J., & Warner, J. B. (1985). Using daily stock returns – the case of event studies. *Journal of Financial Economics, 14*, 3–31.

Cable, J., & Holland, K. (1999). Regression Vs. non-regression models of normal returns – implications for event studies. *Economic Letters, 64*, 81–85.

Campart, S., & Pfister, E. (2003). The Value of Interfirm Cooperation – an Event Study of New Partnership Announcements in the Pharmaceutical Industry, Université du Havre/Université Nancy II Working Paper.

Chan, S. H., Kensinger, J. W., Keown, A., & Martin, J. D. (1997). Do strategic alliances create value? *Journal of Financial Economics, 46*, 199–221.

Chan, S. H., Martin, J. D., & Kensinger, J. W. (1990). Corporate research and development expenditures and share value. *Journal of Financial Economics, 26*, 255–276.

Chaney, P. K., Devinney, T. M., & Winer, R. W. (1991). The impact of new product introductions on the market value of firms. *Journal of Business, 64*, 573–610.

Christensen, R., Davis, J., Muent, G., Ochoa, P., & Schmidt, W. (2002). Biotechnology – an Overview. European Investment Bank (EIB) Sector Papers June 2002.

Coombs, J. E., & Deeds, D. L. (2000). International alliances as sources of capital – evidence from the biotechnology industry. *Journal of High Technology Management Research, 11*(2), 235–253.

Das, S., Sen, P. K., & Sengupta, S. (1998). Impact of strategic alliances on firm valuation. *Academy of Management Journal, 41*(1), 27–41.

DiMasi, J. A. (2000). New drug innovation and pharmaceutical industry structure – trends in the output of pharmaceutical firms. *Drug Information Journal, 34*, 1169–1194.

Fama, E. F., Fisher, L., Jensen, M., & Roll, R. (1969). The adjustment of stock prices of new information. *International Economic Review, 10*, 1–21.

Farag, H. (2009). The Value of Collaboration – How Biotechnology Firms Gain from Alliances, Springer: Contributions to Management Science, Heidelberg, 2009 (also Doctoral Dissertation, European Business School International University 2006).

Fisken, J., & Rutherford, J. (2002). Business models and investment trends in the biotechnology industry in Europe. *Journal of Commercial Biotechnology, 8*(3), 191–199.

Gambanos, L., & Sturchio, J. L. (1998). Pharmaceutical firms and the transition to biotechnology – a study in strategic innovation. *Business History Review, 72*(2), 250–279.

Glaister, K. W. & Buckley, P. J. (1996). Strategic motives for international alliance formation. *Journal of Management Studies, 33*(3), 301–332.

Grossmann, M. (2003). *Entrepreneurship in biotechnology – managing for growth from start-up to initial public offering.* Heidelberg: Physica.

Häussler, C. (2005). *Interfirm collaboration – valuation, contracting, and firm restructuring*, Wiesbaden, also University of Munich Dissertation, Gabler.

Häussler, C. (2006). When does partnering create market value? *European Management Journal, 24*(1), 1–15.

Harrigan, K. R. (1985). *Strategies for joint ventures.* Lexington, MA: Lexington Books.

Höger, T., Fuchs, P., & Bähr, C. (2004). Erfolgskurs Nach Konsolidierung – Biotechnologie – Eine Research Publikation Der Dz Bank Ag, DZ Bank, Frankfurt.

Janney, J. J., & Folta, T. B. (2003). Signaling through private equity placements and its impact on the valuation of biotechnology firms. *Journal of Business Venturing, 18*, 361–380.

Jones, G. R., & Hill, C. W. L. (1988). Transaction cost analysis of strategy-structure choice. *Strategic Management Journal, 9*(2), 159–172.

Kale, P., Dyer, J. H., & Singh, H. (2001). Value creation and success in strategic alliances – alliancing skills and the role of alliance structure and systems. *European Management Journal, 19*(5), 463–471.

Kale P., Dyer J. H., & Singh H. (2002). Alliance capability, stock market response, and long-term alliance success – the role of the alliance function. *Strategic Management Journal, 23*, 747–767.

Karamanos, A. (2002). Market Networks and the Value in Knowledge Exchanges – Evidence from Biotechnology Strategic Alliances, University of Cambridge ESRC Centre for Business Research Working Paper 240.

Kogut, B. (1991). Joint ventures and the option to expand and acquire. *Management Science, 37*(1), 19–33.

Koh, J., & Venkatraman, N. (1991). Joint venture formations and stock market reations – an assessment in the information technology sector. *Academy of Management Journal, 34*(4), 869–892.

Lerner, J., Shane, H., & Tsai, A. (2003). Do equity financing cycles matter? Evidence from biotechnology alliances. *Journal of Financial Economics, 67*, 411–446.

Liu, Q. (2000). How Good Is Good News? Technology Depth, Book-to-Market Ratios, and Innovative Events, UCLA Working Paper, Los Angeles, CA.

MacKinlay, A. C. (1997). Event studies in economics and finance. *Journal of Economic Literature, 35*, 13–39.

Madhavan, R., Koka, B. R., & Prescott, J. E. (1998). Networks in transition – how industry events (re)shape interfirm relationships. *Strategic Management Journal, 19*, 439–459.

Malerba, F., & Orsenigo, L. (2002). Innovation and market structure in the dynamics of the pharmaceutical industry and biotechnology – towards a history-friendly model. *Industrial and Corporate Change, 11*(4), 667–703.

Mallik, A., Zbar, B., & Zemmel, R. W. (2004). Making pharma alliances work. *McKinsey Quarterly, No. 1*, 16–19.

McWilliams, A., & Siegel, D. (1997). Event studies in management research – theoretical and empirical issues. *Academy of Management Journal, 40*(3), 626–657.

National Science Board. (2006). Science and Engineering Indicators 2006, National Science Foundation – Division of Science Resources Statistics, Arlington/VA, February 2006.

Neill, J. D., Pfeiffer, G. M., & Young-Ybarra, C. (2001). Technology R&D alliances and firm value. *Journal of High Technology Management Research, 12*, 227–237.

Ng, R. (2004). *Drugs – from discovery to approval*. Hoboken, NJ: John Wiley & Sons.

Nicholson, S., Danzon, P. M., & McCullough, J. (2002). Biotech-Pharmaceutical Alliances as a Signal of Asset and Firm Quality, NBER Working Paper No. 9007, June 2002.

Park, N., & Mezias, J. M. (2005). Before and after the technology sector crash – the effect of environmental munificence on stock market response to alliances of e-commerce firms. *Strategic Management Journal, 26*, 987–1007.

Park, N., Mezias, J. M., & Song, J. (2004). A resource-based view of strategic alliances and firm value in the electronic marketplace. *Journal of Management, 30*(1), 7–27.

Peterson, P. P. (1989). Event studies – a review of issues and methodology. *Quarterly Journal of Business and Economics, 28*, 36–66.

PhRMA. (2003). *Pharmaceutical industry profile, pharmaceutical research and manufacturers of America*. Washington, DC.

Pisano, G. P., & Mang, P. Y. (1993). Collaborative product development and the market for know-how – strategies and structures in the biotechnology industry, research on technological innovation. *Management and Policy, 5*, 109–136.

Prevezer, M., & Toker, S. (1996). The degree of integration in strategic alliances in biotechnology. *Technology Analysis & Strategic Management, 8*, 117–133.

Pyka, A., & Saviotti, P. (2001). Innovation Networks in the Biotechnology-Based Sector, University of Augsburg Discussion Paper No. 220.

Reuer, J. J. (2000). Parent firm performance across international joint venture life-cycle stages. *Journal of International Business Studies, 31*(1), 1–20.

Roberts, P. W. (1999). Product innovation, product-market competition and persistent profitability in the U.S. pharmaceutical industry. *Strategic Management Journal, 20*(7), 655–670.

Sharma, A. & Lacey, N. (2004). Linking product development outcomes to market valuation of the firm – the case of the U.S. pharmaceutical industry. *Journal of Product Innovation Management, 21*, 297–308.

Strong, N. (1992). Modelling abnormal returns – a review article. *Journal of Business Finance and Accounting, 19*(4), 533–553.

Tollman, P., Guy, P., Altshuler, J., Flanagan, A., & Steiner, M. (2001). *A revolution in R&D – how genomics and genetics are transforming the biopharmaceutical industry*. Boston: The Boston Consulting Group.

Whittaker, E., & Bower, J. D. (1994). A shift to external alliances for product development in the pharmaceutical industry. *R&D Management, 24*(3), 249–260.

Zucker, L. G., & Darby, M. R. (1995). Present at the Revolution – Transformation of Technical Identity for a Large Incumbent Pharmaceutical Firm after the Biotechnological Breakthrough, NBER Working Paper No. 5243.

Are Family-Owned Businesses Better Innovators?

Katinka Wölfer

> *The World Leaders In Innovation (..) Will Also Be World Leaders In Everything Else.*
> Harold R. McAlindon

Contents

1	Family Businesses in Germany	394
2	The Family Business Framework	394
	2.1 The Family Business: A Definition Challenge	394
	2.2 The Effect of Familyness on Performance	396
	2.3 Theoretical Explanations of the Distinctiveness of Family Businesses	397
3	Innovation Drivers in Family Businesses	399
	3.1 Goals	400
	3.2 Structure	402
	3.3 Strategy	405
	3.4 Business Environment	407
	3.5 Conclusion	408
4	The Management of Innovations in Family Businesses	409
	4.1 The Innovation Process	409
	4.2 Innovation Types and Degrees of Innovation	410
	4.3 Innovation Success	411
	4.4 Conclusion	412
5	Impacts of the Global Financial and Economic Crisis on Innovation Activities	412
	References	413

K. Wölfer (✉)
Strategic Finance Institute, European Business School (EBS), Wiesbaden, Germany
e-mail: katinka.woelfer@ebs.edu

1 Family Businesses in Germany[1]

Family influence in a business is most prevalent in the time following the start-up phase of the business, i.e., when the founder or the founding family is the main source of entrepreneurial drive and capital. As a result, family businesses are often associated with small or medium-sized companies. However, family ownership and management is also a widespread form of organization among well-established businesses at later stages of the corporate life cycle.

Accordingly, Hommel and Wölfer (2009) identified about 650 family businesses in Germany and Austria generating yearly sales of more than €100 million. At the head, this includes for example Metro AG, Sal. Oppenheim jr. & Cie. KGaA, and Schaeffler KG. All in all, approximately nine out of ten companies in Germany can be classified as family businesses (depending on the definition used). These family businesses provide 60% of all jobs, generate more than half of the gross domestic product, and develop almost three quarter of all patents and are therefore commonly considered to be the core of the German economy (Sueddeutsche Zeitung, 10/25/07). The large share in the overall number of patent developments suggests that family businesses exhibit a high degree of innovativeness compared to non-family businesses, i.e., publicly held businesses. Accordingly, many family businesses identify their innovation capabilities as one of the utmost important factors of corporate success (Hommel & Wölfer, 2009). The question then arises: What distinguishes family businesses from non-family businesses in order to allow them to achieve these results?

2 The Family Business Framework

Although family businesses are considered a significant and important part of the German economy, family business research has only existed since the 1970s, and is still in its infancy. Until now, there is no generally accepted and consistent theory to explain why and how family businesses differ from publicly held companies. The very basic problem even concerns defining the term "family business".

2.1 The Family Business: A Definition Challenge

While some researchers rely on the self-reported assessment of family businesses (e.g., Birley, 2001), or try to capture the family culture within an organization (e.g.,

---

[1] Family businesses are often placed on a par with so-called "Mittelstand" businesses. Although there is a great overlap between the two, they differ in a way that the term Mittelstand refers to businesses that according to the Institut für Mittelstandsforschung (IfM) generate a yearly turnover between €2 and €50m and provide jobs for 10–500 employees whereas family businesses are characterized by a substantial family influence (see Sect. 2.1). For a detailed discussion see Picot (2007).

Björnberg & Nicholson, 2007), most definition and classification approaches focus on the family's involvement in the business. It is generally accepted that a family's involvement in the business materially influences the firm's goals, strategies and structure, and should thus be used in order to distinguish family businesses from non-family ones.[2] Most researchers therefore concentrate on some combination of the components of a family's involvement in the business: ownership percentage (e.g., Lansberg, Perrow, & Rogolsky, 1988), presence of family members in the management and/or on the supervisory board (e.g., Handler, 1989), and transgenerational succession (e.g., Churchill & Hatten, 1987).[3] Lansberg et al. (1988), for instance, define a family business as an entity in which the members of a family have legal control over ownership. For Daily and Dollinger (1992), at least two relatives of the owner must serve as managers in order to classify the business as a family business. In contrast, Klein (2000) equally includes family ownership and the presence of family members in the management as well as on the board in her definition of family businesses. Practically all definitions focusing on the components of involvement highlight family ownership as the pivotal criterion. However, no consensus is reached on a reasonable threshold of equity ownership. Whereas Anderson and Reeb (2003b) use a threshold of 5% fractional equity ownership, Gallo and Sveen (1991) require the family to own the majority of the stock, and Donckels and Fröhlich (1991) determine a threshold of 60% equity ownership for businesses to be classified as family businesses.[4] The problem with fixed thresholds of equity ownership is the artificial dichotomization of family versus non-family businesses with no further distinction within the two groups.

In response to this definition dilemma, Astrachan, Klein, and Smyrnios (2002) propose to use a continuous rather than a dichotomous scale in order to differentiate various levels of family influence. According to the so-called Power Subscale (PSC), a family can influence a business in a substantial way through its equity ownership stake and the proportion of family representatives who are members of the supervisory board and the management team. These three channels of influence are viewed as additive and interchangeable. According to Klein (2000), family influence can then be measured along the PSC scale as follows:

$$PSC = \frac{Equity_{FamilyOwership}}{Equity_{Total}} + \frac{BoardSeats_{Family}}{BoardSeats_{Total}} + \frac{ManagementMembers_{Family}}{ManagementMembers_{Total}}$$

---

[2] Proponents of the components-of-involvement approach assume that family involvement is sufficient to make a business a family business. Critics argue that family involvement is only a necessary condition and that family involvement must be directed towards behaviors that produce a certain distinctiveness of a family business. For a critical discussion see Chrisman, Chua, and Sharma (2005).

[3] For a detailed overview of alternative definitions of family businesses in the literature see Chua, Chrisman, and Sharma (1999) and Handler (1989).

[4] Jaskiewicz (2006) attributes the differences in ownership thresholds to country-specific, institutional and cultural differences as well as to different types of family businesses in the research focus.

According to Jaskiewicz (2006), the value of the PSC needs to be at least 0.5 in order for a business to classify as a family business. This means an enterprise qualifies as a family business if a family controls at least 50% of the equity. The same is true if a family, for instance, only holds an equity stake of 20% but fills at least three out of ten seats on the board or management.[5] The level of family influence is the more distinctive the higher the PSC value is, and reaches its maximum at a value of 3.0. In this case, the owning family controls 100% of the equity and fills all seats on the board and management with members of the family.

Following Klein (2000) and Jaskiewicz (2006), a family business can then be defined as an entity that is influenced by a family in a substantial way. This means the family owns at least 50% of the equity. If the family owns less than half of the stock but still more than 20%, the lack of influence in ownership is then balanced through either influence on the board (percentage of seats on the supervisory board or external board occupied by family members), or influence in the management (percentage of family members in the top management team). The main advantage of measuring family influence on a continuous rather than dichotomous scale is the resulting continuum of family business characteristics. Aside from identifying the extent of family influence, the approach also allows the manner of family influence on the business to be determined. Depending on whether this influence is derived from ownership, control or management, it yields different types of family businesses: The family-owned business, the family-controlled business and the family-managed business.

2.2 The Effect of Familyness on Performance

Researchers believe that family influence makes a family business distinct from a non-family one in terms of strategy, structure and goals. However, to be cogent, theses differences must have an ultimate effect on the economic performance. Consequently, it is important to analyze differences in performance between family businesses and non-family businesses in order to determine whether family influence does affect performance.

The most widely cited article on the relationship between familyness and corporate performance was published by Anderson and Reeb (2003a). Based on a panel study of S&P 500 firms, their results indicate that family businesses perform significantly better than non-family ones in terms of ROA (6.65% higher in family businesses) and market-based performance measures such as Tobin's q (10% higher in family businesses). Additional analysis reveals that when family members serve as the CEO, performance is better than with outside CEOs. The results point in the same direction as findings by Villalonga and Amit (2004), who show that family businesses are about 25% more valuable than non-family businesses – provided that

---

[5] According to Jaskiewicz (2006), a family must hold at least 20% of the company's cash flow rights as a necessary condition to the PSC approach.

the founder serves as the CEO. Contrary to Anderson and Reeb, they find that corporate value even declines when descendants serve as the CEO. Barontini and Caprio (2005) provide similar evidence for European family businesses. Turning the focus to Germany, Andres (2007) investigates 275 exchange-listed companies and finds that family businesses are not only more profitable than widely held businesses but also outperform companies with other types of blockholders. However, the performance of family businesses is only better in firms where the founding family is still active in the management and/or the supervisory board. In line with prior research, the positive effect of familyness is found to be strongest when the founder serves as the CEO. A possible interpretation for these findings is that owner families might have a deeper relationship with their business or might even feel themselves responsible for other shareholders as long as they serve in the management or supervisory board. If the family is just a large shareholder without active involvement in the business, the performance of their business is statistically not distinguishable from non-family businesses.

In conclusion, empirical evidence suggests that familyness tends to have a positive effect on the value creation in family businesses.[6] The observed outperformance of family businesses then leads researchers to the conclusion that differences in strategy, structure and goals must exist between family businesses and non-family ones (e.g., Chrisman et al., 2005). Further research is needed to identify the nature of these distinctions (see Sect. 3) and to determine how these distinctions result from family involvement; the latter is the topic of the next section.

2.3 Theoretical Explanations of the Distinctiveness of Family Businesses

Researchers basically rely on two theoretical perspectives in explaining the differences between family and non-family businesses: The resource-based view (RBV) of the business and the principal agency theory. According to Jensen and Meckling (1976), agency costs arise because of conflicts of interest and asymmetric information between two (groups of) stakeholders. Owners have therefore an incentive to monitor and discipline the management. This is especially true for shareholders with a large percentage stake in the business – in this case: the family. Agency costs related to the conflict between owners and related managers should therefore be lower in family businesses than in non-family businesses which in turn should ultimately result in a higher corporate value.[7] However, large shareholders

---

[6] In contrast, Chrisman, Chua, and Litz (2004) find little difference between family businesses and non-family ones. Gallo, Tapies, and Cappuyns (2000) even show that family businesses perform less well than publicly held businesses. Reasons for this conflicting evidence might be differing underlying definitions of the term "family business", differing performance measures, varying sample sizes and sample periods as well as differing country and industry focuses.

[7] Agency costs do not exist when an individual owner manages the business by himself (founder-run businesses).

(the family) also have an incentive to use their controlling position to extract private benefits at the expense of minority owners.[8] Villalonga and Amit (2004) argue that control-enhancing mechanisms such as dual share classes with differential voting rights, pyramids, cross-holdings, and voting agreements entitle majority owners to a fraction of the total votes outstanding greater than their share ownership fraction, and find that families make more frequent use of these mechanisms than do other large shareholders in non-family businesses. Corporate wealth can thus be moved towards family control. Agency costs in regard to the relationship between owner families and minority owners are therefore higher in family businesses than in non-family ones. In countervailing reduced agency costs relating to the owner-manager conflict against increased agency costs arising from the majority-minority owner conflict, Villalonga and Amit (2004) show that the positive effects of family ownership prevail. Family ownership thus adds value (especially when the founder serves as CEO), and minority shareholders are likely to be better off or at least no worse off in a family business than they would have been in a non-family one.

Scholars of the behavioral economics school of thought criticize the traditional agency theory view of the family business for assuming pure self-interest, and failing to take into account behavioral aspects emanating from familial bonds: altruism (stewardship) and entrenchment. As altruism improves cooperation and fosters commitment to the family, owners are expected to have an advantage in monitoring and disciplining related managers, and thus benefit from lower agency costs compared to a situation in non-family businesses. Moreover, Carney (2005) suggests that altruism leads to family members' willingness to accept short-term deprivation for the long-term continuity and survival of the business. This should then constitute competitive advantages over non-family businesses in encouraging innovation-based business strategies. At the same time, management entrenchment may occur as managers have the incentive to extract private benefits from the business. Gomez-Mejia, Nuñez-Nickel, and Gutierrez (2001) show that agency problems caused by management entrenchment may even be worse in family businesses than in non-family businesses. Further, ownership entrenchment occurs when succeeding family generations may use their wealth and influence to obtain competitive advantages through political rent seeking rather than through entrepreneurship and innovation. Ownership entrenchment gets even worse in the case of a pyramidal corporate ownership structure. Because innovation can cannibalize existing businesses, pyramidal family ownership can then create disincentives to innovate, if innovation activities occur at levels in the corporate structure where a family's stake in the profit is lower and threatens businesses at levels where its stakes are higher (Chrisman et al., 2005). Altogether, altruism and entrenchment have positive and negative effects on agency costs. Villalonga and

---

[8] If the large shareholder is an institution such as an investment fund, the private benefits of control are diluted among several independent owners. As a result, incentives for expropriating minority shareholders are small, but so are incentives for monitoring the management. If in contrast, the large shareholder is an individual or a family, incentives are greater both for expropriation and monitoring (Villalonga & Amit, 2004).

Amit (2004) argue that contingencies such as the generation managing the business, the extent of ownership control, corporate strategy, and industry affiliation appear to have some impact on whether the influence is positive or negative.

While the principal agency theory explains the distinctiveness of family businesses based on altruism and entrenchment, scholars of the resource-based view in contrast suggest that valuable, imperfectly imitable and non-substitutable resources can lead to a sustainable competitive advantage and hence superior business performance. In order to develop such advantages, businesses have to invest in their core capabilities and resources such as knowledge capital, corporate culture, exceptional infrastructure and business models, and enduring relationships with employees, clients, suppliers or the community partners (Le Breton-Miller & Miller, 2006). Sirmon and Hitt (2003) argue that family businesses identify, acquire, accumulate and leverage their resources in ways that are different from those of non-family businesses, which in turn allows them to enhance a competitive advantage. Carney (2005) for instance shows that family businesses may enjoy long-term relationships with internal and external stakeholders, and through them develop and accumulate social capital which then facilitates relational contracting with partners in external networks. While the fixed costs of creating and maintaining social capital such as filling institutional voids and negotiating ambiguous regulatory environments are high, social capital can contribute to economies of scope because the different units of a large diversified family business can use it advantageously. This in turn might give the family business a competitive advantage in expanding its scope compared to a non-family business. Zahra, Hayton, and Salvato (2004) test whether organizational culture which has also been proposed as an inimitable resource, affects entrepreneurship in family businesses, and find a positive relationship between cultural dimensions such as familyness and the entrepreneurial attitude of the business which in turn encourages innovation activities.

In summary, applying mainstream theories such as the agency theory and the resource-based view to family businesses has shown that family businesses possess distinctive resources and capabilities and most likely tend to experience lower agency costs than non-family businesses. Le Breton-Miller and Miller (2006) further add to the discussion by combining insights from the resource-based view with those of the agency theory and argue that in the case of reduced agency costs relating to the owner-manager conflict potential, resources could be freed up and plowed back into the business, which in turn would generate a resource surplus – a circle of long-term competitive advantages for family businesses.

3 Innovation Drivers in Family Businesses

It appears reasonable that the particular agency problems as well as distinctive resources and capabilities in family businesses consequently lead to different behavior in terms of their goals, structures, and corporate and functional strategies. In addressing these differences, the following section will place a special focus on the driving forces for innovativeness in family businesses.

3.1 Goals

3.1.1 Business Goals

Common economic business goals include financial measures such as market value, profitability and operating efficiency. According to the shareholder value concept, the primary goal for a company is to maximize the wealth of its shareholders (owners) by paying dividends and/or causing the stock price to increase. The outside pressures and incentives for managers to outperform the market are so strong that even the business itself may be sold at an auction to the highest bidder. Eventually, when profits collapse, speculation even tempts managers to make such risky decisions that they may jeopardize the very survival of the business itself. Hence, in their drive to maximize shareholder value, publicly held businesses seek constant growth through short-term profit maximization. In response to the much criticized shareholder value view, most businesses nowadays commit themselves to follow a stakeholder approach by combining economic performance with non-economic success such as usefulness to society, and satisfaction of employees, customers and suppliers. However, this concept is hard to implement in practice because of the difficulty to balance and integrate multiple relationships and mostly conflicting objectives.[9]

In general, goals in family businesses do not differ much from those in publicly held businesses. However, family businesses do not necessarily prioritize economic goals. Instead, Hommel and Wölfer (2009) find the aim to keep the long-term ownership of the business within the family as the topmost goal among German and Austrian family businesses. As a consequence, business strategies in family firms should be geared towards the security of the long-term continuity and survival of the business. Hence, it comes as no surprise that family businesses name steady relationships with customers and suppliers as well as loyal employees as the most important factors for the success of their business (Hommel & Wölfer, 2009). It thus appears that family businesses favor a stable and steady business environment as a consequence of their inherent long-term orientation. Le Breton-Miller and Miller (2006) define long-term orientation as a set of goals – and most of all investments – that pay off over an extended time period of five years or more, and believe that this long-term orientation helps family businesses to develop valuable resources and capabilities that rival businesses cannot imitate or substitute (see Sect. 2.3). In order to create such long-term advantages, family businesses consequently must foster long-term investments such as expenditures in knowledge capital, corporate culture, exceptional infrastructure and business models, enduring relationships with employees, clients, suppliers or the community as well as expenditures in R&D projects (Le Breton-Miller & Miller, 2006). Compared with non-family businesses, this distinctive long-term orientation makes it easier for family businesses to accept short-term

---

[9]See Sundaram and Inkpen (2004) for a detailed discussion on corporate objectives.

losses or profit cuts due to investments that only pay off in the more distant future.[10] This is especially true of investments in R&D projects that, according to the German Commercial Code ("Handelsgesetzbuch"), have to be accounted for as immediate expenses in the financial statements, thereby reducing short-term profit.[11]

3.1.2 Family Goals

As a consequence of their long-term orientation, family businesses differ from non-family businesses in their inherent potential to engage in innovation activities. They further differ from non-family businesses in the sense that not only the business itself formulates goals but also the family – even though these goals might not always coincide. A major goal for the family is, among many others, the long-term wealth accumulation for family members which goes along with the level of dividend payments (Leenen, 2005). It appears reasonable that owner families, like shareholders of publicly held businesses, expect a dividend payment as reward for their capital investments. At the same time, excessive profit distribution may however limit a business' potential to meet its financial performance goals and secure an adequate availability of funds to invest in future growth. This also affects investments in R&D projects. A family-oriented goal setting may thus reduce the potential to foster innovation activities in family businesses. Hommel and Wölfer (2009) however refute the argument by providing evidence that less than 20% of German and Austrian family businesses in the sample gear their dividend payout decisions towards the interests of owner families. Instead, the majority of family businesses value corporate goals such as reinforcing the capital base and securing future growth opportunities clearly higher than meeting owners' expectations on dividend distributions. As a result, family businesses typically prefer the retention of earnings, and hence pay out dividends to a much lesser extent than what would conform to an adequate rate of return on the capital employed. This can be seen as a de facto subsidy of the business by its owners. Compared to publicly held businesses, the consequentially strong capital base then allows family businesses greater financial latitude in financing growth opportunities such as investments in R&D.

A further peculiarity of family businesses is the usually large overlap of business and family wealth. Anderson and Reeb (2003b) find, for example, that large shareholders in S&P 500 businesses have around 70% of their wealth invested in the business they control. Given such a poorly diversified portfolio, a family has then an incentive to reduce the business risk below the level that would be considered optimal for well-diversified outside shareholders in order to reduce the risk

---

[10] Hoskissen, Hitt, Johnson, and Grassman (2002) argue that rates of return based on long-term expectations go along with different innovation strategies compared to short-term expectations on rates of return, and find evidence that pension funds encourage intensive internal R&D activities whereas investment funds prefer acquisitions of externally developed innovations.

[11] Amendments to the accounting standards under the German Commercial Code newly allow a capitalization of costs of development under certain circumstances. Research costs must still be expensed ("Gesetz zur Modernisierung des Bilanzrechts, BilMoG" as of 3/26/09).

level of their personal portfolios. Excessive risk aversion or even risk avoidance can however impose huge costs on the business by, for instance, omitting costly R&D projects. Family ownership can therefore limit innovativeness. Donckels and Fröhlich (1991) add to this concern by finding a higher risk aversion in family businesses than in non-family businesses. Their survey among European family and non-family businesses reveals that family businesses are more inclined to state that innovation involves too much risk, and to disagree to a larger extent with the statement that a manager should encourage even risky innovations. There are however ways of reducing the business risk by means other than omitting R&D expenditures. Family businesses may, for instance, implement a corporate risk management, increase corporate diversification, or decrease leverage.

To conclude, possible interests of family members to extract profits from the business as well as the risk aversion of owner families due to the typically large overlap between business and family wealth may limit the potential for family businesses to extensively invest in R&D. However, the distinctive long-term orientation of owner families and their care for future generations limit the potential as well as the incentives for far excessive dividend payouts, and also demonstrate the need to overcome the risk aversion of the family by means other than an omission of R&D activities, which in turn set the stage for the long-term continuity and survival of the business.

3.2 Structure

3.2.1 Corporate Governance Structures

Owner families may exert their institutional influence through positions held within the management and/or within the supervisory board (see Sect. 2.1). While the management (especially in large companies) is typically dominated by non-family members, owner families exert their influence mainly through their presence on the supervisory board (Hommel & Wölfer, 2009). Nine out of ten family businesses in Germany and Austria have therefore installed one or more supervisory boards or comparable bodies. For businesses with the legal form of a stock corporation ("Aktiengesellschaft") or an association limited by shares ("Kommanditgesellschaft auf Aktien"), the installment of a supervisory board is required by the German stock companies law ("Aktiengesetz"). Above a threshold of 500 or 2,000 employees, family influence within these supervisory boards is however, limited due to the legal requirement to implement codetermination, i.e. one third or half of the seats on the supervisory board must have been filled with employees' representatives ("Drittelbeteiligungsgesetz" or "Mitbestimmungsgesetz").[12] This means for family businesses that employees who do not belong to the family have the legal basis to

---

[12]The installment of a supervisory board along with the implementation of codetermination is also required for limited liability companies ("Gesellschaft mit beschränkter Haftung") with more than 500 employees.

codetermine corporate strategies which also include the control of innovation activities. As a consequence of employees' primary goal to secure employment, and their probably less long-term time horizon (compared to the one of the owner family), employees may have an incentive to impede economically reasonable innovation activities.[13] This is especially true of process innovations. Due to codetermination on the level of the supervisory board, wage agreements ("Tarifvereinbarungen") or the hearing of works councils ("Betriebsräte"), employees in German businesses have many options to refuse their consent to (process) innovations. This innovation adverseness would then result in a lower future corporate performance compared to the performance of (family) businesses for which the codetermination law does not apply. The situation gets even more serious if prospective R&D budgets depend on the financial performance, which would then lead to a long-term downward spiral (Brockhoff, 2006). Nevertheless, the strong involvement of employees also gives the opportunity to articulate suggestions for improvement, and may furthermore improve the employees' acceptance of innovation activities. In support of this argument, Dilgert (2002) finds a positive relationship between the degree of works councils' active involvement and the number of product innovations. Brockhoff (2006) eventually concludes that codetermination encourages (product) innovations. Large family businesses should therefore not experience disadvantages in implementing their innovation strategy.

Beside their institutional presence on the supervisory boards, owner families may also exert influence through the management. Because owner families ultimately aim to foster the family business' long-term continuity, management is expected to implement a long-term orientated perspective which consequently includes an active innovation strategy. However, in analyzing the typical composition of management bodies in German and Austrian family businesses, Hommel and Wölfer (2009) find in about half of the family businesses surveyed that management positions are not entirely occupied by members of the owner family and instead are dominated by non-family members. This is especially true in comparatively large family businesses (see also Klein, 2000).[14] Assuming a more short-term oriented planning horizon for non-family members, it is questionable whether the predominance of external managers leads management to lose sight of the owner family's long-term perspective and thus to refrain from economically important R&D activities (see Sect. 3.1). However, external managers may also, to a certain degree, be incentivized to adopt a more far-sighted perspective in order to eventually realign

---

[13]This argument weakens in the case of family businesses, since the employees' goal to secure long-term employment does not conflict a bit with the goals of the business itself (see Sect. 3.1). In contrast, having bankers on the board could have worse impacts on innovation because external creditors tend to prefer a risk-averse corporate strategy. However, Hommel and Wölfer (2009) find that bankers are rarely represented on supervisory boards of family businesses.

[14]Likely reasons for the predominance of non-family members in the management are the lack of management talents within a family in the course of succession, or the intentional decision to retreat from the operational management.

management objectives with owner family's goals through for instance long-term oriented compensation plans or long-term employment relationships.

Comparing the length of CEO tenures in family businesses and publicly held businesses reveals that CEOs of German publicly traded businesses stay at the job on average 4.7 years whereas CEO tenures in German family businesses typically last 20 years, and in the histories of businesses like Haribo, Dräger, and Hitschler, tenures often have even exceeded 50 years (Frankfurter Allgemeine Zeitung, 5/23/07; Simon, 2007). Especially founder CEOs remain in office for such long periods because they have an incentive to hold office until the next generation is ready to take over. These lengthy tenures may drive CEOs to take a long-sighted, steward-like perspective of the business, and thus may encourage investments in long-term projects such as R&D (Le Breton-Miller & Miller, 2006). In contrast, Zahra (2005) finds that long CEO tenures may create a setting in which strategic simplicity and inertia take hold of the organization and therefore limit innovation activities.

The CEO may not only influence the attitude towards innovation through the length of her tenure but also by means of the management style. It appears plausible that a participative CEO or senior management team who further a two-way communication add to a much larger extent to an innovation-friendly environment compared with more individualistic and entrepreneurial management (Trott, 2005). This is especially true of the interaction with the R&D department. However, no predominant management style can be identified among German so-called hidden champions (Simon, 2007).[15] Nevertheless, the distinctive entrepreneurial orientation of family businesses (see Sect. 3.3) might be evidence of a more participative and hence innovation-supporting leadership style among family businesses.

3.2.2 Organizational Structures

A participative management style usually emerges with the increasing age and size of the company and hence a steadily growing organizational complexity. It is unclear how the age of a business adds to innovation capacity. On the one hand, emerging businesses such as high tech enterprises are per se very innovative. Mature companies, on the other hand, possess the organizational and structural as well as the financial resources and flexibilities required for innovation. However, they may possibly also suffer from strategic rigidity and innovation inertia which would then conflict with the overall goal of long-term continuity and survival of the business. In regard to the level of organizational formalization, Craig and Moores (2006) find that family businesses with a greater innovational posture have a less formal and more decentralized organizational structure.[16] In conclusion, it remains

---

[15] Simon (2007) defines hidden champions as companies which are international leaders in their industries but little known to the general public. For the vast majority, these businesses are family-owned.

[16] The choice of a decentralization strategy empirically depends on the size of the family businesses but also on the individual corporate culture and company history (AlphaZirkel, 2007).

unclear whether the structure of a family business alone can affect innovativeness either positively or negatively, and since structure should ideally follow goals and strategies, a closer investigation of typical strategies in family businesses is required.

3.3 Strategy

3.3.1 Corporate and Business Strategies

Given the typically large overlap of business and family wealth, and the consequent aversion of owner families to take risks in the business, family businesses have an incentive to engage in much greater levels of corporate diversification than non-family businesses (see Sect. 3.1). However, families may also have incentives to forego corporate diversification because of the negative effects of diversification on shareholder value (Amihud & Lev, 1999). Furthermore, family businesses traditionally develop out of a one-entrepreneur company with a one-market focus. Strong relations to customers and suppliers and the inherent long-term orientation then let many family businesses hold on to their established business strategy (Simon, 2007). In other cases, it is simply the lack of expertise or management capacities that prevent firms from diversifying their business portfolio. Indeed, Hommel and Wölfer (2009) find that approximately 75% of German and Austrian family businesses predominantly follow a focused business strategy. Compared with non-family businesses, Anderson and Reeb (2003b) confirm that family businesses exhibit on average about 15% less diversification. Within the business unit(s), family businesses predominantly differentiate from competitors through superior product characteristics (Hommel & Wölfer, 2009). This differentiation strategy requires a steady willingness to innovate in order to extend one's lead over competitors. Alternatively to differentiating from competitors through product characteristics, many family businesses concentrate their activities on servicing market niches. Since the market risk for businesses focusing on one or few business units is relatively high, innovativeness becomes more important in order to counter steadily shortening product life cycles and to stay ahead of competitors as well as potential market entrants. This is especially true of internationally operating businesses as they are forced to adapt and constantly adjust their product portfolio to changing customer needs in global markets.

3.3.2 Functional Strategies

Besides the decisions on corporate and business strategies, the capacity to innovate mainly depends on the financial resources available to a business.[17] A sufficiently

---

[17] For a comprehensive overview of financing particularities in family businesses see Sieger (2007).

large R&D budget allows an innovation-oriented business to develop new products, services and processes, or at least to keep fundamental research on an adequate level. Extensive innovation activities ideally spawn successful innovations which in turn generate cash flows which then can be used to increase the R&D budget in order to fuel further innovation activities. Thus, financial resources channeled into R&D ideally have the potential to start a self-accelerating growth process. Vahs and Burmester (1999) argue that R&D is much easier to realize assuming a business has a strong equity base, since debt financing would burden innovation projects through interest and principal repayments. This is especially true of family businesses where equity owners typically do not receive a dividend payout adequate to the return on their capital brought in, which makes equity even more attractive, and hence relatively cheap (Finance-Studien, 2004). Further, due to the strong preference for family businesses to remain independent of external creditors in order to secure existing ownership structures, family businesses typically finance their operations preferably through the retention of earnings and bank loans, and refrain from extensive external financing. Furthermore, decreasing leverage is also a way for owner families to reduce the likelihood of default, which thereby decreases the risk level of their personal wealth (Hagelin, Holmén, & Pramborg, 2006). Family businesses thus exhibit a relatively high average equity ratio of around 40% compared to an average ratio of 23% for the German "Mittelstand" (Finance-Studien, 2004; Hommel & Wölfer, 2009; KfW, 2008). To conclude, the usually solid equity base gives family businesses a distinctive advantage over non-family businesses in facilitating innovation activities. However, this holds only as long as the strong preference for self-financing does not limit the business' potentials for financing investments compared to non-family businesses, which make use of various internal as well as external financing options.

A further functional strategy to drive innovativeness focuses on the management of human resources. Family businesses emphasize an enduring relationship with their employees (see Sect. 3.1). The number of employees is therefore – to a certain degree – less dependent on decisions on products and services, and decreases less during economic downturns, which in turn ideally leads to more intense loyalty and a greater identification of employees with the family business compared to a situation in non-family businesses. Nevertheless, the comparably lower turnover of employees and hence the hiring of fewer new employees with fresh ideas might also create a situation of walling-off and hence innovation inertia. However, at the same time, family businesses may aim at creating a work environment which supports entrepreneurial orientation. Hommel and Wölfer (2009) find evidence that family businesses in Austria and Germany are confident that entrepreneurial behavior among their employees would be much more distinctive than in comparable non-family counterparts. Assuming that employees in family businesses are loyal and stay with the business for a prolonged period, and therefore possess a distinctive knowledge of the market, and are also incited to act entrepreneurially, family businesses appear to have a substantial advantage over non-family businesses in bearing innovation activities.

3.4 Business Environment

The ability to innovate successfully depends not only on goals, corporate structures and strategies but also on the environment the business operates in. In contrast to internal innovation drivers, external factors of influence can (hardly) be managed by the business.

3.4.1 Market Dynamics and Competition

Market dynamics are driven by ever increasing global competition with high rates of innovation and thus steadily shortening product life cycles. As a consequence, periods of amortization for product innovations are increasingly truncated, which consequently requires products to be produced in large quantities and/or to be sold at a premium price (Vahs & Burmester, 1999). As family businesses typically differentiate themselves from their competitors by means of unique product characteristics (see Sect. 3.3), which in turn might involve relatively higher selling prices, family businesses should have a competitive advantage in operating in dynamic markets or should at least not be worse off than publicly held businesses. Market dynamics therefore have only a limited effect on innovation activities in family businesses with distinguished product characteristics. Alternatively to a differentiation strategy, many family businesses focus their operations on niche markets with few market rivals (see Sect. 3.3). Due to the oligopolistic structures in niche markets, family businesses consequently exhibit relatively large market shares (Simon, 2007). As a result, market power should then reduce market dynamics, and thus increase periods of amortization. Operating in such an innovation-friendly environment allows innovative family businesses to optimally exploit their innovation capabilities. Nevertheless, a competitive advantage can only be realized if market shares are large enough not only in relative but also in absolute terms.

3.4.2 Market Size

A large market in terms of sales volume is important to the successful market launch of product innovations, especially with antecedent cost-intensive R&D efforts. Too narrow a market might consequently make it difficult for innovating businesses to reach break-even (Vahs & Burmester, 1999). Especially family businesses which predominately operate in small market niches might suffer a serious disadvantage with respect to their possibilities to innovate. This argument might lead family businesses to favor incremental and thus less costly product innovations over radical innovations.

3.4.3 R&D Networks

After all, innovation success also depends on the possibility of networking with external partners. Most businesses nowadays work closely together with suppliers and customers, or build strategic alliances and joint ventures with corporate partners. In addition, many businesses cooperate with universities or other research institutions. R&D partnerships usually aim at pooling expertise or resources. This is especially true of ambitious and challenging research intentions or projects that are very cost intensive and thereby require R&D partners to share the financial burden. R&D networks are therefore common practice, for instance, in the automobile industry. Aside from pooling resources, network partners also share the risks of failure for R&D projects with a low probability of success (Vahs & Burmester, 1999). Due to the potentially limited financial resources in family businesses (see Sect. 3.3) and their inherent risk aversion (see Sect. 3.1), family businesses should therefore benefit from the access to R&D networks.

3.5 Conclusion

Apart from potential financial constraints due to the strong preference for internal financing, family businesses may bank on a multitude of competitive advantages in their ability to innovate compared to non-family businesses: Their overall goal to secure the survival and transgenerational continuity of the business leads owner families to accept short-term losses or profit cuts in favor of investments such as in R&D projects which only pay off in the more distant future but establish a long-term competitive advantage over less innovating competitors. The usually high level of involvement of family members on the supervisory board and, even though to a lesser extent, in the management team therefore fuels an active innovation strategy. Corporate governance structures which establish codetermination of employees also encourage the acceptance of corporate innovation activities among employees and furthermore foster entrepreneurial behavior. Finally, the corporate strategy to focus on few business units and to differentiate through superior product characteristics further requires family businesses to steadily engage in innovation activities in order to stay ahead of competitors as well as potential market entrants. Low market dynamics in niche markets and the possibility of engaging in R&D networks typically also establish an innovation-friendly market environment. Overall, family businesses possess a multitude of competitive advantages in order to be better innovators compared to non-family businesses.

However, it appears intuitive that innovativeness alone does not necessarily lead to innovation success. Various studies show that family businesses face difficulties in converting their innovation capabilities into a structured innovation process and instead act on a gut level (e.g., Frankfurter Allgemeine Zeitung, 8/11/08). This is especially true of small and medium-sized family businesses where the founder CEO runs the business. Although the often cited quick decisionmaking and (seeming) flexibility in founder-managed businesses has enabled the success of many family businesses in the past, long-term innovation success inevitably requires a professionally structured management of innovations which is based on strategic foresight as well as a professional evaluation, realization and control of innovation activities. A systematic innovation management is thus a substantial prerequisite for successfully leveraging a family business' competitive advantages in being better innovators.

4 The Management of Innovations in Family Businesses

The successful management of innovations relies on a structured innovation process which translates innovation capabilities into concrete innovation outcomes which (in the case of product innovations) eventually are launched into the market.

4.1 The Innovation Process

Managing innovations first of all requires a clearly stated innovation strategy. Most family businesses rank innovativeness as one of the major aspects of their corporate strategy, and consider innovation capacity as an utmost important factor contributing to their corporate success (Hommel & Wölfer, 2009). Accordingly, one should assume that family businesses rely on clearly defined innovation objectives as well as on a professionally organized innovation process. Indeed, Muhr and Thum (2006) find that most German family businesses exhibit an institutionalized innovation process with regular cross-functional team meetings reporting to the management. Surprisingly, 38% of family businesses do not formally structure their innovation process. They refrain from using sophisticated analyses and strategic planning tools to generate new ideas, and instead mostly rely on their employee suggestion system. Generally, market opportunities and actions are not planned in advance. Innovating then appears to happen rather accidentally on a gut level, or at best, by being reactive to changing market conditions. On the one hand, acting on instinct might grant flexibility advantages. On the other hand, an insufficient innovation strategy and a barely structured process of managing innovations prevent these family businesses from optimally exploiting their resources and capabilities. In other words: Although family businesses usually possess distinctive resources and capabilities required to lay the proper groundwork for innovativeness, this can by no means be understood as a guarantor for actual innovation outcomes. A badly managed innovation process can easily undo the distinctive innovation potential, and thereby puts the continuity and long-term survival of the business clearly at risk.

However, Muhr and Thum (2006) also find that the remaining 62% of family businesses have indeed implemented a professionally structured as well as proactive process to manage their innovations. It appears that customers are the most important external influencers (far ahead of competitors or suppliers) whereas the CEO or the senior management team as well as the R&D department act as major internal influencers of innovations. In contrast to other businesses, hidden champions further prefer neither a pure market-driven nor a pure technology-driven innovation strategy but instead follow a combination thereof (Simon, 2007). This strategy view integrates internal resources and competencies with external market opportunities, and thereby brings together some of family businesses' core competencies: innovation capability, flexibility, good market knowledge as well as steady customer relationships. Researchers believe that such a balanced strategy should ultimately be more successful than a one-sided market or technology-driven strategy (e.g., Vahs & Burmester, 1999). Due to their distinctive resources and capabilities and at the same time their strong market focus, family businesses should experience a competitive advantage over non-family businesses in regard to their innovation strategy.

4.2 Innovation Types and Degrees of Innovation

Innovation basically relates to products and services as well as processes. Process innovations are achieved through a continued, progressive effort of process management techniques such as total quality management (TQM) or Six Sigma programs, and aim at raising productivity, reducing costs, and ultimately increasing profitability. They are hardly observable for outsiders, as process innovations are merely related to the delivery of outcomes, rather than being the outcomes themselves. Family businesses as well as non-family businesses cannot afford to neglect steady and continuous improvements of their operational (and administrative) processes in order to reduce the cost basis and to remain competitive. Concerning product innovations, businesses aim at increasing sales by introducing new products and services, or by providing new features to existing products and services. Compared to publicly held businesses, family businesses are expected to set a distinctively higher value on fostering product innovations than on generating process innovations because family businesses usually operate in market niches or differentiate themselves from competitors by means of unique product characteristics (see Sect. 3.3). With such a strategy, the regular market introduction of new or improved products and services is then a key precondition to long-term continuity and survival. A slowdown in the innovativeness of family businesses would otherwise weaken the competitive advantage developed through their differentiation strategy, or even open a niche market to potential new market entrants. In line with these theoretical arguments, Simon (2007) confirms that hidden champions mainly focus on the innovation of products and services, and less on the development and refinement of processes.

In regard to the innovation degree, product as well as process innovations occur either incrementally or radically. On the one hand, radical innovations have the potential for generating enormous returns that can exceed the returns of incremental innovations by far (Kock, 2007). Moreover, radical improvements can be a good means of differentiation from competitors as new or significantly improved products and services ideally are viewed as more attractive from a customer's perspective since they are connected with a higher relative advantage. In this sense, family businesses could be assumed to follow a strategy of radical innovations as this would be in an optimal way suited to achieving their overall differentiation strategy (see Sect.

3.3). On the other hand, radical innovations are accompanied by an increasing level of complexity and risk. Complexity in radical innovation projects emerges from high market, technology, resource and organizational uncertainties (Kock, 2007). Coping with complexity then demands the application of a lot of resources, thereby making the development of radical innovations costly. At the same time, the risk of technical, market-related or financial failure increases. The high risk of failure, especially of financial failure, contradicts the typically distinctive risk aversion in family businesses (see Sect. 3.1). This is then probably the main reason why most family firms do not innovate radically, and instead prefer the incremental development of new products and services (Muhr & Thum, 2006). Furthermore, family businesses often operate in oligopolistic markets, which impose less pressure for radical innovations compared to atomistic markets with high market dynamics (see Sect. 3.4).

4.3 Innovation Success

Innovation outcomes can, for instance, be determined through input measures such as R&D intensity (R&D expenditures in relation to yearly sales). Despite the usually good data availability, investments in R&D reflect only formal expenditures and are therefore less suitable for determining actual innovation outcomes. Alternative indicators for the innovation success are output measures such as the number of patent applications. In this regard, family businesses clearly outperform non-family businesses. 42 of the 50 most active corporate patent applicants in Germany are German businesses (DPMA, 2008). Families hold a substantial ownership stake in 17 of these businesses (e.g. Schaeffler KG) and a minor stake in 5 businesses (e.g. Robert Bosch GmbH). The remaining 20 businesses are non-family businesses. Although the number of family businesses among the most active patent applicants is comparable to the one for non-family businesses, the absolute number of patents is distinctively higher: Non-family businesses accounted for 5,507 applications in 2007 whereas family businesses filed almost twice as many patent applications. This indicates that family businesses are significantly more successful in regard to innovation outcomes. However, the number of patent filings does not appear to be a reliable proxy for innovation success as many innovative businesses flinch from filing patent applications due to high costs as well as time-consuming and complex administrative requirements. Moreover, many businesses prefer secrecy as an alternative way to protect their intellectual property. Businesses may also file patents not for their own commercial use but rather to legally block out competitors from imitating their innovations. As a consequence, the number of patent filings may be a proxy for the technical results of R&D efforts but is not a good measure of the actual innovation success.

Alternatively, the success of innovations can be measured by means of commercial success, i.e. profitability and market success. While profitability refers to the return on investment, net present value, or profit margins, the dimension market success captures the impact of a new product on revenue, sales volume, or market share (Kock, 2007). Since for most (family) businesses such facts are not available to outsiders and thus do not allow a deeper analysis, researchers instead focus on the composition of corporate product portfolios. A survey conducted on behalf of "Institut der deutschen Wirtschaft" (Koppel, 2006) finds that German businesses realize on average 23% of their annual sales by means of products which are less than five years in the market. For family businesses, Hommel and Wölfer (2009) calculate a slightly higher percentage of 27% of annual sales deriving from new products, whereas Simon (2007) reports new product ratios of 80% and more. The higher percentage of sales deriving from product innovations leads to the conclusion that family businesses are more innovative than their non-family counterparts – or at least as successful in launching innovations as non-family businesses. Regarding the timing of market launches, Hommel and Wölfer (2009) show that family businesses follow in 68% of their launches either a follower or an adopter strategy. This might relate to the preference in family businesses for incremental rather than radical innovations (see Sect. 4.2). Furthermore, a late entry also reduces the risk of premature market introduction, and thereby limits the risk of new product failure (Lilien & Yoon, 1990). This eventually conforms to the owner family's overall risk aversion in order to secure the long-term continuity and survival of the business – the topmost goal in family-owned businesses.

4.4 Conclusion

Although family businesses possess a multitude of competitive advantages in being better innovators than publicly held businesses, these innovation potentials can only be successfully exploited if professionally managed. Studies show that four out of ten family businesses have not implemented a professionally structured innovation process and instead generate innovations rather accidentally on a gut level, or at best, by being reactive to changing market conditions. However, long-term innovation success inevitably requires a professionally structured management of innovations which is based on a professional evaluation, realization and control of innovation activities. The fact that six out of ten family businesses have implemented such an innovation management is evidence that most family businesses have realized that they can only leverage their competitive advantages in being superior innovators by professionally managing them. Thus, the answer to the question whether family businesses are better innovators mainly depends on their attitude towards innovation management.

5 Impacts of the Global Financial and Economic Crisis on Innovation Activities

Through product and process innovations, family businesses aim at increasing sales, raising productivity, reducing costs, and thereby increasing profitability. The ultimate goal of innovation is thus to improve corporate performance and secure the long-term continuity and survival of the firm. Innovations can therefore create long-term competitive advantages. However, generating successful future innovation outcomes requires the initial allocation of sufficient financial resources. It will therefore be interesting to see what impact the current worldwide financial and economic crisis has on the level of corporate R&D expenditures (WirtschaftsWoche, 11/3/08).

The global recession affects family businesses and non-family businesses alike. However, since family businesses historically focus their business activities on trade and industry – which are considered particularly recession-sensitive sectors – family businesses are affected much more by the heavy decline in demand than (non-family) businesses operating in sectors which are less sensitive to economic downturns. Due to the relatively broad exposure to highly recession-stricken sectors, family businesses are expected to feel strong pressure to cut costs that are not economically essential and therewith reduce the level of investments in R&D.

While family businesses are greatly affected by the operational consequences of the worldwide economic downturn, they clearly benefit with regard to financial matters. Due to their strong preference for remaining independent of external creditors, family businesses primarily finance themselves through the retention of earnings and to a lesser extent through outside financing. As a result, family businesses exhibit an average equity ratio of about 40% which exceeds by far the typical equity ratio of other types of businesses. While non-family businesses increasingly complain about tight financial resources due to the often mentioned credit crunch, family businesses benefit from their solid equity base which allows family businesses relatively broad latitude in financing growth opportunities such as investments in R&D – even in times of crisis. The solid equity base then enables family businesses to absorb the operational effects of recessions much better than non-family businesses, and furthermore decreases the pressure to reduce R&D expenditures. Indeed, many family businesses currently possess sufficient financial resources in order to keep their R&D expenditures on a steady or even increasing level (Die Zeit, 2/29/09; Handelsblatt, 2/19/09; ManagerMagazin 12/19/08; WirtschaftsWoche, 3/2/09). These (counter-cyclical) innovation activities will then help to overcome the recession-induced decline in demand, and moreover aim at improving the business' position in the competitive arena as soon as economic activities start to recover. Innovation is thus a means to its end: Securing the long-term continuity and survival of the family business.

References

Amihud, Y., & Lev, B. (1999). Does corporate ownership structure affect its strategy towards diversification? *Strategic Management Journal, 20,* 1063–1069.

Anderson, R. C., & Reeb, D. M. (2003a). Founding-family ownership and firm performance: Evidence from the S&P500. *Journal of Finance, 58*(3), 1301–1328.

Anderson, R. C., & Reeb, D. M. (2003b). Founding-family ownership, corporate diversification, and firm leverage. *Journal of Law and Economics, 46*(2), 653–684.

Andres, C. (2007). Family Ownership as the optimal organizational structure? Working paper, University of Bonn. Retrieved from http://ssrn.com/abstract=903710 (downloaded 2/2/08).

Astrachan, J. H., Klein, S. B., & Smyrnios, K. X. (2002). The F-PEC scale of family influence: A proposal for solving the family business definition problem. *Family Business Review, 15*(1), 45–58.

Barontini, R, & Caprio, L. (2005). The Effect of family control on firm value and performance – Evidence from continental Europe. Working paper No. 88, European Corporate Governance Institute.

Birley, S. (2001). Owner-manager attitudes to family and business issues: A 16 country study. *Entrepreneur Theory and Practice, 26*(2), 63–76.

Björnberg, A., & Nicholson, N. (2007), The family climate scales – development of a new measure for use in family business research. *Family Business Review, 20*(3), 229–246.

Brockhoff, K. (2006). Technologischer Wandel und Corporate Governance. *ZFBF Sonderheft, 54*, 7–31.

Carney, M. (2005). Corporate governance and competitive advantage in family-controlled firms. *Entrepreneur Theory and Practice, 29*(3), 249–265.

Chrisman, J., Chua, J. H., & Litz, R. A. (2004). Comparing the agency cost of family and non-family firms. *Entrepreneur Theory and Practice, 28*(4), 335–354.

Chrisman, J., Chua, J. H., & Sharma, P. (2005). Trends and directions in the development of a strategic management theory of the family firm. *Entrepreneur Theory and Practice, 29*(5), 555–575.

Chua, J. H., Chrisman, J., & Sharma, P. (1999). Defining the family business by behavior. *Entrepreneur Theory and Practice, 23*(4), 19–39.

Churchill, N, C., & Hatten, K. J. (1987). Non-market based transfers of wealth and power: A research framework for family businesses. *American Journal of Small Business, 11*(3), 51–64.

Craig, J. B. L., & Moores, K. (2006). Research note – A 10-year longitudinal investigation of strategy, systems, and environment on innovation in family firms. *Family Business Review, 19*(1), 1–10.

Daily, C. M., & Dollinger, M. J. (1992). An empircial examination of ownership structure in family and professionally managed firms. *Family Business Review, 5*(2), 117–136.

Deutsches Patent- und Markenamt. (2008). Jahresbericht 2007. München.

Die Zeit. (2/26/09). Die stillen Stützen.

Dilgert, A. (2002). Betriebsräte und Innovationen. In: Kahle, E. (Ed.), *Organisatorische Veränderung und Corporate Governance – Aktuelle Themen der Organisationstheorie*, Wiesbaden: Deutscher Universitäts-Verlag.

Donckels, R., & Fröhlich, E. (1991). Are family businesses really different? – European experiences from STRATOS. *Family Business Review, 40*(2), 149–159.

Finance-Studien. (2004). Unabhängig erfolgreich – Wachstums- und Finanzierungsstrategien von großen Familienunternehmen. Frankfurt/Main: F.A.Z.-Institut für Management-, Markt- und Medieninformationen.

Frankfurter Allgemeine Zeitung. (8/11/08). Innovationen in Familienunternehmen.

Frankfurter Allgemeine Zeitung. (5/23/07). Heiße Vorstandsstühle.

Gallo, M. A., & Sveen, J. (1991). Internationalizing the family business – facilitating and restraining factors. *Family Business Review, 4*(2), 181–187.

Gallo, M. A., Tapies, J., & Cappuyns, K. (2000). Comparison of family and non-family business: Financial logic and personal preference. IESE Working Paper No. D/406. IESE Business School.

Gomez-Mejia, L. R., Nuñez-Nickel, M., & Gutierrez, I. (2001), The role of family ties in agency contracts. *Academy of Management of Journal, 44*(1), 81–95.

Hagelin, N., Holmén, M., & Pramborg, B. (2006). Family ownership, dual-class shares, and risk management. *Global Finance Journal, 16*, 283–301.

Handelsblatt. (2/19/09). Mit Innovation durch die Krise.

Handler, W. C. (1989). Methodological issues and considerations in studying family businesses. *Family Business Review, 2*(3), 257–276.

Hommel, U., & Wölfer, K. (2009).*Wertsteigerungsstrategien von Familienunternehmen – Eine empirische Analyse*. Work in Progress.

Hoskissen, R. E., Hitt, M. A., Johnson, R. A., & Grossman, W. (2002). Conflicting voices: The effects of institutional ownership heterogeneity and internal governance on corporate innovation strategies. *Academy of Management Journal, 45*(4), 697–716.

Jaskiewicz, P. (2006). Performance-Studie börsennotierter Familienunternehmen in Deutschland, Frankreich und Spanien. Dissertation, European Business School. Lohmar: Josef Eul Verlag.

Jensen, M., & Meckling, W. (1976). Theory of the firm: Managerial behavior, agency costs, and ownership structure. *Journal of Finance and Economics, 3*, 305–360.

KfW. (2008). Die Entwicklung der Eigenkapitalausstattung kleiner und mittlerer Unternehmen 2002 bis 2005. WirtschaftsObserver 36.

Klein, S. B. (2000). Family businesses in Germany: Significance and structure. *Family Business Review, 13*(3), 157–181.

Kock, A. (2007). Innovativeness and innovation success – a meta-analysis. *ZFBF Special Issue 2*, 1–21.

Koppel, O. (2006). Innovationsverhalten der tecknikaffinen Branchen. Köln: IdW-Verlag.

Lansberg, I. S., Perrow, E. L., & Rogolsky, S. (1988), Family business as an emerging field. *Family Business Review, 1*(1), 1–8.

Le Breton-Miller, I., & Miller, D. (2006). Why do some family businesses out-compete? – governance, long-term orientations, and sustainable capability. *Entrepreneur Theory and Practice, 30*(6), 731–746.

Leenen, S. (2005). Innovation in family businesses – a conceptual framework with case studies of industrial family firms in the German "Mittelstand". Dissertation no. 3096, University of St. Gallen, Difo-Druck, Bamberg.

Lilien, G. L., & Yoon, E. (1990). The timing of competitive market entry: An explanatory study of new industrial products. *Management Science, 36*(5), 568–585.

Mach, A. E. (2007), Alphazirkel Studie: Die Unternehmenskultur im Familienunternehmen. In: Mach, A. E. (Ed.), *Die Unternehmenskultur im Familienunternehmen – Was trägt sie, was prägt sie, was bedeutet sie für den Unternehmenserfolg*. München: Alphazirkel.

ManagerMagazin. (12/19/08). Forsche Geister.

Muhr, D. B., & Thum, G. F. (2006). Tradition verpflichtet? – Innovationsmanagement in Familienunternehmen. *OrganisationsEntwicklung, 4*, 70–77.

Picot, G. (2007). Familien- und Mittelstandsunternehmen im globalen Wandel von Wirtschaft und Gesellschaft. In: Picot, G. (Ed.), *Handbuch für Familien- und Mittelstandsunternehmen – Strategie, Gestaltung, Zukunftssicherung*. Stuttgart: Schäffer-Poeschel.

Sieger, G. (2007). Finanzierung. In: Picot, G. (Ed.), *Handbuch für Familien- und Mittelstandsunternehmen – Strategie, Gestaltung, Zukunftssicherung*. Stuttgart: Schäffer-Poeschel.

Simon, H. (2007). Hidden Champions des 21. Jahrhunderts – Die Erfolgsstrategien unbekannter Weltmarktführer. Frankfurt/Main: Campus Verlag.

Sirmon, D. G., & Hitt, M. H. (2003). Managing resources: Linking unique resources, management, and wealth creation in family firms. *Entrepreneur Theory and Practice, 27*(4), 339–358.

Sueddeutsche Zeitung. (10/25/07). Die Helden im Hintergrund.

Sundaram, A. K., & Inkpen, A. C. (2004). The corporate objective revisited. *Organic Science, 15*(3), 350–363.

Trott, P. (2005). Innovation management and new product development. Prentice Hall: Harlow et al.

Vahs, D., & Burmester, R. (1999). *Innovationsmanagement – Von der Produktidee zur erfolgreichen Vermarktung*. Stuttgart: Schäffer-Poeschel.

Villalonga, B., & Amit, R. (2004). How do family ownership, control, and management affect firm value? *Journal of Finance and Economics, 8*(2), 385–417.

WirtschaftsWoche. (3/2/09). Ich nenne das positives Denken – Ludwig Georg Braun im Interview.

WirtschaftsWoche. (11/3/08). Milberg: Finanzkrise bedroht auch die Forschungslandschaft.

Zahra, S. A. (2005). Entrepreneurial risk taking in family firms. *Family Business Review, 18*(1), 23–40.

Zahra, S. A., Hayton, J. C., & Salvato, C. (2004). Entrepreneurship in family vs. non-family firms: A resource-based analysis of the effect of organizational culture. *Entrepreneur Theory and Practice, 28*(4), 363–381.

Bibliography

Part I: Innovation and International Strategy

Abelshauser, W. (Ed.). (2003). *Die BASF. Eine Unternehmensgeschichte*. München: C.H. Beck.
ADL. (2005a). *Global innovation excellence study 2005. Innovation as strategic lever to drive profitability and growth*. Rotterdam: Arthur D. Little.
ADL. (2005b, March). *Global innovation excellence survey*, Arthur D. Little in Collaboration with VNONCW, ADL Rotterdam.
ADL. (2006, October). *Innovation Excellence. Erfahrungen in Innovation Management*. Wiesbaden.
Aharoni, Y. (1966). *The foreign investment decision process. Graduate school of business*. Boston, MA: Harvard University.
Ambos, B. (2005). Foreign direct investment in industrial research and development: A study of German MNCs. *Research Policy, 34*, 395–410.
Andrew, J., Haanæs, K., Michael, D. C., Sirkin, H. L., & Taylor, A. (2009, April). *Innovation 2009: Making Hard Decisions in the Downturn*.
Barba-Navaretti, G., & Venables, A. J. (2004). *Multinational firms in the world economy*. Princeton, NJ: Princeton University Press.
BASF. (2009). *BASF Presents Sustainable Construction Concepts at the World Future Energy Summit in Abu Dhabi*.
BCG. (2007). *Innovation 2007: A BCG senior management survey*. The Boston Consulting Group, Boston, MA, July 2007.
BCG. (2008, August). *Innovation 2008: Is the tide turning? A BCG senior management survey*. Boston, MA: The Boston Consulting Group.
BCG. (2009a, April). *Innovation 2009: Making hard decisions in the downtown*. Boston, MA: The Boston Consulting Group.
BCG. (2009b, April). *Measuring innovation: The need for action*. Boston, MA: The Boston Consulting Group.
Belitz, H., & Beise, M. (1999). Internationalization of R&D in multinational enterprises: The German perspective. In: Barrell, R. & Pain, N. (Eds.), *Innovation, investment and diffusion of technology in Europe. German direct investment and economic growth in postwar Europe* (pp. 89–119). Cambridge, UK: Cambridge University Press.
Belitz, H., Edler, J., & Grenzmann, C. (2006). Internationalization of industrial R&D. In: Schmoch, U., Rammer, C., & Legler, H. (Eds.), *National systems of innovation in comparison* (pp. 47–66). Dordrecht: Springer.
Belitz, H., Schmidt-Ehmcke, J., & Zloczysti, P. (2008). *Technological and regional patterns in R&D internationalization by German companies*, Weekly Report, Vol. 4, No. 15/2008, pp. 94–101.

Bernstein, M. A., & Bernstein, M. A. (Eds.). (1989). *The great depression: Delayed recovery and economic change in America, 1929–1939, Studies in economic history and policy*. Cambridge, UK: Cambridge University Press.

BERR. (2008). *The 2008 R&D Scoreboard, Department for Business, Enterprise and Regulatory Reform* (BERR), London.

Bloningen, B. A., Davies, R. B., & Head, K. (2002). *Estimating the knowledge-capital model of the multinational enterprise: Comment*, NBER Working Paper No. 8929, May 2002.

Brakman, S., Garretsen, H., & van Marrewijk, C. (2006). Comparative advantage, cross-border mergers and merger waves: International economics meets industrial organization. *CESifo Forum 1/2006*, 22–26.

Bruland, K., & Mowery, D. (2005). *Innovation through time*. The Oxford Handbook of Innovation. Oxford: Oxford University Press.

Chesbrough, H., Vanhaverbeke, W., & West, J. (Eds.). (2006). *Open innovation: Researching a new paradigm*. Oxford: Oxford University Press.

Cooper, R. G. (2009, March–April). How companies are reinventing their idea-to-launch methodologies. *Research Technology Management, 52*(2), 47–57.

Cotterman, R., Fusfeld, A., Henderson, P., Leder, J., Loweth, C., & Metoyer, A. (2009, September–October). Aligning marketing and technology to drive innovation. *Research Technology Management*, 14–20.

Davis, N. (2009). *INSIGHT: R&D will suffer despite stimulus packages*, ICIS Chemical Business 10.06.2009.

Dittrich, K., Duysters, G., & Deman, A. P. (2008, December). Strategic repositioning by means of alliance networks: The case of IBM. *Research Policy, 36*(10), 1496–1511.

Duerand, D. (2009). *Bilanz mit Makel, Wirtschaftswoche*, 34, 19.08.09.

Dunning, H., & Narula, R. (1995). The R&D activities of foreign firms in the United States. *International Studies of Management & Organization, 25*(1–2), 39–73.

ECB. (2005). *Competitiveness and the Export Performance of the Euro Area*. Taskforce of the Monetary Policy Committee of the European System of Central Banks, Occasional Paper Series of the ECB, No. 30, June 2005.

Eisenhardt, K. M., & Martin, J. A. (2000). Dynamic capabilities: What are they? *Strategic Management Journal, 21*, 1105–1121.

Eriksson, K., Johanson, J., Majkgard, A., & Sharma, D. D. (2000). Time and experience in the internationalization process. *Zeitschrift für Betriebswirtschaft, 71*, 21–43.

Field, A. (2003, November). The most technologically progressive decade of the century. *American Economic Review, 4*, 1399–1413.

Fortune Magazine. (2008). America's Most Admired Companies, http://money.cnn.com/magazines/fortune/mostadmired/2008/index.html

Gallecker, G., & Hesse, K. (2009). *Crisis management at BASF. Interview with Dr. Hans-Ulrich Engel*, BASF-Online Reporter 15, June 2009.

Gerybadze, A. (2004). *Technologie- und Innovationsmanagement. Strategie, Organisation und Implementierung*. München: Vahlen.

Gerybadze, A., & Reger, G. (1999). Globalization of R&D: Recent changes in the management of innovation in transnational corporations. *Research Policy, 28*(2–3), 251–274.

Ghoshal, S. (1987) Global strategy: An organizing framework. *Strategic Management Journal, 8*, 425–440.

Grandinetti, R., & Rullani, E. (1994). Sunk internationalisation: Small firms and global knowledge. *Revue DÉconomie Industrielle, 67*, 238–254.

Granstrand, O. (1999). Internationalization of corporate R&D: A study of Japanese and Swedish corporations. *Research Policy, 28*, 275–302.

Granstrand, O., Patel, P., & Pavitt, K. (1997). Multi-technology corporations: Why they have 'distributed' rather than 'distinctive' core competencies. *California Management Review, 39*(4): 8–25.

Hargadon, A. (2003). *How breakthroughs happen*. Boston, MA: Harvard Business School Press.

Hegde, D., & Hicks, D. (2008). The maturation of global corporate R&D: Evidence from the activity of U.S. foreign subsidiaries. *Research Policy, 37*, 390–406.

Hüther, M. (2009). *Gute Diagnose, schlechte Therapie. Steinmeiers "Deutschlandplan" bleibt trotz einzelner guter Ansätze insgesamt enttäuschend, Handelsblatt*, 14.08.2009.

ICIS. (2008). History recommends innovating toward recovery, *ICIS Chemical Business*, 29.12.2008.

Innovativeness: Percentage share of new and substantially improved products/services introduced over the last 5 years in total sales or total contribution over innovation expenditures as percentage of total sales or total contribution.

Jaruzelski, B., & Dehoff, K. (2007). The Customer Connection: The Global Innovation 1000. Strategy and Business, Winter 2008.

Jaruzelski, B., & Dehoff, K. (2008). *Beyond Borders: The Booz & Company Global Innovation 1000*. Strategy and Business, November 2008.

Jennewein, K. (2005). *Intellectual property management. The role of technology-brands in the appropriation of technological innovation*. Heidelberg, New York: Physica.

Jennewein, K., Durand, T., & Gerybadze, A. (2007). Marier Technologies et Marques pour un Cycle de Vie: Le Cas de Routeurs de Cisco. *Revue Française de Gestion, 177*, 57–82.

Johanson, J., & Vahlne, J.-E. (1977). The internationalization process of the firm: A model of knowledge development and increasing foreign market commitments. *Journal of International Business Studies, 8*(1): 23–32.

Johanson, J., & Vahlne, J.-E. (1990). The mechanism of internationalisation. *International Marketing Review, 7*(4): 11–24.

Johanson, J., & Wiedersheim-Paul, F. (1975). The internationalization of the firm: Four Swedish cases. *Journal of Management Studies, 12*(3): 305–322.

Jonash, R. S., & Sommerlatte, T. (1999). *The Innovation Premium*. Random House Business Books, London.

Knott. A. M. (2008). R&D/Returns causality: Absorptive capacity or organizational IQ. *Management Science, 54*, 2054–2067.

Knott, A. M. (2009, September–October). New hope for measuring R&D effectiveness. *Research Technology Management*, 9–13.

Kuemmerle, W. (1997). *Building Effective R&D Capabilities Abroad*, Harvard Business Review, March–April, pp. 61–70.

Lang, M. (2008). *Tradition der Ideen: Reppe-Chemie, BASF-Online Reporter*, 04.10.2008.

Larsen, G. Why megatrends matter, *Futureorientation Nr. 5*, 2006.

Les Bas, C., & Sierra, C. (2002). Location versus home country advantages in R&D activities: Some further results on multinationals locational strategies. *Research Policy, 31*, 589–609.

Malthus, T. (1798). *Essay on the Principle of Population*.

Markusen, J. R. (2001/2002). *Integrating multinational firms into international economics*, NBER Reporter, Winter 2001/2002, pp. 5–7.

Metzger, G., Heger, D., Höwer, D., Licht, G., & Sofka, W. (2009). *High-Tech-Gründungen in Deutschland – Optimismus trotz Krise*, Zentrum für Europäische Wirtschaftsforschung (ZEW).

Nonaka, I. (1991). The knowledge-creating company. *Harvard Business Review, 69*(6): 96–109.

Nonaka, I., & Teece, D. (Eds.). (2001). *Managing industrial knowledge. Creation, transfer and utilization*. London, Thousand Oaks, CA: Sage.

OECD. (2008). *The internationalization of business R&D: Evidence, impacts and implications*. Paris.

Osterhammel, J., & Petersson, N. P. (2003). *Geschichte der Globalisierung: Dimensionen, Prozesse, Epochen*. München: C.H. Beck.

Oviatt, B. M., & McDougall, P. P. (1994). Toward a theory of international new ventures. *Journal of International Business Studies, 25*(1), 45–64.

Oviatt, B. M., & McDougall, P. P. (1995). Global start-ups: Entrepreneurs on a worldwide stage. *Academy of Management Executive, 9*(2), 30–44.

Oviatt, B. M., & McDougall P. P. (1997). Challenges for internationalization process theory: The ease of international new ventures. *Management International Review, 37*(2), 85–99.

Patel, P., & Vega, M. (1999). Patterns of internationalization of corporate technology: Location versus home country advantages. *Research Policy, 28*, 145–155.

Rammer, C., Schmiele, A., & Sofka, W., Innovationsmotor Chemie (2007). Die deutsche Chemieindustrie im globalen Wettbewerb. Studie im Auftrag des Verbands der Chemischen Industrie e. V., Technical Report, Zentrum für Europäische Wirtschaftsforschung (ZEW), Mannheim.

Roberts, E. B. (2001, March–April). Benchmarking global strategic management of technology. *Research Technology Management*, 25–36.

Schwens, C. (2008). *Early Internationalizers: Specificity, Learning and Performance Implications*, Dissertation. Hampp.

Sommerlatte, T. (2004). Capital market orientation in innovation management. *International Journal of Product Development*, *1*(1), 1–11.

Sommerlatte, T., & Krautter, J. (2005). Innovationsportfolio-Management. In: Albers, S., & Gassmann, O. (Eds.), *Handbuch technologie- und innovationsmanagement*. Wiesbaden: Gabler.

Strube, J. (2004). Vom Status Quo zum Innovationsparadies Deutschland – eine entfesselnde Zeitreise, acatech.

Surowiecki, J. (2004). The wisdom of crowds: Why the many are smarter than the few and how collective wisdom shapes business, Economies, Societies and Nations. Little, Brown.

Taleb, N. N. (2007). *The black swan: The impact of the highly improbable*.

Teece, D. J. (1986). Profiting from technological innovation. *Research Policy, 15/6*, 285–305.

Teece, D. J. (2000). *Managing intellectual capital: Organizational, strategic and policy dimensions*. Oxford, UK: Oxford University Press.

Teece, D. J. (2007). Explicating dynamic capabilities: The nature and microfoundations of (sustainable) enterprise performance. *Strategic Management Journal, 28*, 1319–1350.

Total shareholder return, TSR: increase in share value plus dividends.

UNCTAD. (2005). UNCTAD Survey on the Internationalization of R&D, Current Patterns and Prospects on the Internationalization of R&D, Occasional Note, United Nations, New York and Geneva, December 2005.

UNCTAD. (2008). *World Investment Report 2008*, New York and Geneva.

Von Zedtwitz, M., & Gassmann, O. (2002). Market versus technology drive in R&D internationalization: four different patterns of managing research and development. *Research Policy, 31*, 569–588.

Wentz, R. C. (2008). *Die Innovationsmaschine. Wie die weltbesten Unternehmen Innovationen managen*. Berlin, Heidelberg: Springer.

Werner, K., Grabbe, H., & Oden, M. (2009). *Kleine Zukunft für kleine Teilchen, Financial Times Deutschland* (FTD), 26.08.09.

Zahra, S. A. (2005). A theory of international new ventures: A decade of research. *Journal of International Business Studies, 36*, 20–28.

Part II: Efficiency of Innovation Processes in International Enterprises

Abel, W. (1978). *Agrarkrisen und Agrarkonjunktur*. Hamburg.

Al-Ani, A., & Gattermeyer, W. (2001). Entwicklung und Umsetzung von Change Management-Programmen. In: Gattermeyer, W. & Al-Ani, A. (Eds.), *Change Management und Unternehmenserfolg – Grundlagen, Methoden* (pp. 13–40). Gabler Verlag: Praxisbeispiele.

Allen, J., Jimmieson, N. L., Bordia, P., & Irmer, B. E. (2007). Uncertainty during organizational change: Managing perceptions through communication. *Journal of Change Management, 7*(2), 187–210.

Anderson, N. R., De Dreu, C. K. W., & Nijstad, B. A. (2004). The routinization of innovation research: A constructively critical review of the state-of-the-science. *Journal of Organizational Behavior, 25*(2), 147–173.

Ansoff, H. I. (1965). *Corporate strategy*. New York.
Ansoff, H. I. (1977). Strategy formulation as a learning process: An applied managerial theory of strategic behavior. *International Studies of Management & Organization, 7*(2), 58–77.
Antoniou, P. H., & Ansoff, H. I. (2004). Strategic management of technology. *Technology Analysis & Strategic Management, 16*(2), 275–291.
Apenburg, E. (2006). Innovation – wo ist das problem? *Wissenschaftsmanagement, 6*, 14–20.
Argyris, C., & Schön, D. (1978). *Organizational learning: A theory of action perspective*. Reading, MA: Addison Wesley.
Ariely, D. (2008). *Predictably irrational: The hidden forces that shape our decisions*. Harper Collins.
Asenso-Okyere, K., & Davis, K. (2009). *Knowledge and innovation for agricultural development*. Washington: IFPRI.
Ashford, S. J., Lee, C., & Bobko, P. (1989). Content, causes, and consequences of job insecurity: A theory-based measure and substantive test. *Academy of Management Journal, 32*(4), 803–829.
Audi. (2009). Audi/Spore Challenge 2025, http://microsites.audi.com/ea_spore_onlinespecial/. Retrieved 2009-03-16.
Baer, M., & Frese, M. (2003., Innovation is not enough: Climates for initiative and psychological safety, process innovations and firm performance. *Journal of Organizational Behavior, 24*(1), 45–68.
Balmer, R., Invesini, S., Planta, A. & von, Semmer, N. (2000). *Innovation im Unternehmen. Leitfaden zur Selbstbewertung von KMU*. Zürich: vdf Verlag.
Barras, R. (1986). Towards a theory of innovation in services. *Research Policy, 15*, 161–173.
BASF. (1990). *Chemie der Zukunft*. Ludwigshafen.
Bass, B. M. (1999). Two decades of research, and development in transformational leadership. *European Journal of Work and Organizational Psychology, 8*(1), 9–32.
Bauer, H. Albrecht, C. -M., & Kühnl, Ch. (2006). Aspekte der Einführungsstrategien als Erfolgsfaktoren von Produktinnovationen. Reihe Wissenschaftl. Arbeitspapiere W109 des Institutes für Marktorientierte Unternehmensführung, Universität Mannheim.
Bayer. (2008a). *Policy: Future, goals, strategy, values* – The Mission Statement of the Bayer Group.
Bayer. (2008b). *Policy: values and leadership principles* – Living our Values.
Bayer. (2009a). *BSP today: Our behaviors*. Bayer Intranet.
Bayer. (2009b). *Making Acquisitions Work: The Bayer/Schering Case*. Presentation at University of Wuppertal, January 2009.
Bayer HealthCare. (2007). *Towards One BSP Corporate Culture: Creating the Driving Force*. Internal Presentation.
Behlendorf, B. (1999). Open source as a business strategy. In: *Open Sources: Voices from the Open Source Revolution*, O'Reilly.
Benkler, Y. (2002). Coase's Penguin, or, Linux and the nature of the firm. *The Yale Law Journal, 112*, 369–446.
Benner, M. J., & Tushman, M. L. (2003). Exploitation, exploration, and process management: The productivity dilemma revisited. *Academy of Management Review, 28*(2), 238–256.
Bessen, J. E., & Meurer, M. J. (2007). The private costs of patent litigation, law and economics. Working Paper No. 07-08, Boston University School of Law, http://bu.edu/law/faculty/scholarship/workingpapers/documents/BessenJMeurerM050107REV.pdf. Retrieved 2009-02-27.
Birkenmeier, B., & Brodbeck, H. (2005). Marktleistungsentwicklung. In: Hugentobler. W., Schaufelbühl, K., & Blattner, M. (Eds.), *Integrale Betriebswirtschaftslehre*. Zürich: Orell Füssli.
Bledow, R., Frese, M., Anderson, N. R., Erez, M., & Farr, J. L. (2009). A dialectic perspective on innovation: Conflicting demands, multiple pathways, and ambidexterity. *Industrial and Organizational Psychology: Perspectives on Science and Practice, 2*(3), 305–337.

Bohnemeyer, A. (1996) *Innovationsadaption und –diffusion in der Landwirtschaft*, Münster.
Bonner, J. M., Ruekert, R. W., & Walker, O. C. (2001). Upper management control of new product development projects and project performance. *The Journal of Product Innovation Management, 19*, 233–245.
Borsook, P. (1995). *How Anarchy works: On location with the masters of the metaverse.* The Internet Engineering Task Force, Wired 3.10.
Brooks, F. P. (1995). *The mythical man month: Essays on software engineering.* Anniversary Edition 1995/1975. Boston: Addison Wesley.
By, R. T. (2005). Organizational change management: A critical revue. *Journal of Change Management, 5*(4), 369-380.
Cacaci, A. (2006). *Change Management – Widerstände gegen Wandel, Plädoyer für ein System der Prävention.* Wiesbaden: Deutscher Universitäts-Verlag.
Cameron, K. S., & Freeman, S. J. (1991). Cultural congruence, strength and type: relationships to effectiveness. In: Woodman, R. W., & Passmore, W. A. (Eds.), *Research in Organizational Change and Development, 5*, 23–58.
Chesbrough, H. W. (2003). *Open innovation. The new imperative for creating and profiting from technology.* Boston: Harvard Business School Press.
Chesbrough, H. W. (2006). Open innovation: A new paradigm for understanding industrial innovation. In: Chesbrough, H., Vanhaverbeke, W., & West, J. (Eds.), *Open innovation: Researching a new paradigm.* Oxford University Press.
Chmielewski, D. C., & Guynn, J. (2009). Apple CEO Steve Jobs Takes Medical Leave, in: Los Angeles Times, 15 January 2009. http://www.latimes.com/business/la-fi-stevejobs15-2009jan15,0,7042254.story, Accessed 28 January 2009.
Christen, O. (2006). *Entwicklungen, Probleme und Konzepte für den Pflanzenbau von morgen.* ILU-Heft 12, Bonn.
Christensen, C. M. (2006). *The Innovator's Dilemma: The revolutionary book that will change the way you do business.* New York: Harper Collins.
Christensen, C., Stevensen, H., & Marx, M. (2006). Die richtigen Instrumente für den Wandel. In: *Harvard business manager, 12*, 26–38.
Coase, R. H. (1988). *The firm, the market and the law.* University of Chicago Press.
Coenenberg, A. G., & Salfeld, R. (2003). *Wertorientierte Unternehmensführung.* Stuttgart.
Columella, L. I. M. (1981). *Zwölf Bücher über Landwirtschaft (De Re Rustica Libri Duodecim).* Reprint, Vol. 1, München.
COMPO. (2005). *Wachstum braucht Sicherheit.* Münster.
Cooper, R. G. (1998). *Winning at new products.* Reading, MA: Perseus Books.
Cooper, R. G. (1999, April). From experience: The invisible success factors. *Journal of Product Innovation Management, 16*(2), 115–133.
Cooper, R. G. (2008). Perspective: The stage-gate idea-to-launch process – update, Whats' New, and NexGenSystems. *The Journal of Product Innovation Management, 25*, 213–232.
Cooper, R. G., & Kleinschmidt, E. J. (1993). Major new products: What distinguishes the winners in the chemical industry. *The Journal of Product Innovation Management, 10*, 90–111.
Cooper, R. G., & Kleinschmidt, E. J. (2007). Winning businesses in product development: The critical success factors. *Research Technology Management, 50*, 52–66.
Courth, L., Marschmann, B., Kämper, M., & Moscho, A. (2008). Spannungsfeld zwischen Geschwindigkeit und Best-in-Class-Ansätzen – PMI am Beispiel der Bayer-Schering-Übernahme. *M&A Mergers and Aquisitions Review, 1*, 8–14.
Czinkota, M. R., & Ronkainen, I. A. (2005). A forecast of globalization. International business and trade: Report from a Delphi study. *Journal of World Business, 40*(2), 111–123.
Daft, R. L. (2004). *Organization theory and design.* Mason, OH: Thomson South-Western.
DBV. (2009). *Situationsbericht 2009.* Berlin.
Demsetz, H. (1967). Towards a theory of property rights. *The American Economic Review, 57*(2), 347–359.

Desouza, K. C., Awazu, Y., Jha, S., Dombrowski, C., Papagari, S., Baloh, P., et al. (2008). Customer-driven innovation. *Research Technology Management, 51*, 35–44.

DiBella, A. J. (2007). Critical perceptions of organisational change. *Journal of Change Management, 7*(3/4), 231–242.

DiMasi, J. A., Hansen, R. W., & Grabowski, H. G. (2003). The price of innovation: New estimates of drug development costs. *Journal of Health Economics, 22*(2), 151–185.

DLG. (2009). *Landwirtschaft 2020*. Frankfurt.

Elliott, M. S., & Scacchi, W. (2002). Communicating and Mitigating Conflict in Open Source Software Development Projects, http://ics.uci.edu/~melliott/commossd.htm (retrieved 2009-03-16).

Encyclopædia Britannica. (2006). Fatally Flawed: Refuting the Recent Study on Encyclopedic Accuracy by the Journal Nature, http://corporate.britannica.com/britannica_nature_response.pdf. Retrieved 2009-02-28.

Encyclopædia Britannica. (2008). Britannica's New Site: More Participation, Collaboration from Experts and Readers. June 3rd, 2008, http://britannica.com/blogs/2008/06/britannicas-new-site-more-participation-collaboration-from-experts-and-readers/. Retrieved 2009-02-27.

Encyclopædia Britannica. (2009). History of Encyclopædia Britannica and Britannica. Taking a Great Legacy into the 21st Century, http://corporate.britannica.com/company_info.html. Retrieved 2009-02-16.

Erdmann, U. (2001). *Regional essen? Wert und Authentizität der Regionalität von Nahrungsmitteln*, AGEV Tagung.

Ernst, H. (2003) Unternehmenskultur und Innovationserfolg – Eine empirische Analyse, *Zfbf, 55*, 23–44.

Fenwick, S. (1999). Will the mirror crack? In: *Drug Discovery Today, 4*(1), 3.

Fleishman, E. A. (1953). The description of supervisory behavior. *Journal of Applied Psychology, 37*(1), 1–6.

Focus. (2008). Interview with Werner Wenning, Focus, 02/2008, pp. 5–15.

Fraunhofer Gesellschaft. (2005). *Food Chain Management*.

Frese, M., Teng, E., & Wijnen, C. J. (1999). Helping to improve suggestion systems: Predictors of making suggestions in companies. *Journal of Organizational Behavior, 20*(7), 1139–1155.

Fuchs, C., Blachfellner, S., & Bichler, R. (2007). The Urgent Need For Change: Rethinking Knowledge Management, in: Knowledge Management: Innovation, Technology and Cultures, http://icts.sbg.ac.at/media/pdf/pdf1487.pdf. Retrieved 2009-03-16.

Fugate, M., Kinicki, A. J., & Prussia, G. E. (2008). Employee coping with organizational change: An examination of alternative theoretical perspectives and models. *Personnel Psychology, 61*, 1–36.

Galpin, R. J., & Robinson, D. E. (1997). Merger integration: The ultimate change management challenge. *Mergers & Acquisitions, 32*(1/2), 24–28.

Gaubinger, K., Weran, T., & Rabl, M. (2009). *Praxisorientiertes Innovations- und Produktmanagement*. Wiesbaden.

Gibson, C. B., & Birkinshaw, J. (2004). The antecedents, consequences, and mediating role of organizational ambidexterity. *Academy of Management Journal, 47*(2), 209–226.

Giles, J. (2005). Internet encyclopaedias go head to head. *Nature, 438*(7070), 900–901.

Gloor, P., & Cooper, S. (2007). The New Principles of a Swarm Business, MIT Sloan Management Review, April 1, http://sloanreview.mit.edu/the-magazine/articles/2007/spring/48312/the-new-principles-of-a-swarm-business/. Retrieved 2009-03-16.

Goldbrunner, T. (2006). Mehr hilft nicht mehr! Hohe F&E Ausgaben sind kein Garant für Efolg. *Wissenschaftsmanagement, 1*, 30–33.

Graen, G. B., & Uhl-Bien, M. (1995). Relationship-based approach to leadership: Development of Leader-member Exchange (lmx) Theory of Leadership Over 25 Years: Applying a multi-level multi-domain perspective. *The Leadership Quarterly, 6*(2), 219–247.

Griffin, A., & Page, A. L. (1993). An interim report on measuring product development success and failure. *Journal of Product Innovation Management, 10*, 291–308.

Grudowski, S. (1998). *Informationsmanagement und Unternehmenskultur – Untersuchung der wechselseitigen Beziehung des betrieblichen Informationsmanagements und der Unternehmenskultur*, Stuttgart.

Grün, O. (2005). Risiken und Erfolgsfaktoren von Großprojekten: Lehren für das Innovationsmanagement. *zfo, 74*, 207–210.

Gupta, A. K., Smith, K. G., & Shalley, C. E. (2006). The interplay between exploration and exploitation. *Academy of Management Journal, 49*(4), 693–706.

Hamel, G. (2006). The why, what, and how of management innovation. *Harvard Business Review, 84*(6), 140–140.

Hart, S., Hultink, E. J., Tzokas, N., & Commandeur, H. G. (2003). Industrial companies' evaluation criteria in new product development gates. *Journal of Product Innovation Management, 20*, 22–36.

Hauschildt, J. (1997). *Innovationsmanagement*. München.

Hauschildt, J., & Salomo, S. (2007). *Innovationsmanagement*. München.

Hauser, J. R., & Zettelmeyer, F. (1997). Metrics to evaluate R,D&E. *Research and Technology Management, 40*, 32–38.

He, Z.-L., & Wong, P.-K. (2004). Exploration vs. exploitation: An empirical test of the ambidexterity hypothesis. *Organization Science, 15*(4), 481–494.

Heberle, K., & Stolzenberg, K. (2006). *Change Management. Veränderungsprozesse erfolgreich gestalten – Mitarbeiter mobilisieren*. Heidelberg: Springer.

Hecker, F. (2005). Asymmetric Competition, http://blog.hecker.org/2005/09/09/asymmetric-competition/. Retrieved 2009-03-07.

Herstatt, C., & von Hippel, E. (1992). From experience: Developing new product concepts via the lead user method: A case study in a "low-tech" field. *Journal of Product Innovation Management, 9*(3), 213–221.

Hofstede, G. (1981). Culture and organizations. *International Studies of Management and Organization, X*(4), 15–41.

Homburg, C., & Krohmer, H. (2003). *Marketingmanagement*. Wiesbaden: Gabler.

Hopkins, H. (2009). Britannica 2.0: Wikipedia Gets 97% of Encyclopedia Visits, http://weblogs.hitwise.com/us-heather-hopkins/2009/01/britannica_20_wikipedia_gets_9.html. Retrieved 2009-02-16.

Howells, J., & Tether, B. S. (2004). Innovation in Services: Issues at Stake and Trends, http://europe-innova.org/servlet/Doc?cid=6372&lg=EN. Retrieved 2009-02-27.

IVA. (2005). *Pflanzenschutz heute*, "Journalist", April 2005.

Janner, T., Schroth, C., & Schmid, B. (2008). Modelling service systems for collaborative innovation in the enterprise software industry. *IEEE International Conference on Service Computing, 2*, 145–152.

Jansen, J. J. P., Vera, D., & Crossan, M. (2009). Strategic leadership for exploration and exploitation: The moderating role of environmental dynamism. *Leadership Quarterly, 20*(1), 5–18.

Jaworski, J., & Zurlino, F. (2007). *Innovationskultur: Vom Leidensdruck zur Leidenschaft*. Frankfurt a. Main.

Jensen, M., & Meckling, W. (1992). Specific and general knowledge, and organizational structure. In: Werin, L. & Hijkander (Eds.), *Contrast economics*. Cambridge, MA: Basil Blackwell.

Johnson-Laird, P. N. (2004). The history of mental models. In: Manktelow, K. & Chung, M. C. (Eds.), *Psychology of reasoning: Theoretical and historical perspectives* (pp. 179–212). New York: Psychology Press,.

Jørgensen, H. H., Owen, L., & Neus, A. (2008). Making Change Work – Continuing the Enterprise of the Future Conversation, IBM Institute for Business Value.

Keller, R. T. (1992). Transformational leadership and the performance of research and development project groups. *Journal of Management, 18*(3), 489–501.

Keller, R. T. (2006). Transformational leadership, initiating structure, and substitutes for leadership: A longitudinal study of research and development project team performance. *Journal of Applied Psychology, 91*(1), 202–210.

Kelsey, J., & Schneier, B. (1999). The Street Performer Protocol and Digital Copyrights, First Monday, 4, 6–7. http://firstmonday.org/htbin/cgiwrap /bin/ojs/index.php/ fm/article/view/ 673/583. Retrieved 2009-02-27.

Kessler, E. H., & Chakrabarti, A. K. (1996). Innovation speed: A conceptual model of context, antecedents, and outcomes. *Academy of Management Review, 21*(4), 1143–1191.

Kipling, R. (1896). Captains Courageous: A Story of the Grand Banks. http://gutenberg.org/files/ 2186/2186-h/2186-h.htm. Retrieved: 2009-03-03.

Knittel, H., & Albert, E. (2003). *Dünger und Düngung*. Bergen.

Koch, C. (2008). *Die Kunst des Erfolgs: Wie Market-Based Management das weltweit größte Familienunternehmen aufgebaut hat*. Wiley-VCH.

König, M., & Völker, R. (2002). *Innovationsmanagement in der Industrie*. München/Wien.

Kotler, P., Keller, K. L., & Bliemel, F. (2007). *Marketing management*. München.

Krüger, W. (2000). Strategische Erneuerung: Probleme, programme und prozesse. In: Krüger, W. (Ed.), *Excellence in change*. Wiesbaden.

Krüger, W. (2006a). Topmanager als Promotoren und Enabler des Wandels. In: Krüger, W. (Ed.), *Excellence in change, Wege zur strategischen Erneuerung* (pp. 21–46). Wiesbaden: Gabler.

Krüger, W. (2006b). Das 3 W-Modell: Bezugsrahmen für das Wandlungsmanagement. In: Krüger, W. (Ed.), *Excellence in change, Wege zur strategischen Erneuerung* (pp. 125–170). Wiesbaden: Gabler.

K+S. (2006). *Wachstum Erleben*. Kassel.

K+S. (2008). *Unternehmens- und Nachhaltigkeitsbericht 2007*. Kassel.

Kuhn, T. S. (1962). *The structure of scientific revolutions*. University of Chicago Press.

Kummer, K.-F., & Zerulla, W. (2006). *Ablauf und Eigenarten der Wirkstoffforschung in der Pflanzenernährung*, Limburgerhof.

Küster, S., Schuhmacher, M., & Werner, B. (2008). *Open innovation in innovation networks*. Mannheim.

KWS. (2007). *Erfolg kann man säen*. Göttingen.

Langert, M. (2007). *Der Anbau Nachwachsender Rohstoffe in der Landwirtschaft Sachsen-Anhalts und Thüringens*. Halle-Wittenberg.

Langlois, R. N., & Garzarelli, G. (2006). Of Hackers and Hairdressers: Modularity and the Organizational Economics of Open-Source Collaboration, DRUID Summer Conference 2006 on Knowledge, Innovation and Competitiveness. http://2.druid.dk/conferences/viewpaper.php?id=101&cf=8. Retrieved: 2009-03-03.

Lewis, M. W. (2000). Exploring paradox: Toward a more comprehensive guide. *Academy of Management Review, 25*(4), 760–776.

Lühring, N. (2006). *Koordination von Innovationsprojekten*. Wiesbaden: Deutscher Universitäts-Verlag.

Macharzina, K., & Wolf, J. (2008). *Unternehmensführung*. Wiesbaden.

Machiavelli, N. (1532). The Prince, http://gutenberg.org/files/1232/1232.txt Retrieved 2009-03-16.

Manimala, M. J., Jose, P. D., & Thomas, K. R. (2006). Organizational constraints on innovation and intrapreneurship: Insights from public sector. *Vikalpa: Journal for Decision Makers, 31*(1), 49–60.

March, J. G. (1991). Exploration and exploitation in organizational learning. *Organization Science, 2*(1), 71–87.

Marcinowski, S. (2009). *Welternährung 2020*. Berlin: Agrarforum.

Martin, R. L. (2007). *The opposable mind: How successful leaders win through integrative thinking*. Boston, MA: Harvard Business School Press.

Mathieu, J. E., Goodwin, G. F., Heffner, T. S., Salas, E., & Cannon-Bowers, J. A. (2000). The influence of shared mental models on team process and performance. *Journal of Applied Psychology, 85*(2),. 273–283.

Maurer, R. (1996). *Beyond the wall of resistance*. Austin.

McTaggart, R. (1997). Guiding principles for participatory action research. In: McTaggart, R. (Ed.), *Participatory action research: International contexts and consequences*. SUNY Press.

Meffert, H., Burmann, C., & Kirchgeorg, M. (2008). *Marketing*. Wiesbaden.
Mengel, K., & Kirkby, E. A. (1982). *Principles of plant nutrition*. Bern.
Mensch, G. (1977). *Das technologische Patt*. Frankfurt.
Metelmann, K., & Neuwirth, S. (2002). Wachstum und Organisation im Bayer-Konzern. In: Glaum, M., Hommel, U., & Thomaschewski, D. (Eds.), *Wachstumsstrategien internationaler Unternehmungen. Internes vs. externes Wachstum* (pp. 123–157), Stuttgart.
Micic, P. (2006). *Das Zukunftsradar*. Offenbach.
Miles, R. E., & Snow, C. C. (1978). *Organizational strategy, structure, and process*. New York: McGraw-Hill.
Miller, V. D., Johnson, J. R., & Grau, J. (1994). Antecedents to willingness to participate in a planned organizational change. *Journal of Applied Communication Research, 22*, 59–80.
Mintzberg, H. (2007, July–August). Productivity is killing American enterprise. *Harvard Business Review, 85*, 2 25.
Moscho, A., Hodits, R., Friedemann, J., & Leiter J. (2000). Deals that make sense. *Nature Biotechnology, 18*, 719–722.
Mumford, M. D., Scott, G. M., Gaddis, B., & Strange, J. M. (2002). Leading creative people: orchestrating expertise and relationships. *Leadership Quarterly, 13*(6), 705–750.
Murphy, E. (2008). French Publishing Group Sets up Rival to Wikipedia. In: The Independent, 14 May 2008. http://independent.co.uk/news/world/europe/french- group- publishing-sets-up-rival-to-wikipedia-827705.html. Retrieved 2009-02-27.
Nestlé. (2008). *Bericht zur gemeinsamen Wertschöpfung*. Vevey.
Newburger, E. C. (1999). Computer Use in the United States: Population Characteristics, US Census Bureau, http://census.gov/prod/99pubs/p20-522.pdf. Retrieved 2009-03-07.
Newton, I. (1676). Letter to Robert Hooke, Feb. 5th, 1676.
Nielsen. (2008). Nielsen Online Reports Topline U.S. Data for March 2008, http://nielsen-online.com/pr/pr_080414.pdf. Retrieved 2009-03-07.
Noveck, B. S. (2006). "Peer to Patent": Collective intelligence, open review and patent reform. *Harvard Journal of Law & Technology, 20*(1), 123–162.
O'Reilly, T. (2006). Purpose-Driven Media, http://radar.oreilly.com/archives/2006/04/purposedriven-media.html. Retrieved 2009-03-07.
Oakland, J. S., & Tanner, S. (2007). Successful change management. *Total Quality Management, 18*(½), 1–19.
Open Architecture Network. (2009). About the Open Architecture Network, http://openarchitecturenetwork.org/about. Retrieved 2009-03-16.
Ozer, M. (1999). A survey of new product evaluation models. *Journal of Product Innovation Management, 16*, 77–94.
Pardo del Val, M. P., & Fuentes, C. M. (2003). Resistance to change: A literature review and empirical study. *Management Decision, 41*(2), 148–155.
Pendlebury, J., Grouard, B., & Meston, F. (1998). *The ten keys to successful change-management*. Chichester: Wiley.
Pleschak, F., & Sabisch, H. (1996). *Innovationsmanagement*. Stuttgart.
Porter, M. (1985). *Competetive advantage*, London.
PrecisionAg (2009). PrecisionAg in the UK 2009, http://www.precisionag.com.
Raisch, S., & Birkinshaw, J. (2008). Organizational ambidexterity: Antecedents, outcomes, and moderators. *Journal of Management, 34*(3), 375–409.
Raymond, E. (1999). *The Cathedral and the Bazaar*. O'Reilly.
Reasoning, L. L. C. (2004). How Open Source and Commercial Soft-ware Compare: A Quantitative Analysis of Database Implementations in Commercial and in MySQL 4.0.16. http://reasoning.com/pdf/MySQL_White_Paper.pdf. Retrieved 2009-03-16.
Reichwald, R., Meyer, A., Engelmann, M., & Welcher, D. (2007). *Der Kunde als Innovationspartne*. Wiesbaden: Gabler.
Reinwald, R., & Piller, F. (2006). *Interaktive Wertschöpfung*. Wiesbaden.

Rentenbank. (2006). *Landwirtschaftliche Rentenbank, Organisatorische und technische Innovationen in der Landwirtschaft*, vol. 21, Frankfurt.
Rogers, E. (1995). *Diffusion of innovations*. New York.
Rohe, C. (1999). Erfolgreiches Management von Innovation und Wachstum. In: Rohe, C. (Ed.), *Werkzeuge für das Innovations management*, Frankfurt a. Main.
Rösener, W. (1993). *Die Bauern in der Europ*. München: Geschichte.
Rosing, K., & Frese, M. (2009). *Leadership in the Innovation Process: The Importance of Ambidexterity*, Manuscript submitted for publication.
Sackmann, S. (1999). Cultural Change – eigentlich wäre es ja ganz einfach…wenn da nicht die Menschen wären. In: Götz, K. (Eds.), *Cultural change* (vol. 4, pp. 15–37). München/Mering: Managementkonzepte.
Sakkab, N., & Huston, L. (2006). Wie Procter & Gamble zu neuer Kreativität fand. *Harvard Business Manager, 28*, 21–31.
Salter, A., & Tether, B. S. (2006). *Innovation in services: Through the looking glass of innovation studies*.
Satzger, G., & Neus, A. (2009). *Principles of Collaborative Innovation*.
Schein, E .H. (1995). *Unternehmenskultur: Ein Handbuch für Führungskräfte*. New York: Frankfurt am Main.
Schmelzer, H. J., & Sesselmann, W. (2008). Geschäftsprozessmanagement in der Praxis, Kunden zufrieden stellen – Produktivität steigern – Wert erhöhen, Hanser, München.
Schmitz, P. M. (2008). *Bedeutung des AgriFoodBusiness für den Standort Deutschland*. Gießen.
Schoemaker, J. H., & Gunther, R. E. (2006). Machen Sie mehr Fehler: Es lohnt sich! *Harvard Business Manager, 28*, 72–81.
Schreyögg, G. (2000). Neuere Entwicklungen im Bereich des Organisatorischen Wandels. In: Busch, R. (Ed.), *Change Management und Unternehmenskultur: Konzepte in der Praxis, Vol. 20, Forschung und Weiterbildung für die betriebliche Praxis*, München, Mering, pp. 26–44.
Schreyögg, G., & Koch, J. (2007). *Grundlagen des Managements*, Wiesbaden.
Seidenschwarz, W. (2003). *Steuerung unternehmerischen Wandels*. Vahlen, München.
Siemes, J. (2002). Outlook for Potash, "Fertiliser Round Table" Conference, Charleston, SC.
Siguaw, J. A., Simpson, P. M., & Enz, C. A. (2006). Conceptualizing innovation orientation: A framework for study and integration of innovation research. *Journal of Product Innovation Management, 23*, 556–574.
Spath, D., Hirsch-Kreinsen, H., & Kinkel, S. (2008). Organisatorische Wandlungsfähigkeit produzierender Unternehmen, Unternehmenserfahrungen, Forschungs- und Transferbedarfe, Fraunhofer IAO, Stuttgart 2008.
Stern, T., & Jaberg, H. (2007). *Erfolgreiches Innovationsmanagement, Erfolgsfaktoren – Grundmuster – Fallbeispiele*. Wiesbaden: Gabler.
Stewart, T. A., & Raman, A. P. (2007). Lessons from Toyota's long drive. *Harvard Business Review, 85*(7/8), 74–83.
Takeuchi, H., Osono, E., & Shimizu, N. (2008). The contradictions that drive Toyota's success. *Harvard Business Review, 86*(6), 96–104.
Theuvsen, L., Spiller, A., Peupert, M., & Jahn, G. (2007). *Quality management in food chains*. Wageningen.
Thom, N., & Müller, R. (2006). Innovationsmanagement in KMU. In: Bruch, H., Vogel, B., & Krummaker, S. (Ed.), Leadership: Best Practices und Trends. Wiesbaden: Gabler.
Thomaschewski, D. (2002). Strategische Erfolgsfaktoren des internen Wachstums. In: Glaum, M., Hommel, U., & Thomaschewski, D. (Eds.), *Wachstumsstrategien internationaler Unternehmungen*. Stuttgart.
Tidd, J., Bessant, J., & Pavitt, K. (2005). *Managing innovation: Integrating technological, market and organizational change*. West Sussex: John Wiley & Sons.
Time Warner. (2008). Time Warner Annual Report 2007, http://files.shareholder.com/downloads/TWX/574216884x0x172551/CA55EB21-7F45-47BA-BE51-BFF899548A24/2007AR.pdf. Retrieved 2009-03-07.

Top500.org. (2008). Top 500 Supercomputers – Operating System Share, http://top500.org/charts/list/32/os. Retrieved 2009-02-16.
Tversky, A., & Kahnemann, D. (1974). Judgement under un-certainty: Heuristics and biases. *Science, 185*(4157), 1124–1131.
Vahs, D. (2007). *Organisation. Einführung in die Organisationstheorie und –praxis.* Stuttgart: Schäffer-Poeschel.
Vahs, D., & Burmester, R. (2005). *Innovationsmanagement. Von der Produktidee zur erfolgreichen Vermarktung.* Stuttgart: Schäffer-Poeschel.
Van de Ven, A. H., Polley, D. E., Garud, R., & Venkataraman, S. (1999). *The innovation journey.* New York: Oxford University Press.
Van der Penne, G., van Beers, C., & Kleinknecht, A. (2007). Success and failure of innovation: A literature review. *International Journal of Innovation Management, 7*(3),1–30.
Van Dyck, C., Frese, M., Baer, M., & Sonnentag, S. (2005). Organizational error management culture and its impact on performance: A two-study replication. *Journal of Applied Psychology, 90*(6), 1228–1240.
Viitamäki, S. (2007). Crowdsourcing Innovation Principles, http://samiviitamaki.com/2007/03/15/crowdsourcing-innovation-principles/. Retrieved 2009-03-08.
Von Hippel, E. (2005). *Democratizing innovation.* Cambridge, MA: The MIT Press.
Von Hippel, E. (2007). Horizontal innovation networks – by and for users. *Industrial and Corporate Change, 16*(2).
Wagner, J. (2005). *Vertrauen in Netzwerkbeziehungen.* München.
Wahren, K-H. (2003). *Erfolgsfaktor innovation.* Berlin.
Wahren, H. K. (2004). *Erfolgsfaktor Innovation.* Berlin, Heidelberg, New York.
Watson, R. (2008). Celebrate Failure, Fast Company Magazine, July 2008, http://fastcompany.com/resources/innovation/watson/112105.html. Retrieved 2009-03-15.
Welthungerhilfe. (2000). *Jahrbuch Welternährung.* Frankfurt.
Wenger, E., McDermott, R., & Snyder, W. (2002). *Cultivating communities of practice.* Harvard Business School Press.
Wooster, P. D., Simmons, W. L., & Hofstetter, W. K. (2007). Opening Space for Humanity: Applying Open Source Concepts to Human Space Activities, AIAA SPACE 2007 Conference & Exposition, http://wiki.developspace.net/w/images/6/68/AIAA2007-OpeningSpaceforHumanity.pdf. Retrieved 2009-02-27.
YARA. (2009). N-Sensor for site-specific variable application of nitrogen, http://www.yara.com
Zhou, J., & George, J. M. (2003). Awakening employee creativity: The role of leader emotional intelligence. *Leadership Quarterly, 14*(4–5), 545–568.
Zimmer, Y., Nehring, K., Moellmann, T., & Witte, T. (2007). *Agri Benchmark.* Cash Crop Report 2007, FAL, Braunschweig.

Part III: Capital Markets, Finance and Innovation Performance

Adams, R., Bessant, J., & Phelps, R. (2006). Innovation management measurement: A review. *International Journal of Management Reviews, 8,* 21–47.
Albala, A. (1975). Stage approach for the evaluation and selection of R&D projects. *IEEE Transactions on Engineering Management, 23,* 153–164.
Alvarez, L. H., & Stenbacka, R. (2006). Takeover timing, implementation uncertainty and embedded divestment options. *Review of Finance, 10,* 417–441.
Amabile, T. M., Conti, R., Coon, H., Lazenby, J., & Herron, M. (1996). Assessing the work environment for creativity. *Academy of Management Journal, 39,* 1154–1184.
Amihud, Y., & Lev, B. (1999). Does corporate ownership structure affect its strategy towards diversification? *Strategic Management Journal, 20,* 1063–1069.
Anand, B. N., & Khanna, T. (2000a). Do firms learn to create value? The case of alliances. *Strategic Management Journal, 21,* 295–315.

Anderson, R. C., & Reeb, D. M. (2003a). Founding-family ownership and firm performance: Evidence from the S&P500. *Journal of Finance, 58*(3), 1301–1328.

Anderson, R. C., & Reeb, D. M. (2003b). Founding-family ownership, corporate diversification, and firm leverage. *Journal of Law Economy, 46*(2), 653–684.

Andres, C. (2007). Family Ownership as the Optimal Organizational Structure? Working Paper, University of Bonn, Retrieved from http://ssrn.com/abstract=903710. Downloaded 2/2/08.

Andrew, J. P., Haanaes, K., Michael, D. C., Sirkin, H. L., & Taylor, A. (2008, August). Innovation 2008: Is the tide turning? *BCG Senior Management Survey*, 1–31.

Anthony, R. N., & Govindarajan, V. (1998). *Management control systems*, Boston: McGraw-Hill.

Anonymous. (2003). Climbing the Helix Staircase – a Survey of Biotechnology, *The Economist* (March 29th, 2003), pp. 3–18.

Archer, N. P., & Ghasemzadeh, F. (1999). An integrated framework for project portfolio selection. *International Journal of Project Management, 17*(4), 207–216.

Armitage, S. (1995). Event study methods and evidence on their performance. *Journal of Economic Surveys, 8*(4), 25–52.

Arnold, G. (2002). *Corporate financial management*. Pearson Education, Harlow, pp. 385–509.

Arora, A., & Gambarella, A. (1990). Complementarity and external linkages – the strategies of the large firm in biotechnology. *Journal of Industrial Economics, 38*, 361–379.

Astrachan, J. H., Klein, S. B., & Smyrnios, K. X. (2002). The F-PEC scale of family influence: A proposal for solving the family business definition problem. *Family Business Review, 15*(1), 45–58.

Austin, D. H. (1993). An event-study approach to measuring innovative output – the case of biotechnology. *American Economic Review, 83*, 253–258.

Bae, K.-H., Kang, J.-K., & Kim, J.-M. (2002). Tunneling or value added? Evidence from Mergers by Korean Business Groups. *Journal of Finance, 62*(2), 2695–2740.

Baecker, P., Hommel, U., & Lehmann H. (2003). Marktorientierte Investitionsrechnung bei Unsicherheit, Flexibilität und Irreversibilität – Eine Systematik der Bewertungsverfahren. In: Hommel, U., Scholich, M., & Baecker, P. (Eds.), *Reale Optionen – Konzepte, Praxis und Perspektiven strategischer Unternehmens* (pp. 15–35), Berlin, Heidelberg, New York: Springer.

Baghai, M., Coley, S., & White, D. (2000). *The alchemy of growth: Practical insights for building the enduring enterprise*. New York: Perseus.

Balachandra, R., & Raelin, J. A. (1980). How to decide when to abandon a project, *The International Journal of Research Management, 23*, 24–29.

Balagopal, B., & Gilliland, G. (2005). Integrating value and risk in portfolio strategy, *BCG OfA, July 2005*, pp. 1–10.

Barontini, R, & Caprio, L. (2005). The Effect of Family Control on Firm Value and Performance – Evidence from Continental Europe. Working Paper No. 88, European Corporate Governance Institute.

Baum, J. A. C., & Silverman, B. S. (2004). Picking winners or building them? Alliance, intellectual, and human capital as selection criteria in venture financing and performance of biotechnology startups. *Journal of Business Venturing, 19*, 411–436.

Berk, J., & DeMarzo, P. (2007). *Corporate finance, pearson education* (pp. 757–770). Harlow.

Bhagat, S., & Romano, R. (2001). Event Studies and the Law – Part 1 – Technique and Corporate Litigation, Yale International Center for Finance Working Paper No. 00-31/Yale Law School Program for Studies in Law, Economics, and Public Policy Research Paper No. 259.

BIO BIO. (2005). *2005–2006 guide to biotechnology*, Washington, DC.

Birley, S. (2001). Owner-manager attitudes to family and business issues: A 16 country study. *Entrepreneurship Theory and Practice, 26*(2), 63–76.

Bitman, W. R., & Sharif, N. (2008). A conceptual framework for ranking R&D projects. *IEEE Transactions of Engineering Management, 55*(2), 267–278.

Björnberg, Å., & Nicholson, N. (2007). The family climate scales – development of a new measure for use in family business research. *Family Business Review, 20*(3), 229–246.

Black, F., & Scholes, M. (1973). The pricing of options and corporate liabilities. *Journal of Political Economy, 81*(3), 637–654.

Bonduelle, Y., Schmoldt, I., & Scholich, M. (2003). Anwendungsmöglichkeiten der Realoptionsbewertung. In: Hommel, U., Scholich, M., & Baecker, P. (Eds.), *Reale Optionen – Konzepte, Praxis und Perspektiven strategischer Unternehmens* 3–13, Berlin, Heidelberg, New York: Springer.

Boutellier, R., Völker, R., & Voit, E. (1999). *Innovationscontrolling – Forschungs- und Entwicklungsprojekte gezielt planen und steuern*. München: Hanser.

Bowman, R. G. (1983). Understanding and conducting event studies. *Journal of Business Finance and Accounting, 10*(4), 561–584.

Brettel, M., Rudolf, M., & Witt, P. (2005). *Finanzierung von Wachstumsunternehmen*. Wiesbaden.

Brito, N. O. (1977). Marketability restrictions and the valuation of capital assets under uncertainty. *The Journal of Finance, 32*(4), 1109–1123.

Brockhoff, K. (1999). *Forschung und Entwicklung: Planung und Kontrolle*. Munich: Oldenbourg.

Brockhoff, K. K. (1999). Technological progress and the market value of firms. *International Journal of Management Reviews, 1*(4), 485–501.

Brockhoff, K. (2006). Technologischer Wandel und Corporate Governance. *ZFBF Sonderheft, 54*, 7–31.

Brown, D. M., & Laverick, S. (1994, October/November). Is your measurement system well balanced? *Journal for Quality and Participation*, 6–11.

Brown, S. J., & Warner, J. B. (1980). Measuring security price performance. *Journal of Financial Economics, 8*, 205–258.

Brown, S. J., & Warner, J. B. (1985). Using daily stock returns – the case of event studies. *Journal of Financial Economics, 14*, 3–31.

Brunner, M. F., Fahlbusch, N., & Hundertmark, N. (2000). *Corporate Venture Capital-Programme: Die Beziehung zwischen Corporate und Venture*.

Bucher, M., Mondello, E., & Marbacher, S. (2002). Unternehmensbewertung mit Realoptionen. *Der Schweizer Treuhänder, 9*, 779–786.

Buchner, R., & Englert, J. (1994). Die Bewertung von Unternehmen auf der Basis des Unternehmensvergleichs. *Betriebs-Berater, 49*(23), 1573–1580.

Cable, J., & Holland, K. (1999). Regression vs. non-regression models of normal returns – implications for event studies. *Economic Letters, 64*, 81–85.

Calantone, R. J., di Benedetto, C. A., & Schmidt, J. B. (1999). Using the analytic hierarchy process in new product screening. *Journal of Product Innovation Management, 16*, 65–76.

Campart, S., & Pfister, E. (2003). *The Value of Interfirm Cooperation – an Event Study of New Partnership Announcements in the Pharmaceutical Industry*. Université du Havre/Université Nancy II Working Paper.

Carney, M. (2005). Corporate governance and competitive advantage in family-controlled firms. *Entrepreneurship Theory and Practice, 29*(3), 249–265.

Chan, S. H., Martin, J. D., & Kensinger, J. W. (1990). Corporate research and development expenditures and share value. *Journal of Financial Economics, 26*, 255–276.

Chan, V., Musso, C., & Shankar, V. (2008). McKinsey global survey results: Assessing innovation metrics. *McKinsey Quarterly, Oct 2008*, 1–11.

Chaney, P. K., Devinney, T. M., & Winer, R. W. (1991). The impact of new product introductions on the market value of firms. *Journal of Business, 64*, 573–610.

Chien, C. F. (2002). A portfolio-evaluation framework for selecting R&D projects. *R&D Management, 32*(4), 359–368.

Chiva, R., Alegre, J., & Lapiedra, R. (2007). Measuring organizational learning capability among the work force. *International Journal of Manpower, 28*, 224–242.

Chrisman, J., Chua, J. H., & Litz, R. A. (2004). Comparing the agency cost of family and non-family firms. *Entrepreneurship Theory and Practice, 28*(4), 335–354.

Chrisman, J., Chua, J. H., & Sharma, P. (2005). Trends and directions in the development of a strategic management theory of the family firm. *Entrepreneurship Theory and Practice, 29*(5), 555–575.

Christensen, C. M. (1997). *The innovator's Dilemma: When new technologies cause great firms to fail*, ix–xxvii, Harvard Business School Press.

Christensen, R., Davis, J., Muent, G., Ochoa, P., & Schmidt, W. (2002). "Biotechnology – an Overview." European Investment Bank (EIB) Sector Papers June 2002.

Chua, J. H., Chrisman, J., & Sharma, P. (1999). Defining the family business by behavior. *Entrepreneurship Theory and Practice, 23*(4), 19–39.

Churchill, N. C., & Hatten, K. J. (1987). Non-market based transfers of wealth and power: A research framework for family businesses. *American Journal of Small Business, 11*(3), 51–64.

Coombs, J. E., & Deeds, D. L. (2000). International alliances as sources of capital – evidence from the biotechnology industry. *Journal of High Technology Management Research, 11*(2), 235–253.

Cooper, R. G. (1993). *Winning at new products*. Reading, MA: Addison-Wesley.

Cooper, R. G. (1999). The invisible success factors in product innovation. *Journal of Product Innovation Management, 16*, 115–133.

Cooper, R. G. (2002). *Top oder Flop in der Produktentwicklung*. Weinheim.

Cooper, R. G. (2008). Perspective: The Stage-Gate-Idea-to-Launch Process-Update, What's New and NexGen Systems. *Journal of Product Innovation Management, 25*(3), 213–232.

Cooper, R. G., & Edgett, S. J. (2008). Maximizing productivity in product innovation. *Research Technology Management, 51*, 47–58.

Cooper, R. G., Edgett, S. J., & Kleinschmidt, E. J. (1999). New product portfolio management: Practices and performance. *Journal of Product Innovation Management, 16*, 333–351.

Cooper, R. G., Edgett, S. J., & Kleinschmidt, E. J. (2001). Portfolio management for new product development: Results of an industry practices study. *R&D Management, 31*(4), 361–380.

Cooper, R. G., & Kleinschmidt, E. J. (1990). *New products: The key factors in success*. Chicago: American Marketing Association.

Cox, J. C., Ross, S. A., & Rubinstein, M. (1979). Option pricing: A simplified approach. *Journal of Financial Economics, 7*, 229–263.

Craig, J. B. L., & Moores, K. (2006). Research note – A 10-year longitudinal investigation of strategy, systems, and environment on innovation in family firms. *Family Business Review, 19*(1), 1–10.

Daily, C. M., & Dollinger, M. J. (1992). An empircial examination of ownership structure in family and professionally managed firms. *Family Business Review, 5*(2), 117–136.

Damodaran, A. (2008). *Strategic risk taking*. Upper Saddle River, NJ: Wharton School Publishing.

Das, S., Sen, P. K., & Sengupta, S. (1998). Impact of strategic alliances on firm valuation. *Academy of Management Journal, 41*(1), 27–41.

Davila, T., Epstein, M. J., & Matusik, S. F. (2004). Innovation strategy and the use of performance measures. *Advances in Management Accounting, 13*, 27–58.

Day, G. S. (2007, December). Is It Real? Can We Win? Is It Worth Doing? *Harvard Business Review*, 110–120.

Deutsches Patent- und Markenamt. (2008). *Jahresbericht 2007*. München.

Diedrich, R. (2003). Die Sicherheitsäquivalentmethode der Unternehmensbewertung: Ein (auch) entscheidungstheoretisches wohlbegründbares Verfahren. *Zeitschrift für betriebswirtschaftliche Forschung, 55*(3), 281–286.

Die Zeit (2/26/09). *Die stillen Stützen*.

Dilgert, A. (2002). Betriebsräte und Innovationen. In: Kahle, E. (Ed.), *Organisatorische Veränderung und Corporate Governance – Aktuelle Themen der Organisationstheorie*. Wiesbaden, Deutscher Universitäts-Verlag.

DiMasi, J. A. (2000). New drug innovation and pharmaceutical industry structure – trends in the output of pharmaceutical firms. *Drug Information Journal, 34*, pp. 1169–1194.

Dixit, A., & Pindyck, R. (1994). *Investment under uncertainty*. Princeton: Princeton University Press.

Donckels, R., & Fröhlich, E. (1991). Are family businesses really different? – European experiences from STRATOS. *Family Business Review, 40*(2), 149–159.

Drukarczyk, J., & Schüler, A. (2007). *Unternehmensbewertung*. Munich: Vahlen.

Engel, K., & Diedrichs, E. (2006). Die Kunst, Innovationen erfolgreich zu Management. In: Garn, M. & Kalt, G. (Eds.), *Innovationstreiber am Standort Deutschland* (pp. 72–78). Unternehmen übernehmen Verantwortung.

Eisenführ, F., & Weber, M. (2003). *Rationales entscheiden.* Berlin: Springer.

Farag, H. (2009). The Value of Collaboration – How Biotechnology Firms Gain from Alliances. Springer: Contributions to Management Science, Heidelberg, 2009 (also Doctoral Dissertation, European Business School International University 2006).

Faulkner, T. (1996). Applying "options thinking" to R&D valuation. *Research Technology Management, 39,* 50–57.

F. A.Z.-Institut für Management-Markt- und Medieninformationen, Frankfurt am Main.

Finance-Studien. (2004). Unabhängig erfolgreich – Wachstums- und Finanzierungsstrategien von großen Familienunternehmen. F.A.Z.-Institut für Management-, Markt- und Medieninformationen et al., Frankfurt/Main et al.

Fisken, J., & Rutherford J (2002). Business models and investment trends in the biotechnology industry in Europe. *Journal of Commercial Biotechnology, 8*(3), 191–199.

Flamholtz, E. G. (1996). Effective organizational control: A framework, applications and implications. *European Management Journal, 14,* 596–611.

Floricel, S., & Ibanescu, M. (2008). Using R&D portfolio management to deal with dynamic risk. *R&D Management, 38*(5), 452–467.

Frankfurter Allgemeine Zeitung (8/11/08) Innovationen in Familienunternehmen.

Frankfurter Allgemeine Zeitung (5/23/07) Heiße Vorstandsstühle.

Gallo, M. A., & Sveen, J. (1991). Internationalizing the family business – facilitating and restraining factors. *Family Business Review, 4*(2), 181–187.

Gambanos, L., & Sturchio, J. L. (1998). Pharmaceutical firms and the transition to biotechnology – a study in strategic innovation. *Business History Review, 72*(2), 250–279.

Garvin, D., & Levesque, L. (2004). Emerging Business Opportunities at IBM, Harvard Business School case, No. 304-075.

Gassmann, O., & Enkel, E. (2006). Open innovation: Externe Hebeleffekte in der Innovation erzielen. *Zeitschrift Führung und Organisation, 3,* 132–138.

Gino, F., & Pisano, G. (2006). Do Managers' Heuristics Affect R&D Performance Volatility? Working Paper HBS Division of Research.

Girotra, K., Terwiesch, C., & Ulrich, K. T. (2007). Valuing R&D projects in a portfolio: Evidence from the pharmaceutical industry. *Management Science, 53*(9), 1452–1466.

Glaister/Buckley. (1996). Strategic motives for international alliance formation. *Journal of Management Studies, 33*(3), 301–332.

Gleich, R. (1997). Stichwort performance measurement. *Die Betriebswirtschaft, 57,* 114–117.

Gleich, R. (2001). *Das System des Performance Measurement – Theoretisches Grundkonzept, Entwicklungs- und Anwendungsstand.* München: Vahlen.

Gleich, R., Nestle, V., & Sommer, L. (2009a). Innovationsorientiertes Performance Measurement. Entwicklung und Implementierung einer Innovation Scorecard am Beispiel der Festo AG & Co. KG. In: Fisch, J. H. & Roß, J. M. (Hsrg.) *Fallstudien zum Innovationsmanagement – Methodengestützte Lösung von Problemen aus der Unternehmenspraxis,* Wiesbaden, pp. 187–204.

Gleich, R., Henke, M., Quitt, A., & Sommer, L. (2009b), New approaches in performance measurement: Methods for specification and operationalisation within the context of supply management. *International Journal of Business Excellence, 2,* 105–123.

Gomez-Mejia, L. R., Nuñez-Nickel, M., & Gutierrez, I. (2001). The role of family ties in agency contracts. *Academy of Management Journal, 44*(1), 81–95.

Gompers, P. A., & Lerner, J. (1998). The Determinants of Corporate Venture Capital Success: Organizational Structure, Incentives and Complementarities, Working Paper, No. 6725, National Bureau of Economic Research, NBER Working Paper Series, Cambridge.

Granig, P. (2007). *Innovationsbewertung. Potentialprognose und -steuerung durch Ertrags- und Risikosimulation.* DUV, Wiesbaden.

Greiner, O., Römer, S., & Russo, P. (2009). *Innovationsstudie 2009: Das verschwendete Innovationspotenzial – geniale Ideen im Unternehmen finden und nutzen*. Stuttgart: Horváth & Partners Management Consultants.

Grossmann, M. (2003). *Entrepreneurship in biotechnology – managing for growth from start-up to initial public offering*. Heidelberg: Physica.

Günther, T., & Grüning, M. (2002). Performance measurement-systeme im praktischen Einsatz. *Controlling, 14*, 5–13.

Gustafsson, J., & Salo, A. (2005). Contingent portfolio programming for the management of risky projects. *Operations Research, 53*(6), 946–956.

Hagelin, N., Holmén, M., & Pramborg, B. (2006). Family ownership, dual-class shares, and risk management. *Global Finance Journal, 16*, 283–301.

Hall, B. H., & Mairesse, J. (1995). Exploring the relationship between R&D and productivity in french manufacturing firms. *Journal of Econometrics, 65*, 263–293.

Handelsblatt (2/19/09). Mit Innovation durch die Krise.

Handler, W. C. (1989). Methodological issues and considerations in studying family businesses. *Family Business Review, 2*(3), 257–276.

Harrigan, K. R. (1985). *Strategies for joint ventures*. Lexington, MA: Lexington Books.

Hartmann, M. (2006). Realoptionen als Bewertungsinstrument für frühe Phasen der Forschung und Entwicklung in der pharmazeutischen Industrie, Dissertation, Technische Universität Berlin, Berlin.

Häussler, C. (2005). Interfirm Collaboration – Valuation, Contracting, and Firm Restructuring, Wiesbaden, also University of Munich Dissertation, Gabler.

Häussler, C. (2006). When does partnering create market value? *European Management Journal, 24*(1), 1–15.

Hauschildt, J. (2004). *Innovationsmanagement*. München: Vahlen.

Heidenberger, K., & Stummer, C. (1999). Research and development project selection and resource allocation: A review of quantitative modelling approaches. *International Journal of Management Review, 1*, 197–224.

Hofmann, J. (2005). *Value Intangibles: Deutsche Bank Research*, International Topics, Frankfurt.

Höger, T., Fuchs, P., & Bähr, C. (2004). Erfolgskurs Nach Konsolidierung – Biotechnologie – Eine Research Publikation Der Dz Bank Ag, DZ Bank, Frankfurt.

Hommel, U., & Pritsch, G. (1999). Investitionsbewertung und Unternehmensführung mit dem Realoptionsansatz. In: Achleitner, A. K. & Thoma, G. F. (Eds), *Handbuch corporate finance*, Cologne: Verlag Deutscher Wirtschaftsdienst.

Hommel, U., & Wölfer, K. (2009). Wertsteigerungsstrategien von Familienunternehmen – Eine empirische Analyse. Work in Progress.

Hoskissen, R. E., Hitt, M. A., Johnson, R. A., & Grossman, W. (2002). Conflicting voices: The effects of institutional ownership heterogeneity and internal governance on corporate innovation strategies. *Academy of Management Journal, 45*(4), 697–716.

Hughes, G. D., & Chain, D. C. (1996). Turning new product development into a continuous learning process. *Journal of Product Innovation Management, 13*, 89–104.

Jackson, B. (1983). Decision methods for selecting a portfolio of R&D projects. *Research Management*, 21–26.

Jäger, R., & Himmel, H. (2003). Die Fair Value-Bewertung immaterieller Vermögenswerte vor dem Hintergrund der Umsetzung internationaler Rechnungslegungsstandards. *Betriebswirtschaftliche Forschung und Praxis, 55*(4), 417–440.

Janney, J. J., & Folta, T. B. (2003). Signaling through private equity placements and its impact on the valuation of biotechnology firms. *Journal of Business Venturing, 18*, 361–380.

Jaskiewicz, P. (2006). Performance-Studie börsennotierter Familienunternehmen in Deutschland, Frankreich und Spanien, Dissertation, European Business School, Josef Eul Verlag, Lohmar.

Jensen, M., & Meckling, W. (1976). Theory of the firm: Managerial behavior, agency costs, and ownership structure. *Journal of Finance and Economy, 3*, 305–360.

Jerez-Gómez, P., Céspedes-Lorente, J., & Valle-Cabrera, R. (2005). Organizational learning capability: A proposal of measurement. *Journal of Business Research, 58*, 715–726.
Jesch, T. A. (2004). *Private equity beteiligungen* (pp. 21–24). Wiesbaden: Gabler.
Jones, G. R., & Hill, C. W.L. (1988). Transaction cost analysis of strategy-structure choice. *Strategic Management Journal, 9*(2), 159–172.
KfW. (2008). Die Entwicklung der Eigenkapitalausstattung kleiner und mittlerer Unternehmen 2002 bis 2005. WirtschaftsObserver 36.
Kahl, M., Liu, J., & Longstaff, F. A. (2003). Paper millionaires: How valuable is stock to a stockholder who is restricted from selling it? *Journal of Financial Economics, 67*(3), 385–410.
Kale, P., Dyer, J. H., & Singh, H. (2001). Value creation and success in strategic alliances – alliancing skills and the role of alliance structure and systems. *European Management Journal, 19*(5), 463–471.
Kale P., Dyer J. H., & Singh H. (2002). Alliance capability, stock market response, and long-term alliance success – the role of the alliance function. *Strategic Management Journal, 23*, 747–767.
Kaplan, R. S., & Norton, D. P. (1992). The balanced scorecard – measures that drive performance. *Harvard Business Review, 70*, 71–79.
Kaplan, R. S., & Norton, D. P. (2003). *The strategy-focused organization – how balanced scorecard companies thrive in the new business environment*. Boston: Harvard Business School Press.
Karamanos, A. (2002). Market Networks and the Value in Knowledge Exchanges – Evidence from Biotechnology Strategic Alliances, University of Cambridge ESRC Centre for Business Research Working Paper 240.
Kianto, A. (2008). Development and validation of a survey instrument for measuring organisational renewal capability. *International Journal of Technology Management, 42*, 69–88.
Killen, C. P., Hunt, R. A., & Kleinschmidt, E. J. (2008). Project portfolio management for product innovation. *International Journal of Quality and Reliability Management, 25*, 24–38.
Kim, W. C., & Mauborgne, R. (1999). Strategy, value innovation and the knowledge economy. *Sloan Management Review, 40*(3), 41–54.
Klein, S. B. (2000). Family businesses in Germany: Significance and structure. *Family Business Review, 13*(3), 157–181.
Kock, A. (2007). Innovativeness and Innovation Success – A Meta-Analysis. *ZFBF Special Issue 2*, pp. 1–21.
Kogut, B. (1991). Joint ventures and the option to expand and acquire. *Management Science, 37*(1), 19–33.
Koh, J., & Venkatraman, N. (1991). Joint venture formations and stock market reations – an assessment in the information technology sector. *Academy of Management Journal, 34*(4), 869–892.
Koppel, O. (2006). *Innovationsverhalten der tecknikaffinen Branchen*. Köln: IdW-Verlag.
Krimphove, D., & Tytko, K. (2002). *Praktiker-Handbuch Unternehmensfinanzierung*, pp. 879–890.
Krostewitz, A. (2008). Unternehmensbewertung im Risikoverbund – Neue Methoden der Unternehmensbewertung bei Mergers & Acquisitions, Dissertation, Friedrich-Schiller-Universität Jena, Jena.
Kürsten, W. (2002). "Unternehmensbewertung unter Unsicherheit", oder: Theoriedefizit einer künstlichen Diskussion über Sicherheitsäquivalent- und Risikozuschlagsmethode. *Zeitschrift für betriebswirtschaftliche Forschung, 54*(2), 128–144.
Kürsten, W. (2003). Grenzen und Reformbedarfe der Sicherheitsäquivalentmethode in der (traditionellen) Unternehmensbewertung. *Zeitschrift für betriebswirtschaftliche Forschung, 55*(3), 306–314.
Kürsten, W. (2007). Kontextadäquates Risikomanagement: Eine komprimierte Einführung. *Jenaer Schriften zur Wirtschaftswissenschaft, 18*, 1–4.
Kürsten W., & Straßberger M. (2004). Risikomessung, Risikomaße und Value-at-Risk. *Das Wirtschaftsstudium, 33*(2), 202–207.
Kuhner, C., & Maltry, H. (2006). *Unternehmensbewertung*. Berlin, Heidelberg: Springer.

Lansberg, I. S., Perrow, E. L., & Rogolsky, S. (1988). Family business as an emerging field. *Family Business Review, 1*(1), 1–8.

Lawson, B., & Samson, D. (2001). Developing innovation capability in organisations: A dynamic capabilities approach. *International Journal of Innovation Management, 5*, 377–400.

Learned, E., Christensen, C. R., Andrews, K. R., & Guth, W. (1969). *Business policy – text and cases.* New York, NY: Richard D. Irwin.

Le Breton-Miller, I., & Miller, D. (2006). Why do some family businesses out-compete? – Governance, long-term orientations, and sustainable capability. *Entrepreneurship Theory and Practice, 30*(6), 731–746.

Leenen, S. (2005). Innovation in Family Businesses – A Conceptual Framework with Case Studies of Industrial Family Firms in the German "Mittelstand", Dissertation No. 3096, University of St. Gallen, Difo-Druck, Bamberg.

Leitner, K.-H., Sammer, M., Graggober, M., Schartinger, D., & Zielowski, Ch. (2001). Wissensbilanzierung für Universitäten (IC Statements for Universities), Seibersdorf Research Report ARC-S-0145, www.systemforschung.arcs.ac.at/ Publikationen/21.pdf. Access 10 April 2004.

Lev, B. (2001). *Intangibles – management, measurement, and reporting.* Washington, DC: Brookings.

Lev, B., & Zarowin, P. (1999). The boundaries of financial reporting and how to extend them. *Journal of Accounting Research, 37*(3), 353–378.

Liberatore, M. J., & Titus, G. J. (1983). The practice of management science on R&D project management. *Management Science, 29*, 62–974.

Lilien, G. L., & Yoon, E. (1990). The timing of competitive market entry: An explanatory study of new industrial products. *Management Science, 36*(5), 568–585.

Littekmann, J. (Ed.). (2005). *Innovationscontrolling.* München: Vahlen.

Littkemann, J., & Holtrup, M. (2008). Evaluation von Dienstleistungsinnovationen – Möglichkeiten und Grenzen aus Sicht des Controllings. Der Controlling-Berater, pp. 261-284.

Liu, Q. (2000). How Good Is Good News? Technology Depth, Book-to-Market Ratios, and Innovative Events, UCLA Working Paper, Los Angeles, CA.

Loch, CH., & Kavadias, S. (2002). Dynamic portfolio selection of NPD programs using marginal returns. *Management Science, 48*(10), 1227–1241.

Löhnert, P., & Böckmann, U. J. (2005). Multiplikatorverfahren in der Unternehmensbewertung. In: Peemöller, V. (Eds), *Praxishandbuch der Unternehmensbewertung* (pp. 403–428). Herne & Berlin: Verlag Neue Wirtschaftsbriefe.

Longstaff, F. A. (1995). How much can marketability affect security values? *The Journal of Finance, 50*(5), 1767–1774.

Longstaff, F. A. (2001). Optimal portfolio choice and the valuation of illiquid securities. *The Review of Financial Studies, 14*(2), 407–431.

Mach, A. E. (2007). Alphazirkel Studie: Die Unternehmenskultur im Familienunternehmen. In: Mach, A. E. (Ed.), *Die Unternehmenskultur im Familienunternehmen – Was trägt sie, was prägt sie, was bedeutet sie für den Unternehmenserfolg.* München: Alphazirkel.

MacKinlay, A. C. (1997). Event studies in economics and finance. *Journal of Economic Literature, 35*, 13–39.

MacMillan, I. C., & McGrath, R. G. (2002). Crafting R&D project portfolios. *Research Technology Management, 45*(5), 48–59.

Mackenstedt, A., Fladung, H. D., & Himmel, H. (2006). Ausgewählte Aspekte bei der Bestimmung beizulegender Zeitwerte nach IFRS 3 – Anmerkungen zu IDW RS HFA 16. *Die Wirtschaftsprüfung, 16*, 1097–1048.

Madhavan, R., Koka, B. R., & Prescott, J. E. (1998). Networks in transition – how industry events (re)shape interfirm relationships. *Strategic Management Journal, 19*, 439–459.

Malerba, F., & Orsenigo, L. (2002). Innovation and market structure in the dynamics of the pharmaceutical industry and biotechnology – towards a history-friendly model. *Industrial and Corporate Change, 11*(4), 667–703.

Mallik, A., Zbar, B., & Zemmel, R. W. (2004). Making pharma alliances work. *McKinsey Quarterly, 1,* 16–19.
ManagerMagazin (12/19/08), Forsche Geister.
Mandl, G., & Rabel, K. (2005). Methoden der Unternehmensbewertung (Überblick). In: Peemöller, V. H. (Eds), *Praxishandbuch der Unternehmensbewertung.* Herne: Verlag Neue Wirtschaftsbriefe.
Markowitz, H. M. (1952). Portfolio selection. *Journal of Finance, 7,* 77–91.
Martino, J. P. (1995). *R&D project selection.* New York, NY: Wiley.
Matschke, M. J., & Brösel, G. (2007). *Unternehmensbewertung: Funktionen –Methoden – Grundsätze.* Wiesbaden: Gabler.
Mayers, D. (1972). Nonmarketable assets and capital market equilibrium under uncertainty. In: Jensen, M. (Eds.), *Studies in the theory of capital markets: Papers of the conference on modern capital theory* (pp. 223–248). New York: Praeger Publishers.
Mayers, D. (1973). Nonmarketable assets and the determination of capital asset prices in the absence of a riskless asset. *The Journal of Business, 46*(2), 258–267.
McWilliams, A., & Siegel, D. (1997). Event studies in management research – theoretical and empirical issues. *Academy of Management Journal, 40*(3), 626–657.
Miao, J., & Wang, N. (2007). Investment, consumption, and hedging under incomplete markets. *Journal of Financial Economics, 86*(3), 608–642.
Mikkola, J. H. (2001). Portfolio management of R&D projects: Implications for innovation management, *Technovation, 21,* 423–435.
Möller, K., & Janssen, S. (2009). Performance Measurement von Produktinnovationen. Konzepte, Instrumente und Kennzahlen des Innovationscontrollings. *Controlling, 21,* 89–96.
Montoya-Weiss, M. M., & Calantone, R. (1994). Determinants of new product performance: A review and meta-analysis. *Journal of Product Innovation Management, 11*(5), 397–417.
Morck, R., Stangeland, D. A., & Yeung, B. (2000). Inherited wealth, corporate control and economic growth: The Canadian disease. In: Morck, R. (Ed.), *Concentrated corporate ownership.* Chicago: University of Chicago Press.
Mouritsen, J., Bukh, P. N., Larsen, H. T., & Johansen, M. R. (2002). Developing and managing knowledge through intellectual capital statements. *Journal of Intellectual Capital, 3*(1), 10–29.
Muhr, D. B., & Thum, G. F. (2006). Tradition verpflichtet? – Innovationsmanagement in Familienunternehmen. *OrganisationsEntwicklung, 4,* 70–77.
Müller-Stewens, G. (1998). Performance Measurement im Lichte eines Stakeholderansatzes. In: Reinecke, S., Tomczak, T., & Dittrich, S. (Eds.), *Marketingcontrolling* (pp. 34–43). Thexis, St. Gallen.
Nakamura, L. (2003). A trillion dollars a year in intangible investments and new economy. In: Hand, J. & Lev, B. (Eds.), *Intangible assets – values, measures and risks* (pp. 19–47). Oxford University Press.
National Science Board. (2006). Science and Engineering Indicators 2006, National Science Foundation – Division of Science Resources Statistics, Arlington/VA, February 2006.
Neely, A. (2005). The evolution of performance measurement research. Developments in the last decade and a research agenda for the next. *International Journal of Operations & Production Management, 25,* 1264–1277.
Neely, A., Gregory, M., & Platts, K. (1995). Performance measurement system design. *International Journal of Operations & Production Management, 15,* 80–116.
Neely, A., Mills, J., Platts, K., Gregory, M., & Richards, H. (1996). Performance measurement system design: should process based approaches be adopted. *International Journal of Production Economics,* 46–47/423–431.
Neill, J. D., Pfeiffer, G. M., & Young-Ybarra, C. (2001). Technology R&D alliances and firm value. *Journal of High Technology Management Research, 12,* 227–237.
Ng, R. (2004). *Drugs – from Discovery to Approval.* Hoboken/NJ: John Wiley & Sons.
Nicholson, S., Danzon, P. M., & McCullough, J. (2002). Biotech-Pharmaceutical Alliances as a Signal of Asset and Firm Quality, NBER Working Paper No. 9007, June 2002.

Noor, I., Martin, R., & Bowman, D. (2005). Implementation of successful risk-based portfolio management. *AACE International Transactions RISK, 02,* 1–6.

Ordóñez De Pablos, P. (2003). Intellectual capital reporting in Spain – a comparative view. *Journal of Intellectual Capital, 4*(1), 61–81.

O'Reilly, B., & Rao, R. M. (1997). The secrets of America's most admired corporations: New ideas, new products. *Fortune, 135,* 60–64.

Orens, R., & Lybaert, N. (2007). Does the financial analysts' usage of non-financial information influence the analysts' forecast accuracy – some evidence from the Belgian sell-side financial analyst. *The International Journal of Accounting, 42*(3), 237–271.

Organisation for Economic Co-Operation and Development (OECD) (2005). Oslo Manual – Guidelines for collecting and interpreting innovation data. http://www.oecd.org/document/23/0,3343,en_2649_37417_35595607_1_1_1_37417,00.html. Accessed 13 March 2009.

Pakes, A. (1985). On patents, R&D, and the stock market rate of return. *Journal of Political Economy, 93*(2), 390–409.

Park, N., & Mezias, J. M. (2005). Before and after the technology sector crash – the effect of environmental munificence on stock market response to alliances of e-commerce firms. *Strategic Management Journal, 26,* 987–1007.

Park, N., Mezias, J. M., & Song, J. (2004). A resource-based view of strategic alliances and firm value in the electronic marketplace. *Journal of Management, 30*(1), 7–27.

Pellens, B., Fülbier, R. U., Gassen, J., & Sellhorn, T. (2008). *Internationale Rechnungslegung.* Stuttgart: Schäffer-Poeschel.

Penrose, E. T. (1959). *The theory of the growth of the firm.* Oxford: Blackwell Publishers.

Peterson, P. P. (1989). Event studies – a review of issues and methodology. *Quarterly Journal of Business and Economics, 28,* 36–66.

PhRMA. (2003). *Pharmaceutical industry profile, pharmaceutical research and manufacturers of America.* Washington, DC.

Picot, G. (2007). Familien- und Mittelstandsunternehmen im globalen Wandel von Wirtschaft und Gesellschaft. In: Picot, G. (Ed.), *Handbuch für Familien- und Mittelstandsunternehmen – Strategie, Gestaltung, Zukunftssicherung.* Stuttgart: Schäffer-Poeschel.

Pisano, G. P., & Mang, P. Y. (1993). Collaborative product development and the market for know-how – strategies and structures in the biotechnology industry. *Research on Technological Innovation, Management and Policy, 5,* 109–136.

Prevezer, M., & Toker, S. (1996). The degree of integration in strategic alliances in biotechnology. *Technology Analysis & Strategic Management, 8,* 117–133.

Probst, G., Raub, S., & Romhardt, K. (2006). *Wissen managen: Wie Unternehmen ihre wertvollste Ressource optimal nutzen.* Wiesbaden: Gabler.

Pyka, A., & Saviotti, P. (2001). Innovation Networks in the Biotechnology-Based Sector, University of Augsburg Discussion Paper No. 220.

Raake, A. (2008). Strategisches Performance Measurement, Anwendungsstand und Gestaltungsmöglichkeiten am Beispiel des Öffentlichen Personennahverkehrs, Lit, Berlin.

Rams, A. (1999). Realoptionsbasierte Unternehmensbewertung, Finanz Betrieb, No. 11, pp. 349–364.

Rastogi, P. N. (2003). The nature and role of IC – rethinking the process of value creation and sustained enterprise growth. *Journal of Intellectual Capital, 4*(2), 227–248.

Reilly, R., & Schweihs, R. P. (1999). *Valuing intangible assets.* New York: McGraw-Hill.

Remer, D. S., Stokdyk, S. B., & Van Driel, M. (1993). Survey of project evaluation techniques currently used in industry. *International Journal of Production Economics, 32,* 103–115.

Reuer, J. J. (2000). Parent firm performance across international joint venture life-cycle stages. *Journal of International Business Studies, 31*(1), 1–20.

Riegler, C., & Kristandl, G. (2004). Value Reporting in österreichischen Unternehmen – Beobachtung des Berichtverhaltens von ATX und ATX-Prime Unternehmen (Value Reporting in Austrian Companies – Observations of the Disclosure of ATX and ATX Prime Companies),

Seicht G. (Ed.), *Jahrbuch für Controlling und Rechnungswesen 2004*, LexisNexis, Vienna, pp. 245–267.

Rind, K. W. (1981). The role of venture capital in corporate development. *Strategic Management Journal, 2*, 169–180.

Roberts, P. W. (1999). Product innovation, product-market competition and persistent profitability in the U.S. pharmaceutical industry. *Strategic Management Journal, 20*(7), 655–670.

Ross, S. A., Westerfield, R. W., & Jaffe, J. (2005). *Corporate finance* (pp. 540–568). New York: McGraw Hill.

Rumelt, R. P. (1984). Towards a strategic theory of the firm. In: Lamb, R. (Ed.), *Competitive strategic management* (pp. 556–570). Prentice Hall: Englewood Cliffs.

Rummler, G. A., & Brache, A. P. (1990). *Improving performance. How to manage the white space in the organization chart*. San Francisco: Jossey-Bass.

Saaty, T. L., Rogers, P. C., & Pell, R. (1980). Portfolio selection through hierarchies. *Journal of Portfolio Management, 6*(3), 16–21.

Sakurai, M. (1989). Target costing and how to use it. *Journal of Cost Management, 3*, Summer, 39–50.

Sammerl, N. (2006). *Innovationsfähigkeit und nachhaltiger Wettbewerbsvorteil: Messung, Determinanten, Wirkungen*. Wiesbaden: DUV.

Sanwal, A. (2007). *Optimizing corporate portfolio management*. Hoboken, NJ: John Wiley & Sons.

Schmeisser, W., Kantner, A., Geburtig, A., & Schindler, F. (2006). *Forschungs- und Technologie-Controlling. Wie Unternehmen Innovationen operativ und strategisch steuern*. Stuttgart: Schäffer-Poeschel.

Scholich, M., Mackenstedt, A., & Greinert, M. (2004). Valuation of intangible assets for financial reporting. In: Fandel, G., Backes-Gellner, U., Schlüter, M., & Staufenbiel, J. E. (Eds.), *Modern concepts of the theory of the firm – managing enterprises of the new economy* (pp. 491–504). Berlin: Springer.

Scholich, M., & Robers, D. I. (2007). Vom Beginner zum Professional – Innovation bei PricewaterhouseCoopers. In: Schmidt, K., Gleich, R., & Richter, A. (Eds.), *Innovations management in der Serviceindustrie – Grundlagen, Praxisbeispiele und Perspektiven* (pp. 325–338). Haufe, Freiburg, Berlin, Munich.

Scholich, M., & Wulff, C. (2002). Ansätze und Methoden zur Bewertung von Wachstumsunternehmen. In: Hommel, U., & Knecht, T. (Eds), *Wertorientiertes Start-Up-Management* (pp. 563–579). Munich: Vahlen.

Scholich, M., Gleich, R., & Grobusch, H., (2006). *Innovation Performance – Das Erfolgsgeheimnis innovativer Dienstleister*. PricewaterhouseCoopers, Deutsches Zentrum für Luft- und Raumfahrt (DLR), European Business School (EBS).

Schultze, W. (2003). Kombinationsverfahren und Residualgewinnmethode in der Unternehmensbewertung: konzeptioneller Zusammenhang. *Kapitalmarktorientierte Rechnungslegung, 3*(10), 458–464.

Schumpeter, J. (1942). *Capitalism, socialism, and democracy*. New York: Harper and Row.

Schumpeter, J. A. (1943). *Capitalism, socialism, and democracy*. London: Allen & Unwin.

Schwartz, E. S., & Trigeorgis, L. (2001). *Real options and investment under uncertainty: Classical readings and recent contributions*. Cambridge, MA: MIT Press.

Schween, K. (1996). *Corporate Venture Capital: Risikokapitalfinanzierung deutscher Industrieunternehmen*. Wiesbaden.

Schwetzler, B. (2000). Unternehmensbewertung unter Unsicherheit- Sicherheitsäquivalent- oder Risikozuschlagsmethode? *Zeitschrift für betriebswirtschaftliche Forschung, 52*(8), 469–486.

Schwetzler, B. (2002). Das Ende des Ertragswertverfahrens? – Replik zu den Anmerkungen von Wolfgang Kürsten zu meinem Beitrag in der zfbf. *Zeitschrift für betriebswirtschaftliche Forschung, 54*, 145–158.

Sharma, A., & Lacey, N. (2004). Linking product development outcomes to market valuation of the firm – the case of the U.S. pharmaceutical industry. *Journal of Product Innovation Management, 21*, 297–308.

Sieger, G. (2007). Finanzierung. In: Picot, G. (Ed.), *Handbuch für Familien- und Mittelstandsunternehmen – Strategie, Gestaltung, Zukunftssicherung*, Stuttgart: Schäffer-Poeschel.

Simon, H. (2007). *Hidden Champions des 21. Jahrhunderts – Die Erfolgsstrategien unbekannter Weltmarktführer*. Frankfurt/Main: Campus Verlag.

Sirmon, D. G., & Hitt, M. H. (2003). Managing resources: Linking unique resources, management, and wealth creation in family firms. *Entrepreneurship Theory and Practice, 27*(4), 339–358.

Souder, W. E. (1984). *Project selection and economic appraisal*. New York, NY: Van Nostrand Reinhold.

Speckbacher, G., Güldenberg, S., & Ruthner, R. (2004). Externes Reporting über immaterielle Vermögenswerte (External Reporting on Intangible Assets). In: Horváth, P. & Möller, K. (Eds.), *Intangibles in der Unternehmenssteuerung* (pp. 435–455). Munich: Vahlen.

Spinler, S., & Huchzermeier, A. (2004). Realoptionen: Eine marktbasierte Bewertungsmethodik für dynamische Investitionsentscheidungen unter Unsicherheit. *Zeitschrift für Controlling & Management, Sonderheft 1*, 66–71.

Spremann, K. (2002). *Finanzanalyse und Unternehmensbewertung, IMF – International Management and Finance*. Munich, Vienna: Oldenbourg.

Stewart, T. (1997). *Intellectual capital – The new wealth of organizations*. London: Broadway.

Storey, C., & Kelly, D. (2001). Measuring the performance of new service development activities. *The Service Industries Journal, 21*, 71–90.

Strong, N. (1992). Modelling abnormal returns – a review article. *Journal of Business Finance and Accounting, 19*(4), 533–553.

Stummer, C., & Heidenberger, K. (2003). Interactive R&D portfolio analysis with project interdependencies and time profiles of multiple objectives. *IEEE Transactions of Engineering Management, 50*(2), 175–183.

Sueddeutsche Zeitung (10/25/07), Die Helden im Hintergrund.

Sull, D. N., & Houlder, D. (2006). How companies can avoid a midlife crisis. *MIT Sloan Management Review, 48*(1), 26–34.

Sundaram, A. K., & Inkpen, A. C. (2004). The corporate objective revisited. *Organic Scienc, 15*(3), 350–363.

Sveiby, K.-E. (1997). *The new organizational wealth – managing and measuring knowledge-based assets*. San Francisco: Berrett-Koehler.

Tasker, S. (1998). Technology company conference calls: A small sample study. *Journal of Financial Statement Analysis, 4*(1), 6–14.

Thom, N. (1992). *Innovationsmanagement*. Bern: Schweizerische Volksbank.

Tinsley, R. (2000). *Advanced Project Financing, Euromoney Books*, London, pp. 1–14.

Tobin, J. (1958). Liquidity preferences as behavior towards risk. *Review of Economic Studies, 25*(2), 65–86.

Tollman, P., Guy, P., Altshuler, J., Flanagan, A., & Steiner, M. (2001). *A revolution in R&D – how genomics and genetics are transforming the biopharmaceutical industry*. Boston: The Boston Consulting Group.

Triantis, A., & Borison, A. (2001). Real options: state of the practice. *Journal of Applied Corporate Finance, 14*(2), 8–24.

Trigeorgis, L. (1996). *Real options, managerial flexibility and strategy in resource allocation*. Cambridge, MA: MIT Press.

Trott, P. (2005). *Innovation management and new product development*, Harlow et al.: Prentice Hall.

Ulrich, K. T., & Eppinger, S. D. (1995). *Product design and development*. New York: McGraw-Hill.

Vahs, D., & Burmester, R. (1999). *Innovationsmanagement – Von der Produktidee zur erfolgreichen Vermarktung*. Stuttgart: Schäffer-Poeschel.

Vahs, D., & Burmester, R. (2002). *Innovationsmanagement: von der Produktidee zur erfolgreichen Vermarktung*. Stuttgart: Schäffer-Poeschel.

Vandemaele, S., Vergauwen, P., & Smits, A. (2005). Intellectual capital disclosure in The Netherlands, Sweden and the UK. *Journal of Intellectual Capital, 6*(3), 417–426.

Vassolo, R. S., Anand, J., & Folta, T. B. (2004). Non-additivity in portfolios of exploration activities: A real options-based analysis of equity alliances in biotechnology, *Strategic Management Jounal, 25*, 1045–1061.

Ven, A.v.d. (1986). Central problems in the management of innovation. *Management Science, 32*(5), 590–607.

Verma, D., & Sinha, K. K. (2002). Toward a theory of project interdependencies in high tech R&D environments. *Journal of Operations Management, 20*, 451–468.

Villalonga, B., & Amit, R. (2004). How do family ownership, control, and management affect firm value? *Journal of Finance Economy, 8*(2), 385–417.

Wernerfeldt, B. (1984). A resource-based view of the firm. *Strategic Management Journal, 5*(2), 171–180.

Weiss, M., Zirkler, B., & Guttenberger, B. (2008). Performance Measurement Systeme und ihre Anwendung in der Praxis. Ergebnisse empirischer Studien, *Controlling, 20*, 139–147.

Whittaker, E., & Bower, J. D. (1994). A shift to external alliances for product development in the pharmaceutical industry. *R&D Management, 24*(3), 249–260.

Wiese, J. (2003). Zur theoretischen Fundierung der Sicherheitsäquivalentmethode und des Begriffs der Risikoauflösung bei der Unternehmensbewertung. *Zeitschrift für betriebswirtschaftliche Forschung, 55*, 287–305.

Wilhelm, J. E. (2005). Unternehmensbewertung – Eine finanzmarkttheoretische Untersuchung. *Zeitschrift für Betriebswirtschaft, 75*(6), 631–665.

WirtschaftsWoche (3/2/09), Ich nenne das positives Denken – Ludwig Georg Braun im Interview.

WirtschaftsWoche (11/3/08), Milberg: Finanzkrise bedroht auch die Forschungslandschaft.

Witt, J. (1996). Grundlagen für die Entwicklung und die Vermarktung neuer Produkte. In: Witt, J. (Eds.), *Produktinnovation*. Munich: Vahlen.

Witt, P. (2005). Corporate venture capital. In: Börner, C., Grichnik, D. (Eds.), *Entrepreneurial Finance, Kompendium der Gründungs- und Wachstumsfinanzierung* (pp. 259–276). Heidelberg.

Witt, P., & Brachtendorf, G. (2002). Gründungsfinanzierung durch Großunternehmen, Die Betriebswirtschaft, pp. 681–692.

Workgroup "Accounting and Reporting of Intangible Assets" (WGARIA) (2001). Kategorisierung und bilanzielle Erfassung immaterieller Werte (Categorization and Recognition of Intangible Assets). *Der Betrieb, 54*(19), 989–995.

Workgroup "Accounting and Reporting of Intangible Assets" (WGARIA) (2005). Corporate Reporting on Intangibles – A Proposal from a German Bachground, Schmalenbach Business Review, Special Issue 2/2005, pp. 65-100.

Workgroup "Accounting and Reporting of Intangible Assets" (WGARIA) (2008). Leitlinien zur Bilanzierung selbstgeschaffener immaterieller Vermögensgegenstände des Anlagevermögens nach dem Regierungsentwurf des BilMoG (Guidelines for the Recognition of Self Created Fixed Intangible Assets Due to Government's Draft for BilMoG). *Der Betrieb, 61*(34), 1813–1821.

Wulf, I. (2008). Immaterielle Vermögenswerte nach IFRS – Ansatz, Bewertung, Goodwill-Bilanzierung (Intangilbe Assets According to IFRS – Recognition, Measurement, Goodwill Accounting), Erich Schmidt, Berlin.

Zahra, S. A. (2005). Entrepreneurial risk taking in family firms. *Family Business Review, 18*(1), 23–40.

Zahra, S. A., Hayton, J. C., & Salvato, C. (2004). Entrepreneurship in family vs. non-family firms: A resource-based analysis of the effect of organizational culture. *Entrepreneurship Theory and Practice, 28*(4), 363–381.

Zucker, L. G., & Darby, M. R. (1995).Present at the Revolution – Transformation of Technical Identity for a Large Incumbent Pharmaceutical Firm after the Biotechnological Breakthrough, NBER Working Paper No. 5243.

Editors

Prof. Dr. Alexander Gerybadze Professor of International Management, University of Hohenheim, Stuttgart. Studies in Economics, Mathematics and Business Administration in Heidelberg (1973–1978) and Stanford University (1979–1980). Ph.D. on Evolutionary Models of Technical Change, Heidelberg University 1980. 1981–1983 Consultant, VDI Technology Center Berlin. 1983–1990 Arthur D. Little International in Wiesbaden, Member of the European Directorate. Professor of Technology Management at St. Gallen Business School, Switzerland 1991–1995. Since 1996 Director, Center for International Management and Innovation, University of Hohenheim, Member of the Executive Board, Center for Innovation and Services, Member of the Expert Commission for Research and Innovation, German Federal Government.

Research on Technology and Innovation Management, R&D Internationalization and Offshoring, R&D and Knowledge Management in Multinational Firms, National Innovation Systems and National Innovation Strategy, Management of Innovation Clusters and Standard-setting Consortia.

Prof. Ulrich Hommel, Ph.D. is a Full Professor of Finance and heads the Endowed Chair of Corporate Finance & Capital Markets at European Business School (EBS). Ulrich Hommel holds a Ph.D. in Economics from the University of Michigan, Ann Arbor, and has completed his habilitation in Business Administration at the WHU, Germany. He was an Assistant Professor of Finance at the WHU from 1994 to 1999 and has subsequently joined the faculty of the EBS. He is the Director of the Strategic Finance Institute at the EBS. In the past, Ulrich Hommel has held visiting appointments at the Stephen M. Ross School of Business (University of Michigan), the Krannert School of Management (Purdue University) and the Bordeaux Business School. His main research interests are corporate risk management, venture capital & private equity, family business finance and corporate restructuring. Ulrich Hommel has been Academic Dean of the Faculty at the EBS from 2000 to 2002 and has subsequently held the position of Rector and Managing Director from 2003 to 2006. Since 2007, he is also an Associate Director of Quality Services at the European Foundation for Management Development (EFMD) in Brussels and, as one of the Directors, is responsible for the EFMD Programme Accreditation System (EPAS).

Prof. Dr. Dieter Thomaschewski Professor of Business Administration/ Management, University of Applied Sciences, Ludwigshafen a.Rh. Studies in Economics and Business Administration in Saarbrücken (1964–1969). Ph.D. on Synergies in Mergers and Acquisitions at the European Business School Oestrich-Winkel (2004). Management Executive BASF SE 1970–2005, amongst others President of BASF Venezolana (1986–1989), of BASF Information Systems in US (1990–1991), of BASF Fertilizer Division (1992–1998) of BASF Region Europe (1999–2005). Faculty Member of the Danubia University in Krems (Austria). Since 2006 Professor at the Ludwigshafen University of Applied Sciences. Founder and Scientific Director of the Middle-East Europe Institute (MOI) in the University (2007). Board Member of various Companies/Institutes.

Research on Corporate Management, Strategic and International Business, Innovation Management particular New Product Marketing, Intercultural behaviour.

Hans W. Reiners President of Performance Chemicals Division, BASF SE, Ludwigshafen, Germany. Studies in Business Administration especially Marketing, Finance, Organization and Metallurgy at the Technical University in Aachen (1982–1987). 1987–1994 starting in various Management positions with BASF in Germany and the United States. 1995–1998 Director of Business Management Agricultural Products for Southern Europe in Barcelona, Spain. 1998–2000 Group Vice President for Global Marketing Agricultural Products and Global Integration of American Cyanamid in 2000. President for Agricultural Products division with headquarters in New Jersey, USA (2001–2005) and for Styrenics Division in Ludwigshafen, Germany (2005–2009). From 2003 to 2005 Vice President and President of CropLife International, representing the Plant Science Industry.

Authors

Part I: Innovation and International Strategy

Dr. Reinhold Achatz Reinhold Achatz is Corporate Vice President of Siemens AG, in this function he is head of Corporate Research and Technologies (since 2006) and head of the Corporate Development Center (since April 2009). He is responsible for the global research activities of Siemens AG with over 2,200 employees. In the Development Center, Dr. Achatz is driving the integration of the product development process with the expertise of 2,600 software engineers worldwide, the majority of whom are from India and Central Eastern Europe. Reinhold Achatz holds a degree in Electrical Engineering (Dipl. Ing.), University of Erlangen-Nuremberg (1979) and a Ph.D. (Dr.-Ing.) in Information Technology in Mechanical Engineering, Technical University of Munich (2009). He joined Siemens AG Automation, as a software engineer in 1980 and held numerous management positions over the years. In addition, Dr. Achatz is a board member and chairman of the EU's NESSI Technology platform. He is a member of the European Research Area Board (ERAB). Also, Dr. Achatz is a member of the Board of the Committee for Research, Innovation and Technology of the German Bundesverband der Deutschen Industrie e.V. (BDI) and chair of the Research and Development Executive Working Group for the German Zentralverband Elektrotechnik- und Elektronikindustrie e.V. (ZVEI).

Dr. Heike Belitz is a senior researcher at the German Institute for Economic Research Berlin (Deutsches Institut für Wirtschaftsforschung DIW), Department Innovation, Manufacturing, Service. She entered the institute in 1991 and has conducted several studies related to technology policy, R&D of multinational companies, innovation systems, innovation indicators, and evaluation of selected R&D promoting programmes. From 2000 to 2002 she worked as secretary of the evaluation commission of the System of Industry-Integrating Research Assistance at the Federal Ministry of Economics and Technology. She studied mathematical economics at the University of Economics in Berlin (1977–1981), where she completed her PhD thesis in 1986.

Gabriela Buchfink Responsible for project management and facilitation techniques in creative workshops in product development at TRUMPF Werkzeugmaschinen

GmbH und Co. KG in Ditzingen. Studied Product Engineering, with focus on Documentation and Communication at Furtwangen University (1999–2004). Author of the TRUMPF books "Fascination of Sheet Metal" and "The Laser as a Tool", published by Vogel Verlag (2004–2006) and also the internal TRUMPF book "Thinking and moving in new directions" (2008). Since 2006 management of groupwide and development-related projects, since 2008 additionally responsible for creativity techniques in innovation management.

Dr. Hans Jörg Heger is Partner and Principal Consultant in the CTO Office of Siemens AG. He studied Physics, in Frankfurt and Hamburg, and obtained his Ph.D. from the Technical University of Munich in 1999. During his Ph.D., he gained international experience at the University of Pretoria, South Africa. After scientific work at the HelmholtzCenter Munich, he joined Siemens AG in 2000. He held the positions of a Senior Research Scientist, Strategic Account Manager and Senior Consultant before becoming a Partner for innovation strategy in the CTO Office. In this role, he is responsible for the client support of Corporate Research and Technologies and is focused on Strategic Planning with special emphasis on technology strategy and portfolio management.

Prof. Dr. Rüdiger Kabst is Professor of Business Administration and Human Resource Management at the University of Giessen and director of the Interdisciplinary Research Unit on Evidence-based Management and Entrepreneurship. He is the German representative of the Cranfield Network on International Human Resource Management (cranet). He had been a visiting research scholar at the University of Illinois/Urbana-Champaign in 1996, at the University of California/Berkeley in 2001, and at EWHA University in Seoul/South Korea in 2006. Rüdiger Kabst is co-editor of the peer-reviewed journal "Management Revue: The International Review of Management Studies", co-editor of the scientific book series "Empirische Personal- und Organisationsforschung" as well as co-editor of the professional HR Journal "Personal". His current research interests include international comparative human resource management, expatriate management, human resource practices between market and hierarchy (e.g. outsourcing, downsizing, interim-management, working time flexibility, etc.), interfirm cooperations (e.g. joint ventures), trust between organizations, young technology start-ups, international entrepreneurship, and internationalization of medium-sized enterprises.

Dr. Christian Koerber Head of New Business Development at TRUMPF GmbH + Co. KG in Ditzingen. Studied Mechanical Engineering, specialized in manufacturing technology at RWTH Aachen University (1988–1993). Training as International Welding Engineer (IWE) at the SLV Duisburg (1993). Awarded PhD in Engineering, thesis on welding with high-powered lasers at the Fraunhofer Institute for Laser Technology in Aachen (1993–1995). Followed by 2 years (1996–1997) as Project Manager for industrial projects at the Coopération Laser Franco-Allemande in Paris. 1998 worked as consultant for production, logistics and integral factory organization with agiplan Aktiengesellschaft in Muelheim. Head of the core market segment

Machinery and Industrial Equipment Manufacturing and later the Production competence center. In 2002 joined Kienbaum Management Consultants in Duesseldorf as Division Manager, later Head of the Industry section. Assumed current position at TRUMPF in 2007.

Dr. Andreas Kreimeyer (Member of the Board of Executive Directors BASF SE.) studied biology at the Universities of Hannover and Hamburg. After being awarded his doctorate, he joined BASF's Main Laboratory in 1986. In 1993, he became personal assistant to the Chairman of the Board of Executive Directors. Dr. Kreimeyer moved to Singapore in 1995. From 1998 to 2002, he was president of several BASF divisions, including the Fertilizer division (1998); Dispersions division (2000) and Functional Polymers division (2001). In 2002, Dr. Kreimeyer was appointed to the Board of Executive Directors of BASF SE effective January 1, 2003. Dr. Andreas Kreimeyer is responsible for Inorganics; Petrochemicals; Intermediates; Chemicals Research & Engineering; Science Relations & Innovation Management and BASF Future Business. In addition, he is Research Executive Director and a Member of the Wintershall Holding AG Supervisory Board.

Dr. Christian Schwens is post-doc (Habilitand) at the University of Giessen, Germany. He studied business administration at the University of Paderborn and at the University of Stockholm. Christian Schwens is a member of the Interdisciplinary Research Unit on Management and Entrepreneurship. In 2006 he was visiting scholar at the Carlson School of Management at the University of Minnesota, Minneapolis, USA. His research interests include the internationalization of technology firms, international entrepreneurship, market entry mode choices of small and medium-sized firms (SMEs), foreign institutions, international staffing, and internationalization of SMEs.

Prof. Dr. Tom Sommerlatte was chairman of the advisory board of Arthur D. Little (until 2008); he studied chemistry and chemical engineering at the Free and the Technical University of Berlin; the University of Rochester, New York; and the University of Paris and obtained PhD in chemical engineering from the University of Paris (1968). He did research on artificial intelligence and knowledge management at Studiengruppe für Systemforschung, Heidelberg. Thereafter, he did his Master of Business Administration (MBA) at the European Institute of Business Administration, INSEAD, Fontainebleau (1970), and joined the international consulting firm Arthur D. Little, Inc., in Brussels. He was transferred to the newly founded branch office in Wiesbaden (1973) and became a member of the European Directorate of the firm (1976) in charge of operations management, later of Telematics consulting. He was the managing director of German, Austrian and Swiss operations (1983) and then of all European consulting activities (1990) and chairman of Arthur D. Little's global consulting activities (1996). He started teaching systems sciences at the University of Kassel (1999), stepped down from executive role at Arthur D. Little (1997) to become chairman of the advisory board (until 2008) and created Osiris MIC GmbH (2002), a firm specializing in counselling services in the area of innovation management. He has authored and co-authored over 20 books on management issues, mainly related to innovation strategy and technology management, e.g. "Innovation Premium", a bestseller in the USA in 1999.

Dr. Holger Steinmetz is Psychologist and works as a post-doc (Habilitand) at the University of Giessen, Germany. His research interests include cross-cultural research, work – and organizational psychology and diverse methodological issues (structural equation modeling, psychometrics, meta-analysis). He conducted studies, for instance, on work stress and subjective health, work-life balance, and measurement invariance.

Dipl.-Oec. Harald Völker (Chief Financial Officer at TRUMPF GmbH + Co. KG in Ditzingen and Head of the Medical Technology Division.) studied Business Administration at Giessen University. Consultant at McKinsey & Co. in Duesseldorf and Frankfurt (1986–1990). At TRUMPF since 1990, initially as Director Controlling, as of 1992, as Director Finance and Controlling. Since 1996, Head of the Electronics and Medical Technology Business Division and Managing Director, HÜTTINGER Elektronik GmbH + Co. KG. Since July 2001, Executive Vice President and Chief Financial Officer, TRUMPF GmbH + Co. KG, responsible for finances, controlling, information technology, and legal affairs as well as head of the Medical Technology Business Division. Since 2007 responsible for acquisition management and since 2008 also for consulting activities.

Part II: Efficiency of Innovation Processes in International Enterprises

Dr. Lydia Bals Senior Consultant at Bayer Business Consulting (since 2008). Studies in Business Administration at European Business School (EBS; Wiesbaden, Germany), at EGADE (Escuela de Graduados en Administration de Empresas; Monterrey, Mexico) at Helsinki University of Technology (Espoo, Finland) (2001–2005). PhD thesis on international sourcing of services at the Supply Chain Management Institute of EBS (2005–2008). Visiting Scholar, University of Pennsylvania/Wharton (Philadelphia, USA) and Columbia University/Columbia Business School (New York, USA) (2008). Recently visiting Scholar at Copenhagen Business School.

Prof. Dr. Michael Frese Professor at National University of Singapore Business School and Leuphana (University of Lueneburg, Germany). Worked at University of Giessen, University of Amsterdam, University of Munich, University of Pennsylvania , Technical University of Berlin. He has been and is visiting professor at various universities in England (London Business School), Africa (Zimbabwe and Markere University Business School, Uganda), China, and USA. Has authored more than 200 articles and belongs to the most frequently cited work and organizational psychologists from Europe. Research on a wide range of topics, for example, psychological effects of unemployment, stress at work, a psychological theory of errors. Michael has worked on innovation, personal initiative, entrepreneurship, and cultural issues (both national culture and organizational culture).

Alexa Hergenröther Head of Corporate Development, M&A, K+S Aktiengesellschaft, Kassel. Studies in Economics and Business Administration at University of Mannheim (1992–1997). Consultant (Audit, Tax and M&A) Deloitte, Hannover (1997–2001). Degree as German Tax Accountant (2002). Senior Manager Tax, K+S Aktiengesellschaft, Kassel (2002–2006). Head of Corporate Development, M&A in K+S Aktiengesellschaft, Kassel (since 2006).

Matthias Kämper Head of the Marketing & Sales Department as well as Post Merger Integration Practice at Bayer Business Consulting. Bank apprenticeship at Commerzbank AG (1985–1987), studies in Business Administration at the University of Cologne (1987–1992). Worked as a project manager at IFEP GmbH, a marketing research institute (1988–1992). Marketing for consumer goods, in charge of the brands Natreen and Satina (1992–1995). Marketing and sales for prescription free drugs (OTC) and the brand Aspirin (1995–1999). Marketing manager for various OTC categories (1999–2002). Regional Brand Manager for the Analgetics business of Bayer Healthcare in Europe/Middle East/Africa (2002–2005). Project Manager of the integration of Roche's OTC business for Europe at Bayer Health Care (2004–2005).

Michael Kielkopf Program Manager of the Master of Science (M.Sc) in Innovation Management and research assistant at the Institute for Change Management and Innovation (CMI) faculty of Management, University of Applied Sciences Esslingen. Studies in Business Administration in Tübingen (1999–2005).

Dr. Verena Koch Research assistant at the faculty of Management, University of Applied Sciences Esslingen. Studies in Business Administration in Regensburg (1999–2004), at the Ecole de Commerce et de Gestion La Rochelle (2000–2001) and the University of Glasgow (2003–2004). Consultant for Corporate Finance, Deloitte & Touche (2004–2006). Ph.D. on interaction work in the service industry at the University of Augsburg.

Dr. Alexander Moscho Head of Business Consulting at Bayer Business Services GmbH, internal management consulting of Bayer AG (since 2006). Studies on biotechnology at TU Braunschweig (1990–1996). Ph.D. on BioSciences at TU Munich (1999–2001). Visiting Scholar at Stanford University (California, USA) (1995–1996), Visiting Professor in Bioentrepreneurship at Danube University Krems (Austria) (since 2005). Associate Principal at McKinsey&Company lastly as a member of the international Pharma/Healthcare-Leadership-Team (1996–2006). Worked in different positions within the branches of Life Sciences as well as Venture Capital.

Andreas Neus Service Innovation Lead, Karlsruhe Service Research Institute, Karlsruhe Institute of Technology (KIT). Studies in psychology, communications research and computer science in Bonn, Trier and Luxembourg. Thesis on Trust and Information Quality in Online Communication (1999). Cofounder, Metabit Service & Consulting, Bonn (1997–1999). Consultant, e-business Innovation Center, IBM Unternehmensberatung GmbH, Hamburg (1999–2001). Senior Consultant, Strategy & Change, IBM Global Business Services, Hamburg (2003–2005). Media & Entertainment EMEA Lead, IBM Institute for Business Value, Amsterdam (2006–2007). Research on Collaborative Innovation in Services, Open Innovation, Innovation Culture, Business Model Innovation and Change Management.

Dr. Stefan Neuwirth Head of the function "Human Resources Strategy" at Bayer AG (since 2009). Head of the Shared Services Department and the Deputy Head of Bayer Business Consulting (2003–2009). Apprenticeship at Siemens AG (1989–1991), studies in Business Administration at TU Berlin (1992–1995). Ph.D. on "Organization Theory" (1995–1998). Visiting Scholar at Duke University (USA) (1997). Held various positions at Bayer in the areas of Organization, Human Resource Management and Inhouse Consulting (since 1998).

Dr. Nina Rosenbusch Research Associate, Faculty of Economics and Business Administration, Friedrich Schiller University, Jena. Studies in Business Administration, University of Frankfurt (2001–2004). Ph.D. on Innovation and Firm Performance, Friedrich Schiller University, Jena, (2008). Junior Marketing Manager, International Business Development, Union Asset Management AG (2001–2004). Since 2004 Member of the Interdisciplinary Research Unit on Management and Entrepreneurship, University of Giessen. Research on Performance Implications of Entrepreneurial Strategies, Corporate Governance and Top Management Teams in Entrepreneurial Firms, Exploration, Exploitation and Ambidexterity, Gender Issues in Entrepreneurship.

Kathrin Rosing Research Associate, Department of Work and Organizational Psychology, Justus Liebig University, Giessen. Studies in Psychology, University of Osnabrueck (2001–2007). Since 2007 Member of the Interdisciplinary Research Unit on Management and Entrepreneurship, Justus Liebig University, Giessen. Research on Creativity, Innovation, Exploration, Exploitation, and Ambidexterity, Leadership, Self-Regulation.

Prof. Dr. Gerhard Satzger Director Karlsruhe Service Research Institute, Professor of Service Innovation and Management, Karlsruhe Institute of Technology (KIT). Master of Business Engineering, University of Karlsruhe (1982–1989); MBA, Oregon State University/USA (1987–1988); PhD on Datacenter Investment Planning, University of Giessen (1993); "Habilitation" on Capital-Intensive Services in Global Markets, University of Augsburg (1998); various positions IBM Germany (1989–1996), Assistant Professor University of Augsburg (1997–1998); Senior finance management IBM Germany (1998–2000) and IBM Europe/Middle East/Africa, Paris (2001–2002); CFO IBM Global Services Central Europe/Germany (2002–2007). Since 2008 Director of the IBM-sponsored Karlsruhe Service Research Institute. Research on Service Innovation,

Economics and Design of Service Relationships, Service Value Networks, Service Management, and Financial Engineering.

Michael Schürle Head of Communications of Voith Paper in Heidenheim, Germany: studies in international business administration at the Aalen University (1997–2001). Gained experience in marketing and controlling while working for Behr America in Detroit, USA. Joined Voith Paper in 2001; worked on the development of its CRM (customer relationship management) software. Marketing activities for Voith Paper (2003–2005). Named Head of Communications (2006).

Dr. Johannes Siemes Head of Marketing Projects and Market Research of K+S KALI GmbH, Kassel (retiered). Studies in Agricultural Economics in Bonn and Göttingen (1968–1973). Ph.D. on Simulated Management Models within Food Markets at Bonn University (1976). From 1977 to 2008 various Management Functions for K+S AG (Kassel) in the field of Public Relations, Marketing and Sales and later at K+S KALI GmbH head of Marketing Projects, Trade Policy and Market Research on Agriculture and Fertilizer Markets. Consultant for agricultural Innovation, Food Markets and Farm Management (since 2008).

Bertram Staudenmaier Board Member of the Voith AG in Heidenheim, Germany, and President of the Voith Paper division Fabric and Roll Systems; studies in production engineering at the Ulm University of Applied Sciences (1984–1987). Management Executive positions at leading suppliers to the paper industry. For example business segments and operating locations at Xerium Technologies Inc., including its Stowe Woodward AG affiliate (1988–2003). President of Xerium North America (2003–2004). Joined the Executive Board of the Voith AG (2005). In charge of Voith Paper's Fabrics and Rolls businesses. Executive Board Member of the industrial Association for Printing & Paper Equipment and Supplies of the German Engineering Federation (VDMA).

Dr. Alexander Tarlatt Managing Director and Co-founder of Santiago Advisors, an international consultancy focused on organizational design. Studies in Economics and Business Administration at the University of Cologne (1993–1998), Ph.D. on Strategy Implementation at the University of Cologne (2001). Consultant at Droege & Comp. GmbH 2002–2008, amongst others Senior Principal of Droege's Organization and Change Practice (2007–2008). Professional Service focus: Large Scale Organizational Change, Post Merger Integration, Business Transformation, Innovation Systems and Organization, R&D-Organization, Shared Service Operations.

Prof. Dr. Dietmar Vahs Professor of Business Administration, Change and Innovation Management at Esslingen University. Studies in Economics and Business Administration in Tübingen (1981–1986) and Charlottesville (University of Virginia, 1984). Ph.D. with a Study on The Development of Controlling in Industrial Firms (1989). Trainee at the Daimler-Benz AG and Management Executive at the Mercedes-Benz Truck Division (1989–1993). Since 1993 Professor at Esslingen University. Director of the Steinbeis-Transferzentrum "Innovative

Corporate Leadership" in Stuttgart (1994–1998). Since 1998 Director of the Institute for Change Management and Innovation (CMI). Research on Corporate Management, Key Success Factors of Change and Innovation Management, Organizational Culture and Organizational Structures.

Part III: Capital Markets, Finance and Innovation Performance

Prof. Dr. Malte Brettel was born in 1967. He studied engineering and business administration at the Technical University of Darmstadt. In 1996 he received a Ph.D. and in 2003 a Habilitation in business administration from the WHU Koblenz. Since 2003, Mr. Brettel has been professor of business administration and entrepreneurship at Aachen University (RWTH). He is a co-founder of a successful e-commerce start up and other companies and has extensive practical experience as a consultant with start-up companies and established firms like Porsche AG or Deutsche Post AG.

Dr. Hady Farag is a consultant in the Frankfurt and Singapore offices of The Boston Consulting Group (BCG). His work focuses on corporate strategy, capital market and M&A-related projects across a variety of industries. Prior to joining BCG, Dr. Farag served as researcher and later Head of University Development at the European Business School (EBS), where he earned a doctorate degree in business administration. He also holds master-level degrees from both EBS and the Joseph M. Katz Graduate School of Business (University of Pittsburgh).

Prof. Dr. Ronald Gleich a former partner of Horváth & Partners Management Consultants, became a faculty member at the European Business School (EBS), Germany as the Head of the Endowed Chair of Industrial Management in 2003. In 2004–2005, as a member of the EBS management board, he was responsible for the financial and commercial development of the programmes of Executive Education. Since 2007, he has been a Managing Director of the EBS Executive Education GmbH and Head of the Strascheg Institute for Innovation and Entrepreneurship (SIIE), which focuses on research and teaching in management accounting, innovation management, entrepreneurship and project management. He has published numerous books and journals as author and editor in the research areas of management accounting and innovation management.

Prof. Dr. Thomas W. Günther 1981–1986 Studies in Business Management at University of Augsburg, 1986–1994 Assistant at the Chair of Auditing and Control of Prof. Dr. A. G. Coenenberg, University of Augsburg, 1994–1996 Lecturer at Chair of Management and Control at Dresden University of Technology, 1996 Habilitation at University of Augsburg, since 1996 Full Professor for Management Accounting and Control at Dresden University of Technology, 1999 Chair offer from Vienna University of Economics and Business, 2007 Chair offer from European Business School, Oestrich-Winkel/Wiesbaden, since 2001 Head of Working Group "Value Based Management in SMEs" and member of Working Group "Intangibles in Accounting" of Schmalenbach Society, 2001–2002 and 2006

Visiting Professor at University of Virginia, Charlottesville, VA, USA, since 2004 Board Member of the Schmalenbach Society, Consulting and Coaching of SMEs, listed companies and NPOs in the design of management control systems.

Dr. Andreas Krostewitz is Consultant with PricewaterhouseCoopers AG, Frankfurt am Main. He studied Business Administration at the Friedrich-Schiller-University of Jena, Germany (Diploma Degree) and International Management at the Anglia Polytechnic University of Cambridge, UK (Bachelor of Arts). Dr. Andreas Krostewitz received his doctorate in Business Administration from the Friedrich-Schiller-University of Jena. In 2008 he joined PwC's service line Advisory – Valuation & Strategy and is specialized in Business Enterprise Valuation, Purchase Price Allocations und Impairment Tests according to international accounting standards.

Dr. Frank Lindner is management consultant at Innovation Navigators, a Horváth & Partners Company. His research activities are focused on innovation and technology management. He received a Master of Science in Technology and Management (Diplom-Ingenieur) from the University of Stuttgart and a Doctorate from the European Business School.

Dr. Ulrich Pidun works as Principal and Global Expert for Corporate Development with The Boston Consulting Group in Frankfurt. His project work and research is focused on corporate strategy and portfolio management, value and risk management, and innovation topics. Ulrich Pidun has studied chemistry and mathematics in Marburg and London and holds a PhD in Theoretical Chemistry and an MBA from INSEAD, Fontainebleau. Since 2006 he is also a lecturer for Strategic Management at the University of Karlsruhe.

Frank Rohwedder studied economics in Kiel and Paris before joining a trainee program for investment specialists at a Hamburg bank. Since 1999 he has been involved in the field of structured investments with a focus on media and private equity. In August 2005 he joined the Asset Finance & Leasing team of the Deutsche Bank in Frankfurt, responsible for developing the asset class Intellectual Property, a role he still occupies. In addition to his banking activities Frank Rohwedder has held an honorary lecturer position at the Hamburg Business Academy (now Hamburg School of Business Administration) teaching economics with a focus on macroeconomics.

Dr. Thomas Rüschen completed his business studies in Mannheim and Frankfurt before joining Deutsche Bank in 1990 in the field of European financial markets. As Senior Relationship Manager in London he was responsible for corporate accounts in continental Europe and then became Country Manager for Italy. He currently heads Asset Finance & Leasing, which is responsible for the financing of long-term economic commodities worldwide, e.g. in the area of renewable energies. He is a member of Global-Banking-Exco as well as the Environmental Steering Committee, which coordinates climate change related proposals and activities of the Deutsche Bank.

Ervin Schellenberg is Joint Chief Executive (Member of the Board) and Founder of EquityGate, Wiesbaden, an independent investment banking advisory firm. He worked for over 19 years at leading investment management and investment banking firms in London and Frankfurt including Duke Street Capital Private Equity Ltd, London, Salomon Smith Barney, London, Dresdner Kleinwort Benson, Frankfurt.

He is a graduate from Johann Wolfgang Goethe-Business School, Frankfurt. He publishes and lectures on various topics such as mergers and acquisitions, private equity and financing as well as valuation for the European Business School, Reichartshausen, School of Applied Sciences, Wiesbaden, EUROFORUM, Euromoney, etc.

He is fluent in German, English and speaks Croatian and Slovenian.

Dr. Peter Schentler is a Postdoctoral Researcher and Head of the Competence Center Management Accounting & Performance Measurement at the Strascheg Institute for Innovation and Entrepreneurship (SIIE) at the European Business School (EBS), International University Schloss Reichartshausen, Germany. He received his Doctorate in Economic Sciences from the University of Rostock/Germany and his Master's degree in Business Administration and Engineering from the FH Joanneum in Kapfenberg, Austria. His research and teaching concentrate on management accounting and performance measurement with a special focus on innovation and procurement.

Martin Scholich is Partner and Member of the Managing Board with PricewaterhouseCoopers AG. He studied Business Administration at the University of Cologne and obtained his MBA degree at the Eastern Illinois University, USA. Martin Scholich joined PricewaterhouseCoopers' Corporate Finance practice in 1991 and has been leading the German Advisory practice since 2004. Since July 2008 he has been responsible for the client support in the Rhine-Main area as Senior Relationship Partner for the Frankfurt office.

Martin Scholich is a Certified Public Accountant and a Certified Tax Advisor. He has significant experience in Business Enterprise Valuation and Advising on Transactions.

Katinka Wölfer is a doctoral candidate and research assistant at the Strategic Finance Institute at European Business School (EBS) International University, Wiesbaden (Germany) since 2007. She received a diploma in business with majors in banking, finance and international accounting from the Justus Liebig University Giessen (Germany) and the University of Kentucky (USA). She also graduated with an MBA from the University of Wisconsin-Milwaukee (USA). In recognition of high scholastic achievements, Katinka Wölfer was selected for membership in Beta Gamma Sigma, the international honor society for collegiate schools of business. Her main research interests concentrate on family business finance.